GREG

MODERN VLSI DESIGN
System-on-Chip Design
Third Edition

Wayne Wolf

Prentice Hall PTR
Upper Saddle River, NJ 07458
www.phptr.com

ISBN 0-13-061970-1

90000

9 790130 619708

Library of Congress Cataloging-in-Publication Data

A CIP catalog record for this book can be obtained from the Library of Congress.

Editorial/production supervision: *Patti Guerrieri*
Acquisitions editor: *Bernard Goodwin*
Marketing manager: *Dan DePasquale*
Manufacturing manager: *Alexis R. Heydt-Long*
Editorial assistant: *Michelle Vincenti*
Cover design director: *Jerry Votta*
Cover designer: *Anthony Gemmellaro*

 © 2002, 1998, 1994 by Prentice Hall PTR
Prentice-Hall, Inc.
Upper Saddle River, NJ 07458

Prentice Hall books are widely used by corporations and government agencies
for training, marketing, and resale.

The publisher offers discounts on this book when ordered in bulk quantities.
For more information, contact: Corporate Sales Department, Phone: 800-382-3419;
Fax: 201-236-7141; E-mail: corpsales@prenhall.com; or write: Prentice Hall PTR,
Corp. Sales Dept., One Lake Street, Upper Saddle River, NJ 07458.

Illustrated and typeset by the author. This book was typeset using FrameMaker. Illustrations were
drawn using Adobe Illustrator, with layout plots generated by cif2ps.

Photo credits: Example 1-1: Intel Corporation; Figure 2-2: IBM; Figure 2-5: Agere; Figure 2-15:
UMC; Figure 2-16: IBM; Figure 6-1: Hewlett-Packard Company; Example 7-1: Tom Way, IBM
Microelectronics, Essex Junction, VT; Figure 7-19: Wayne Wolf; Figure 7-20: Wayne Wolf and
IBM; Example 8-4: Philips.

Second edition published under the title *Modern VLSI Design: Systems on Silicon*.

Printed in the United States of America
10 9 8 7 6 5 4 3 2 1

ISBN 0-13-061970-1

Pearson Education LTD.
Pearson Education Australia PTY, Limited
Pearson Education Singapore, Pte. Ltd.
Pearson Education North Asia Ltd.
Pearson Education Canada, Ltd.
Pearson Educación de Mexico, S.A. de C.V.
Pearson Education — Japan
Pearson Education Malaysia, Pte. Ltd.

for Nancy (as always)

Table of Contents

3 Logic Gates   111

Preface to the Third Edition

This third edition of *Modern VLSI Design* includes both incremental refinements and new topics. All these changes are designed to help keep up with the fast pace of advancements in VLSI technology and design.

The incremental refinements in the book include improvements in the discussion of low power design, the chip project, and the lexicon. Low power design was discussed in the second edition, but has become even more complex due to the higher leakages found at smaller transistor sizes. The PDP-8 used in previous editions has been replaced with a more modern data path design. Designing a complete computer is beyond the scope of most VLSI courses, but a data path makes a good class project. I have also tried to make the lexicon a more comprehensive guide to the terms in the book.

This edition shows more major improvements to the discussions of interconnect and hardware description languages. Interconnect has become increasingly important over the past few years, with interconnect delays often dominating total delay. I decided it was time to fully embrace the importance of interconnect, especially with the advent of copper interconnect. This third edition now talks more thoroughly about interconnect models, crosstalk, and interconnect-centric logic design.

The third editon also incorporates a much more thorough discussion of hardware description languages. Chapter 8, which describes architectural design, now introduces VHDL and Verilog as the major hardware description languages. Though these sections are not meant to be thorough manuals for these languages, they should provide enough information for the reader to understand the major concepts of the languages and to be able to read design examples in those languages.

As with the second edition, you can find additional helpful material on the World Wide Web at http://www.ee.princeton.edu/~wolf/modern-vlsi. This site includes overheads useful either for teaching or for self-paced learning. The site also includes supplementary materials, such as layouts and HDL descriptions. Instructors may request a book of answers to the problems in the book by calling Prentice Hall directly.

I'd like to thank Al Casavant and Ken Shepard for their advice on interconnect analysis and Joerg Henkel for his advice on low power design. I'd also like to thank Fred Rosenberger for his many helpful comments on the book. As always, any mistakes are mine.

Wayne Wolf

Princeton, New Jersey

Preface to the Second Edition

Every chapter in this second edition of *Modern VLSI Design* has been updated to reflect the challenges looming in VLSI system design. Today's VLSI design projects are, in many cases, mega-chips which not only contain tens (and soon hundreds) of millions of transistors, but must also run at very high frequencies. As a result, I have emphasized circuit design in a number of ways: the fabrication chapter spends much more time on transistor characteristics; the chapter on gate design covers a wider variety of gate designs; the combinational logic chapter enhances the description of interconnect delay and adds an important new section on crosstalk; the sequential logic chapter covers clock period determination more thoroughly; the subsystems chapter gives much more detailed descriptions of both multiplication and RAM design; the floorplanning chapter spends much more time on clock distribution.

Beyond being large and fast, modern VLSI systems must frequently be designed for low power consumption. Low-power design is of course critical for battery-operated devices, but the sheer size of these VLSI systems means that excessive power consumption can lead to heat problems. Like testing, low-power design cuts across all levels of abstraction, and you will find new sections on low power throughout the book.

The reader familiar with the first edition of this book will notice that the combinational logic material formerly covered in one chapter (Chapter 3) has been split into two chapters, one of logic gates and another on combinational networks. This split was the result of the great amount of material added on circuit design added to the early chapters of the book. Other, smaller rearrangements have also been made in the book, hopefully aiding clarity.

You can find additional helpful material on the World Wide Web at http://www.ee.princeton.edu/~wolf/modern-vlsi. This site includes overheads useful either for teaching or for self-paced learning. The site also includes supplementary materials, such as layouts and VHDL descriptions. Instructors may request a book of answers to the problems in the book by calling Prentice Hall directly.

I would especially like to thank Derek Beatty, Luc Claesen, John Darringer, Srinivas Devadas, Santanu Dutta, Michaela Guiney, Alex Ishii, Steve Lin, Rob Mathews, Cherrice Traver, and Steve Trimberger for their comments and suggestions on this second edition.

Wayne Wolf

Princeton, New Jersey

Preface

This book was written in the belief that VLSI design is *system* design. Designing fast inverters is fun, but designing a high-performance, cost-effective integrated circuit demands knowledge of all aspects of digital design, from application algorithms to fabrication and packaging. Carver Mead and Lynn Conway dubbed this approach the tall-thin designer approach. Today's hot designer is a little fatter than his or her 1979 ancestor, since we now know a lot more about VLSI design than we did when Mead and Conway first spoke. But the same principle applies: you must be well-versed in both high-level and low-level design skills to make the most of your design opportunities.

Since VLSI has moved from an exotic, expensive curiosity to an everyday necessity, universities have refocused their VLSI design classes away from circuit design and toward advanced logic and system design. Studying VLSI design as a system design discipline requires such a class to consider a somewhat different set of areas than does the study of circuit design. Topics such as ALU and multiplexer design or advanced clocking strategies used to be discussed using TTL and board-level components, with only occasional nods toward VLSI implementations of very large components. However, the push toward higher levels of integration means that most advanced logic design projects will be designed for integrated circuit implementation.

I have tried to include in this book the range of topics required to grow and train today's tall, moderately-chubby IC designer. Traditional logic design topics, such as adders and state machines, are balanced on the one hand by discussions of circuits and layout techniques and on the other hand by the architectural choices implied by scheduling and allocation. Very large ICs are sufficiently complex that we can't tackle them using circuit design techniques alone; the top-notch designer must understand enough about architecture and logic design to know which parts of the circuit and layout require close attention. The integration of system-level design techniques, such as scheduling, with the more traditional logic design topics is essential for a full understanding of VLSI-size systems.

In an effort to systematically cover all the problems encountered while designing digital systems in VLSI, I have organized the material in this book relatively bottom-up, from fabrication to architecture. Though I am a strong fan of top-down design, the technological limitations which drive architecture are best learned starting with fabrication and layout. You can't expect to fully appreciate all the nuances of why a particular design step is formulated in a certain way until you have completed a chip design yourself, but referring to the steps as you proceed on your own chip design should help guide you. As a result of the bottom-up organization, some topics may be broken up in unexpected ways. For example, placement and routing are not treated as a single subject, but separately at each level of abstraction: transistor, cell, and floor plan. In many instances I purposely tried to juxtapose topics in unexpected ways to encourage new ways of thinking about their interrelationships.

This book is designed to emphasize several topics that are essential to the practice of VLSI design as a system design discipline:

- **A systematic design methodology reaching from circuits to architecture.** Modern logic design includes more than the traditional topics of adder design and two-level minimization—register-transfer design, scheduling, and allocation are all essential tools for the design of complex digital systems. Circuit and layout design tell us which logic and architectural designs make the most sense for CMOS VLSI.

- **Emphasis on top-down design starting from high-level models.** While no high-performance chip can be designed completely top-down, it is excellent discipline to start from a complete (hopefully executable) description of what the chip is to do; a number of experts estimate that half the application-specific ICs designed execute their delivery tests but don't work in their target system because the designer didn't work from a complete specification.

- **Testing and design-for-testability.** Today's customers demand both high quality and short design turnaround. Every designer must understand how chips are tested and what makes them hard to test. Relatively small changes to the architecture can make a chip drastically easier to test, while a poorly designed architecture cannot be adequately tested by even the best testing engineer.

- **Design algorithms**. We must use analysis and synthesis tools to design almost any type of chip: large chips, to be able to complete them at all; relatively small ASICs, to meet performance and time-to-market goals. Making the best use of those tools requires understanding how the tools work and exactly what design problem they are intended to solve.

The design methodologies described in this book make heavy use of computer-aided design (CAD) tools of all varieties: synthesis and analysis; layout, circuit, logic, and architecture design. CAD is more than a collection of programs. CAD is a way of thinking, a way of life, like Zen. CAD's greatest contribution to design is breaking the process up into manageable steps. That is a conceptual advance you can apply with no computer in sight. A designer can—and should—formulate a narrow problem and apply well-understood methods to solve that problem. Whether the designer uses CAD tools or solves the problem by hand is much less important than the fact that the chip design isn't a jumble of vaguely competing concerns but a well-understood set of tasks.

I have explicitly avoided talking about the operation of particular CAD tools. Different people have different tools available to them and a textbook should not be a user's guide. More importantly, the details of how a particular program works are a diversion—what counts is the underlying problem formulations used to define the problem and the algorithms used to solve them. Many CAD algorithms are relatively intuitive and I have tried to walk through examples to show how you can think like a CAD algorithm. Some of the less intuitive CAD algorithms have been relegated to a separate chapter; understanding these algorithms helps explain what the tool does, but isn't directly important to manual design.

Both the practicing professional and the advanced undergraduate or graduate student should benefit from this book. Students will probably undertake their most complex logic design project to date in a VLSI class. For a student, the most rewarding aspect of a VLSI design class is to put together previously-learned basics on circuit, logic, and architecture design to understand the tradeoffs between the different levels of abstraction. Professionals who either practice VLSI design or develop VLSI CAD tools can use this book to brush up on parts of the design process with which they have less-frequent involvement. Doing a truly good job of each step of design requires a solid understanding of the big picture.

A number of people have improved this book through their criticism. The students of COS/ELE 420 at Princeton University have been both patient and enthusiastic. Profs. C.-K. Cheng, Andrea La Paugh, Miriam Leeser, and John "Wild Man" Nestor all used drafts in their classes and gave me valuable feedback. Profs. Giovanni De Micheli, Steven Johnson, Sharad Malik, Robert Rutenbar, and James Sturm also gave me detailed and important advice after struggling through early drafts. Profs. Malik and Niraj Jha also patiently answered my questions about the literature. Any errors in this book are, of course, my own.

Thanks to Dr. Mark Pinto and David Boulin of AT&T for the transistor cross section photo and to Chong Hao and Dr. Michael Tong of AT&T for the ASIC photo. Dr. Robert Mathews, formerly of Stanford University and now of Performance Processors, indoctrinated me in pedagogical methods for VLSI design from an impressionable age. John Redford of DEC supplied many of the colorful terms in the lexicon.

Wayne Wolf

Princeton, New Jersey

1

Digital Systems and VLSI

Highlights:

VLSI and Moore's Law.

CMOS vs. other logic families.

Hierarchical design.

1.1 Why Design Integrated Circuits?

This book describes design methods for integrated circuits. That may seem like a specialized topic. But, in fact, integrated circuit (IC) technology is the enabling technology for a whole host of innovative devices and systems that have changed the way we live. Integrated circuits are much smaller and consume less power than the discrete components used to build electronic systems before the 1960s. Integration allows us to build systems with many more transistors, allowing much more computing power to be applied to solving a problem. Integrated circuits are also much easier to design and manufacture and are more reliable than discrete systems; that makes it possible to develop special-purpose systems that are more efficient than general-purpose computers for the task at hand.

Electronic systems now perform a wide variety of tasks in daily life. Electronic systems in some cases have replaced mechanisms that operated mechanically, hydraulically, or by other means; electronics are usually smaller, more flexible, and easier to service. In other cases electronic systems have created totally new applications. Electronic systems perform a variety of tasks, some of them visible, some more hidden:

- Personal entertainment systems such as portable MP3 players and DVD players perform sophisticated algorithms using very low power to save battery life.

- Electronic systems in cars clearly operate stereo systems and displays; they also control fuel injection systems, adjust suspensions to varying terrain, and perform the control functions required for anti-lock braking (ABS) systems.

- Digital electronics perform video compression and decompression on-the-fly in consumer electronics.

- Low-cost terminals for Web browsing still require sophisticated electronics, despite their dedicated function.

- Personal computers and workstations provide word-processing, financial analysis, and games. Computers include both central processing units (CPUs) and special-purpose hardware for disk access, faster screen display, etc.

- Medical electronic systems measure bodily functions and perform complex processing algorithms to warn about unusual conditions. The availability of these complex systems, far from overwhelming consumers, only creates demand for even more complex systems.

The growing sophistication of applications continually pushes the design and manufacturing of integrated circuits and electronic systems to new levels of complexity. And perhaps the most amazing characteristic of this collection of systems is its variety—as systems become more complex, we build not a few general-purpose computers but an ever wider range of special-purpose systems. Our ability to do so is a testament to our growing mastery of both integrated circuit manufacturing and design, but the increasing demands of customers continue to test the limits of design and manufacturing.

While we will concentrate on integrated circuits in this book, the properties of integrated circuits—what we can and cannot efficiently put in an integrated circuit—largely determine the architecture of the entire system. Integrated circuits improve system characteristics in several critical ways. ICs have three key advantages over digital circuits built from discrete components:

- **Size**. Integrated circuits are much smaller—both transistors and wires are shrunk to micrometer sizes, compared to the millimeter or centimeter scales of discrete components. Small size leads to advantages in speed and power consumption, since smaller components have smaller parasitic resistances, capacitances, and inductances.

- **Speed**. Signals can be switched between logic 0 and logic 1 much quicker within a chip than they can between chips. Communication within a chip can occur hundreds of times faster than communication between chips on a printed circuit board. The high speed of circuits on-chip is due to their small size—smaller components and wires have smaller parasitic capacitances to slow down the signal.

- **Power consumption**. Logic operations within a chip also take much less power. Once again, lower power consumption is largely due to the small size of circuits on the chip—smaller parasitic capacitances and resistances require less power to drive them.

These advantages of integrated circuits translate into advantages at the system level:

- **Smaller physical size**. Smallness is often an advantage in itself—consider portable televisions or handheld cellular telephones.

- **Lower power consumption**. Replacing a handful of standard parts with a single chip reduces total power consumption. Reducing power consumption has a ripple effect on the rest of the system: a smaller, cheaper power supply can be used; since less power consumption means less heat, a fan may no longer be necessary; a simpler cabinet with less shielding for electromagnetic shielding may be feasible, too.

- **Reduced cost**. Reducing the number of components, the power supply requirements, cabinet costs, and so on, will inevitably reduce system cost. The ripple effect of integration is such that the cost of a system built from custom ICs can be less, even though the individual ICs cost more than the standard parts they replace.

Understanding why integrated circuit technology has such profound influence on the design of digital systems requires understanding both the technology of IC manufacturing and the economics of ICs and digital systems.

1.2 Integrated Circuit Manufacturing

1.2.1 Technology

Most manufacturing processes are fairly tightly coupled to the item they are manufacturing. An assembly line built to produce Buicks, for example, would have to undergo moderate reorganization to build Chevys—tools like sheet metal molds would have to be replaced, and even some machines would have to be modified. And either assembly line would be far removed from what is required to produce electric drills. Integrated circuit manufacturing technology, on the other hand, is remarkably versatile. While there are several manufacturing processes for different circuit types—CMOS, bipolar, etc.—a manufacturing line can make any circuit of that type simply by changing a few basic tools called masks. For example, a single CMOS manufacturing plant can make both microprocessors and microwave oven controllers by changing the masks that form the patterns of wires and transistors on the chips.

Figure 1-1: A wafer divided into chips.

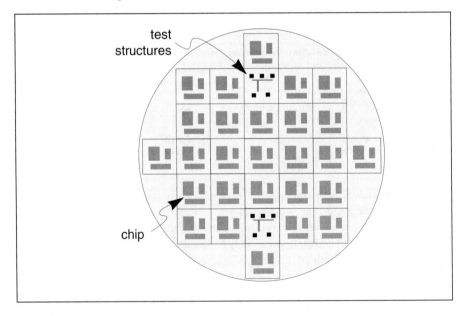

Silicon wafers are the raw material of IC manufacturing. The fabrication process forms patterns on the wafer that create wires and transistors. As shown in Figure 1-1, a series of identical chips are patterned onto the wafer (with some space reserved for test circuit structures which allow manufacturing to measure the results of the manufacturing process). The IC manufacturing process is efficient because we can produce many identical chips by processing a single wafer. By changing the masks that determine what patterns are laid down on the chip, we determine the digital circuit that will be created. The IC fabrication line is a generic manufacturing line—we can

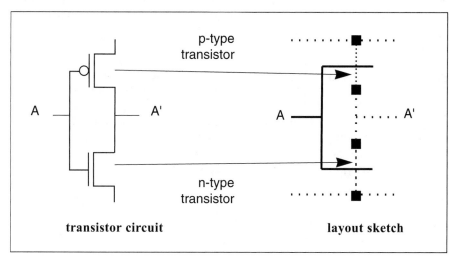

transistor circuit   layout sketch

Figure 1-2: An inverter circuit and a sketch for its layout.

quickly retool the line to make large quantities of a new kind of chip, using the same processing steps used for the line's previous product.

Figure 1-2 shows the schematic for a simple digital circuit. From this description alone we could build a breadboard circuit out of standard parts. To build it on an IC fabrication line, we must go one step further and design the **layout**, or patterns on the masks. The rectangular shapes in the layout (shown here as a sketch called a stick diagram) form transistors and wires which conform to the circuit in the schematic. Creating layouts is very time-consuming and very important—the size of the layout determines the cost to manufacture the circuit, and the shapes of elements in the layout determine the speed of the circuit as well. During manufacturing, a **photolithographic** (photographic printing) process is used to transfer the layout patterns from the masks to the wafer. The patterns left by the mask are used to selectively change the wafer: impurities are added at selected locations in the wafer; insulating and conducting materials are added on top of the wafer as well. These fabrication steps require high temperatures, small amounts of highly toxic chemicals, and extremely clean environments. At the end of processing, the wafer is divided into a number of chips.

Because no manufacturing process is perfect, some of the chips on the wafer may not work. Since at least one defect is almost sure to occur on each wafer, wafers are cut into smaller, working chips; the largest chip that can be reasonably manufactured today is just over 1.5 cm on a side, while a wafer is in the 20-25 cm. Each chip is individually tested; the ones that pass the test are saved after the wafer is diced into chips. The working chips are placed in the packages familiar to digital designers. In some packages, tiny wires connect the chip to the package's pins while the package body protects the chip from handling and the elements; in others, solder bumps directly connect the chip to the package.

Integrated circuit manufacturing is a powerful technology for two reasons: all circuits can be made out of a few types of transistors and wires; and any combination of wires and transistors can be built on a single fabrication line just by changing the masks that determine the pattern of components on the chip. Integrated circuits run very fast because the circuits are very small. Just as important, we are not stuck building a few standard chip types—we can build any function we want. The flexibility given by IC manufacturing lets us build faster, more complex digital systems in ever greater variety.

1.2.2 Economics

Because integrated circuit manufacturing has so much leverage—a great number of parts can be built with a few standard manufacturing procedures—a great deal of effort has gone into improving IC manufacturing.

Figure 1-3:
Moore's Law.

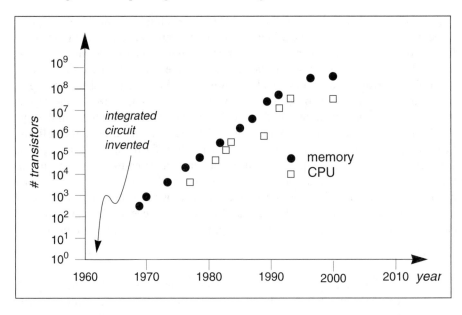

In the 1960s Gordon Moore, an industry pioneer, predicted that the number of transistors that could be manufactured on a chip would grow exponentially. His prediction, now known as **Moore's Law**, was remarkably prescient. Moore's ultimate prediction was that transistor count would double every 18 months, an estimate that has held up remarkably well. Figure 1-3 shows advances in manufacturing capability by charting the introduction dates of key products that pushed the state of the manufacturing art. Over the past thirty years, the number of transistors per chip has doubled about once a year. The circles show various logic circuits, primarily central processing units (CPUs) and digital signal processors (DSPs), while the black dots show random-access memories, primarily dynamic RAMs or DRAMs. At any given time, memory chips have more transistors per unit area than logic chips, but both have obeyed Moore's Law. The next example shows how Moore's Law has held up in one family of microprocessors.

Example 1-1: *Moore's Law and Intel microprocessors*

The Intel microprocessors are one good example in the growth in complexity of integrated circuits. Here are the sizes of several generations of the microprocessors descended from the Intel 8086.

microprocessor	date of introduction	# transistors	feature size (microns)
80286	2/82	134,000	1.5
80386	10/85	275,000	1.5
80486	4/89	1,200,000	1.0
Pentium™	3/93	3,100,000	0.8
Pentium Pro™	11/95	5,500,000	0.6

The photomicrographs of these processors, all courtesy of Intel, vividly show the increase in design complexity implied by this exponential growth in transistor count.

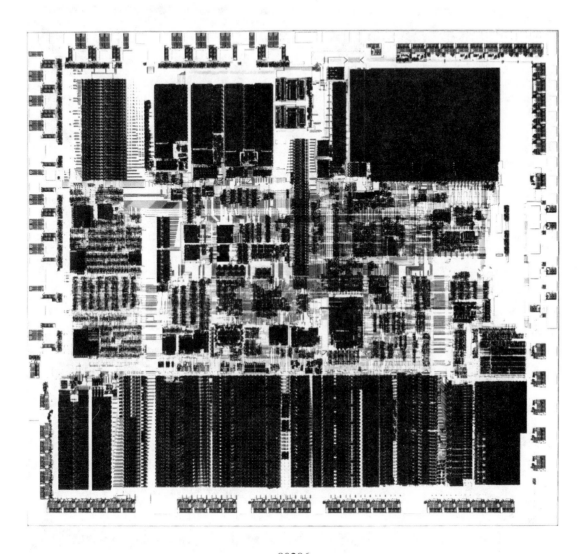

80286

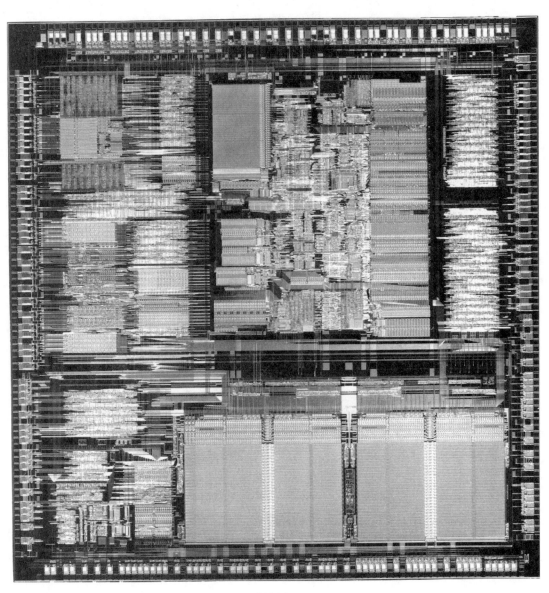

80386

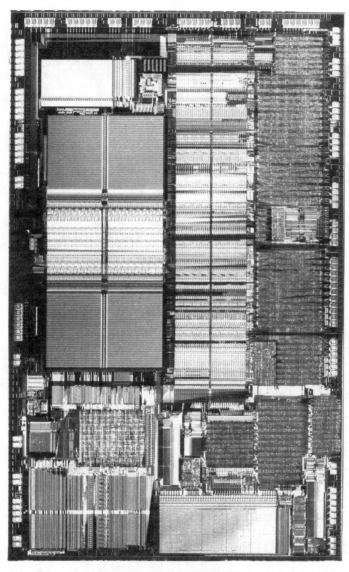

80486

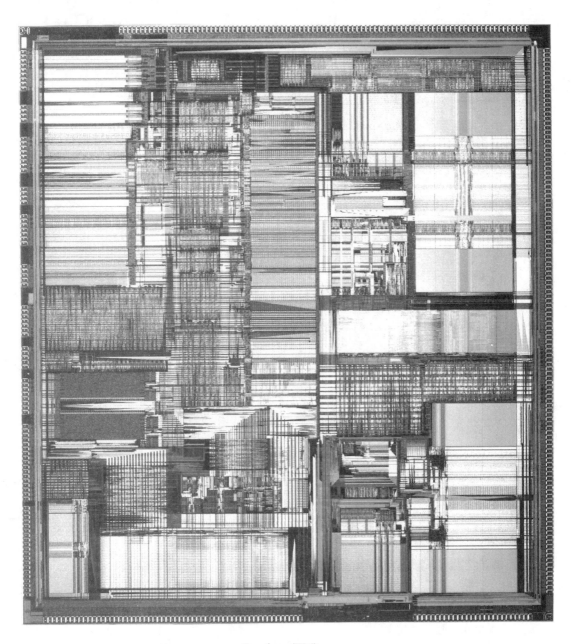

Pentium (TM)

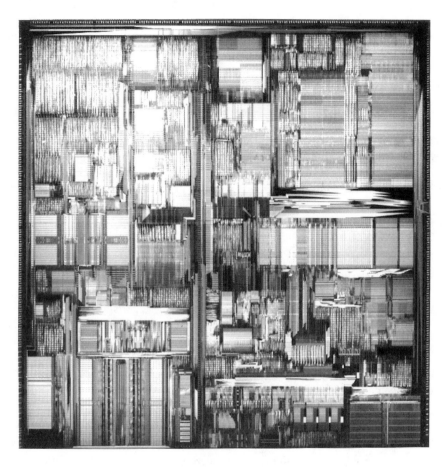

Pentium Pro (TM)

IC manufacturing plants are extremely expensive. A single plant costs $1 billion or more [NYT93]. Given that a new, state-of-the-art manufacturing process is developed every three years, that is a sizeable investment. The investment makes sense because a single plant can manufacture so many chips and can easily be switched to manufacture different types of chips. In the early years of the integrated circuits business, companies focused on building large quantities of a few standard parts. These parts are commodities—one 80 ns, 16Mb dynamic RAM is more or less the same as any other, regardless of the manufacturer. Companies concentrated on commodity parts in part because manufacturing processes were less well understood and manufacturing variations are easier to keep track of when the same part is being fabricated day after day. Standard parts also made sense because designing integrated circuits was hard—not only the circuit, but the layout had to be designed, and there were few computer programs to help automate the design process.

The preponderance of standard parts pushed the problems of building customized systems back to the board-level designers who used the standard parts. Since a function built from standard parts usually requires more components than if the function were built with custom-designed ICs, designers tended to build smaller, simpler systems. The industrial trend, however, is to make available a wider variety of integrated circuits. The greater diversity of chips includes:

- **More specialized standard parts**. In the 1960s, standard parts were logic gates; in the 1970s they were LSI components. Today, standard parts include fairly specialized components: communication network interfaces, graphics accelerators, floating point processors. All these parts are more specialized than microprocessors but are used in enough volume that designing special-purpose chips is worth the effort. In fact, putting a complex, high-performance function on a single chip often makes other applications possible—for example, single-chip floating point processors make high-speed numeric computation available on even inexpensive personal computers.

- **Application-specific integrated circuits (ASICs)**. Rather than build a system out of standard parts, designers can now create a single chip for their particular application. Because the chip is specialized, the functions of several standard parts can often be squeezed into a single chip, reducing system size, power, heat, and cost. Application-specific ICs are possible because of computer tools that help humans design chips much more quickly.

- **Systems-on-chips (SoCs)**. Fabrication technology has advanced to the point that we can put a complete system on a single chip. For example, a single-chip computer can include a CPU, bus, I/O devices, and memory. SoCs allow systems to be made at much lower cost than the equivalent board-level system. SoCs can also be higher performance and lower power than board-level equivalents because on-chip connections are more efficient than chip-to-chip connections.

A wider variety of chips is now available in part because fabrication methods are better understood and more reliable. More importantly, as the number of transistors per chip grows, it becomes easier and cheaper to design special-purpose ICs. When only a few transistors could be put on a chip, careful design was required to ensure that even modest functions could be put on a single chip. Today's VLSI manufacturing processes, which can put millions of carefully-designed transistors on a chip, can also be used to put tens of thousands of less-carefully designed transistors on a chip. Even though the chip could be made smaller or faster with more design effort, the advantages of having a single-chip implementation of a function that can be quickly designed often outweighs the lost potential performance. The problem and the challenge of the ability to manufacture such large chips is design—the ability to make effective use of the millions of transistors on a chip to perform a useful function.

1.3 CMOS Technology

1.3.1 CMOS Circuit Techniques

A number of different IC fabrication technologies are available to us. The most important difference between fabrication technologies is the types of transistors they can produce. Different transistor types require different circuit designs for Boolean logic and memory functions, which have very different speed and power characteristics.

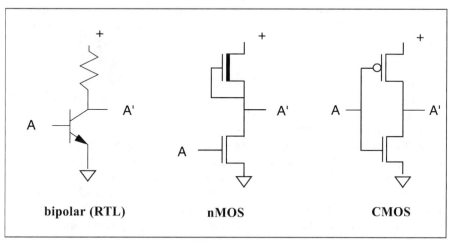

Figure 1-4: An inverter implemented in bipolar, nMOS, and CMOS.

bipolar (RTL)   nMOS   CMOS

Figure 1-4 shows three different circuit topologies for an inverter (logical NOT) using different transistor types. A bipolar transistor circuit can be used along with a resistor to build an inverter (though more sophisticated circuits are generally used in practice). An n-channel enhancement mode MOS transistor can be coupled with an n-channel depletion mode transistor (which is drawn as a resistor with a bar) to create a static nMOS gate. And a pair of p-type and n-type enhancement mode MOS transistors is used to build a static complementary, or CMOS, inverter. The speeds of these circuits types are ranked in decreasing order from left to right: the bipolar circuit is definitely fastest, followed by the nMOS and CMOS circuits. On the other hand, the power consumption of these circuits decreases from left to right: the bipolar circuit requires a great deal of power, the nMOS circuit uses considerably less but still not negligible amounts, while the CMOS circuit requires very little power. The negative correlation between speed and power is partly due to the nature of the universe—changing the value of a physical variable requires more power as you change it more quickly. It is also due to the clever design of the CMOS circuit—while the bipolar and nMOS circuits consume power in steady state, after their outputs have settled to their final values, the CMOS circuit consumes no steady-state power.

1.3.2 Power Consumption

The huge chips that can be fabricated today are possible only because of the extremely low power consumption of CMOS circuits. Power consumption is critical at the chip level because much of the power is dissipated as heat, and chips have limited heat dissipation capacity. Even if the system in which a chip is placed can supply large amounts of power, most chips are packaged to dissipate fewer than 10 to 15 Watts of power before they suffer permanent damage (though some chips dissipate well over 50 Watts thanks to special packaging). The power consumption of a logic circuit limits the number transistors we can effectively put on a single chip.

Limiting the number of transistors per chip changes system design in several ways. Most obviously, it increases the physical size of a system. Using high-powered circuits also increases power supply and cooling requirements. A more subtle effect is caused by the fact that the time required to transmit a signal between chips is much larger than the time required to send the same signal between two transistors on the same chip; as a result, some of the advantage of using a higher-speed circuit family is lost. Another subtle effect of decreasing the level of integration is that the electrical design of multi-chip systems is more complex: microscopic wires on-chip exhibit parasitic resistance and capacitance, while macroscopic wires between chips have capacitance and inductance, which can cause a number of ringing effects that are much harder to analyze.

The close relationship between power consumption and heat makes low-power design techniques important knowledge for every CMOS designer. Of course, low-power design is especially important in battery-operated systems like cellular telephones and personal digital assistants (PDAs). Battery-powered systems require not only low power consumption, but also low energy consumption per computation—power can be reduced by slowing down a computation while still consuming the same total energy. Energy, in contrast, must be saved by avoiding unnecessary work. In general, both types of design techniques are referred to as low-power design. We will see throughout the rest of this book that minimizing power and energy consumption requires careful attention to detail at every level of abstraction, from system architecture down to layout.

1.3.3 Design and Testability

Our ability to build large chips of unlimited variety introduces the problem of checking whether those chips have been manufactured correctly. Designers accept the need to **verify** or **validate** their designs to make sure that the circuits perform the specified function. (Some people use the terms verification and validation interchangeably; a finer distinction reserves verification for formal proofs of correctness, leaving validation to mean any technique which increases confidence in correctness, such as simulation.) Chip designs are simulated to ensure that the chip's circuits compute the proper functions to a sequence of inputs chosen to exercise the chip. But each chip that comes off the manufacturing line must also undergo **manufacturing test**—the chip must be exercised to demonstrate that no manufacturing defects rendered the chip useless. Because IC manufacturing tends to introduce certain types of defects and because we want to minimize the time required to test each chip, we can't just use the input sequences created for design verification to perform manufacturing test. Each chip must be designed to be fully and easily testable. Finding out that a chip is bad only after you have plugged it into a system is annoying at best and dangerous at worst. Customers are unlikely to keep using manufacturers who regularly supply bad chips.

Defects introduced during manufacturing range from the catastrophic—contamination that destroys every transistor on the wafer—to the subtle—a single misfired wire or a crystalline defect that kills only one transistor. While some bad chips can be found very easily, each chip must be thoroughly tested to find even subtle flaws that produce erroneous results only occasionally. Tests designed to exercise functionality and expose design bugs don't always uncover manufacturing defects. We use fault models to identify potential manufacturing problems and determine how they affect the chip's operation. The most common fault model is stuck-at-0/1: the defect causes a logic gate's output to be always 0 (or 1), independent of the gate's input values. We can often determine whether a logic gate's output is stuck even if we can't directly observe its outputs or control its inputs. We can generate a good set of manufacturing tests for the chip by assuming each logic gate's output is stuck at 0 (then 1) and finding an input to the chip which causes different outputs when the fault is present or absent. (Both the stuck-at-0/1 fault model and the assumption that faults occur only one at a time are simplifications, but they often are good enough to give good rejection of faulty chips.)

Unfortunately, not all chip designs are equally testable. Some faults may require long input sequences to expose; other faults may not be testable at all, even though they cause chip malfunctions that aren't covered by the fault model. Traditionally, chip designers have ignored testability problems, leaving them to a separate test engineer who must find a set of inputs to adequately test the chip. If the test engineer can't change the chip design to fix testability problems, his or her job becomes both difficult and unpleasant. The result is often poorly tested chips whose manufacturing problems are found only after the customer has plugged them into a system. Companies now recognize that the only way to deliver high-quality chips to customers is to make the chip designer responsible for testing, just as the designer is responsible for making the chip run at the required speed. Testability problems can often be fixed easily early in the design process at relatively little cost in area and performance. But modern designers must understand testability requirements, analysis techniques which identify hard-to-test sections of the design, and design techniques which improve testability.

1.4 Integrated Circuit Design Techniques

To make use of the flood of transistors given to us by Moore's law, we must design large, complex chips quickly. The obstacle to making large chips work correctly is complexity—many interesting ideas for chips have died in the swamp of details that must be made correct before the chip actually works. Integrated circuit design is hard because designers must juggle several different problems:

- **Multiple levels of abstraction**. IC design requires refining an idea through many levels of detail. Starting from a specification of what the chip must do, the designer must create an architecture which performs the required function, expand the architecture into a logic design, and further expand the logic design into a layout like the one in Figure 1-2. As you will learn by the end of this book, the specification-to-layout design process is a lot of work.

- **Multiple and conflicting costs**. In addition to drawing a design through many levels of detail, the designer must also take into account costs—not dollar costs, but criteria by which the quality of the design is judged. One critical cost is the speed at which the chip runs. Two architectures which execute the same function (multiplication, for example) may run at very different speeds. We will see that chip area is another critical design cost: the cost of manufacturing a chip is exponentially related to its area, and chips much larger than 1 cm^2 cannot be manufactured at all. Furthermore, if multiple cost criteria—such as area and speed requirements—must be satisfied, many design decisions will improve one cost metric at the expense of the other. Design is dominated by the process of balancing conflicting constraints.

- **Short design time**. In an ideal world, a designer would have time to contemplate the effect of a design decision. We do not, however, live in an ideal world. Chips which appear too late may make little or no money because competitors have snatched market share. Therefore, designers are under pressure to design chips as quickly as possible. Design time is especially tight in application-specific IC design, where only a few weeks may be available to turn a concept into a working ASIC.

Designers have developed two techniques to eliminate unnecessary detail: **hierarchical design** and **design abstraction.** Designers also make liberal use of computer-aided design tools to analyze and synthesize the design.

1.4.1 Hierarchical Design

Hierarchical design is commonly used in programming: a procedure is written not as a huge list of primitive statements but as calls to simpler procedures. Each procedure breaks down the task into smaller operations until each step is refined into a procedure simple enough to be written directly. This technique is commonly known as **divide-and-conquer**—the procedure's complexity is conquered by recursively breaking it down into manageable pieces.

Chip designers divide and conquer by breaking the chip into a hierarchy of components. As shown in Figure 1-5, a component consists of a **body** and a number of **pins**—this full adder has pins a, b, cin, cout, and sum. If we consider this full adder the definition of a **type**, we can make many **instances** of this type. Repeating commonly used components is very useful, for example, in building an n-bit adder from n full adders. We typically give each component instance a **name**. Since all components of the same type have the same pins, we refer to the pins on a particular component by giving the component instance name and pin name together; separating the instance and pin names by a dot is common practice. If we have two full adders, *add1* and *add2*, we can refer to *add1.sum* and *add2.sum* as distinct **terminals** (where a terminal is a component-pin pair).

*Figure 1-5: Pins
on a component.*

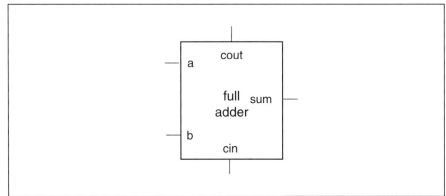

*Figure 1-6: A
hierarchical logic
design.*

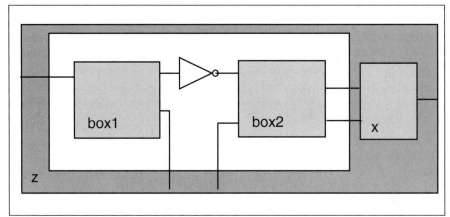

*Figure 1-7: A
component hierar-
chy.*

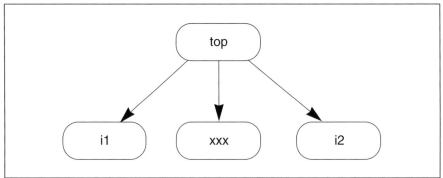

We can list the electrical connections which make up a circuit in either of two equivalent ways: a **net list** or a **component list**. A net list gives, for each net, the terminals connected to that net. Here is a net list for the top component of Figure 1-6:

```
net1: top.in1 i1.in
net2: i1.out xxx.B
topin1: top.n1 xxx.xin1
topin2: top.n2 xxx.xin2
botin1: top.n3 xxx.xin3
net3: xxx.out i2.in
outnet: i2.out top.out
```

A component list gives, for each component, the net attached to each pin. Here is a component list version of the same circuit:

```
top: in1=net1 n1=topin1 n2=topin2 n3=topin3 out=outnet
i1: in=net1 out=net2
xxx: xin1=topin1 xin2=topin2 xin3=botin1 B=net2 out=net3
i2: in=net3 out=outnet
```

Given one form of connectivity description, we can always transform it into the other form. Which format is used depends on the application—some searches are best performed net-by-net and others component-by-component. As an abuse of terminology, any file which describes electrical connectivity is usually called a **netlist file**, even if it is in component list format.

As shown in Figure 1-6, a logic design can be recursively broken into components, each of which is composed of smaller components until the design is described in terms of logic gates and transistors. In this figure, we have shown the type and instance as instance(type); there are two components of type A. Component ownership forms a hierarchy. The component hierarchy of Figure 1-6's example is shown in Figure 1-7. Each rounded box represents a component; an arrow from one box to another shows that the component pointed to is an element in the component which points to it. We may need to refer to several instance names to differentiate components. In this case, we may refer to either *top/i1* or *top/i2*, where we trace the component ownership from the most highest-level component and separate component names by slashes (/). (The resemblance of this naming scheme to UNIX file names is intentional—many design tools use files and directories to model component hierarchies.)

Each component is used as a black box—to understand how the system works, we only have to know each component's input-output behavior, not how that behavior is implemented inside the box. To design each black box, we build it out of smaller, simpler black boxes. The internals of each type define its behavior in terms of the components used to build it. If we know the behavior of our primitive components, such as transistors, we can infer the behavior of any hierarchically-described component.

People can much more easily understand a 10,000,000-transistor hierarchical design than the same design expressed directly as ten million transistors wired together. The hierarchical design helps you organize your thinking—the hierarchy organizes the function of a large number of transistors into a particular, easy-to-summarize function. Hierarchical design also makes it easier to reuse pieces of chips, either by modifying an old design to perform added functions or using by one component for a new purpose.

1.4.2 Design Abstraction

Design abstraction is critical to hardware system design. Hardware designers use multiple levels of design abstraction to manage the design process and ensure that they meet major design goals, such as speed and power consumption. The simplest example of a design abstraction is the logic gate. A logic gate is a simplification of the non-linear circuit used to build the gate: the logic gate accepts binary Boolean values. Some design tasks, such as accurate delay calculation, are hard or impossible when cast in terms of logic gates. However, other design tasks, such as logic optimization, are too cumbersome to be done on the circuit. We choose the design abstraction that is best suited to the design task.

We may also use higher abstractions to make first-cut decisions that are later refined using more detailed models: we often, for example, optimize logic using simple delay calculations, then refine the logic design using detailed circuit information. Design abstraction and hierarchical design aren't the same thing. A design hierarchy uses components at the same level of abstraction—an architecture built from Boolean logic functions, for example—and each level of the hierarchy adds complexity by adding components. The number of components may not change as it is recast to a lower level of abstraction—the added complexity comes from the more sophisticated behavior of those components.

The next example illustrates the large number of abstractions we can create for a very simple circuit.

Example 1-2: Layout and its abstractions

Layout is the lowest level of design abstraction for VLSI. The layout is sent directly to manufacturing to guide the patterning of the circuits. The configuration of rectangles in the layout determines the circuit topology and the characteristics of the components. However, the layout of even a simple circuit is sufficiently complex that we want to introduce more abstract representations that help us concentrate on certain key details.

Here is a layout for a simple circuit known as a dynamic latch:

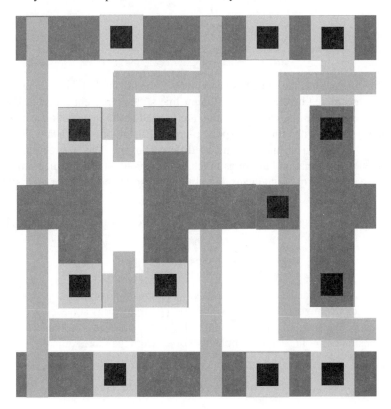

This layout contains rectangles that define the transistors, wires, and vias which connect the wires. The rectangles are drawn on several different layers corresponding to distinct layers of material or process steps in the integrated circuit.

Here is an abstraction for that layout: a stick diagram, which is a sketch of a layout:

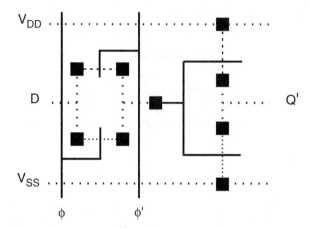

This stick diagram has the same basic structure as the layout, but the rectangles in the layout are abstracted here as lines. Different line styles represent different layers of material: metal, diffusion, etc. Transistors are formed at the intersection a line representing polysilicon with either a n-type or p-type diffusion line. The heavy dots represent vias, which connect material on two different layers. This abstraction conveys some physical information but not as much as the layout—the stick diagram reflects the relative positions of components, but not their absolute positions or their sizes.

Going one more step up the abstraction hierarchy, we can draw a transistor-level schematic:

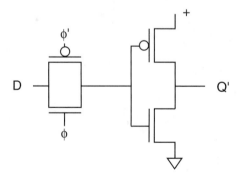

This formulation is not intended to describe the physical layout of the circuit at all—though the placement of transistors may resemble the organization of the transistors in the layout, that is a matter of convenience. The intent of the schematic is to describe the major electrical components and their interconnections.

We can go one step higher in the abstraction hierarchy to draw a **mixed schematic**:

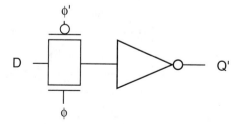

This is called mixed because it is built from components at different levels of abstraction: not only transistors, but also an inverter, which is in turn built from transistors. The added abstraction of the inverter helps to clarify the organization of the circuit.

The next example shows how a slightly more complex hardware design is built up from circuit to complex logic.

Example 1-3: Digital logic abstractions

A transistor circuit for an inverter is relatively small. We can determine its behavior over time, representing input and output values as continuous voltages to accurately determine its delay:

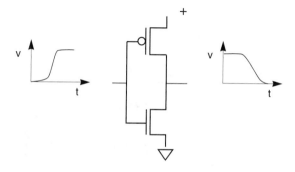

We can use transistors to build more complex functions like the full adder. At this point, we often simplify the circuit behavior to 0 and 1 values which may be delayed in continuous time:

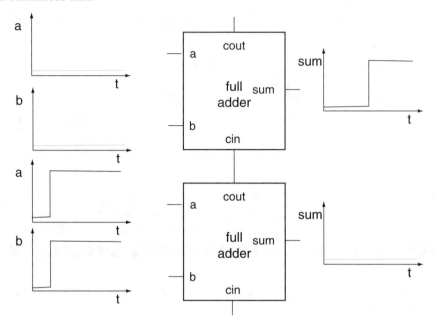

As circuits get bigger, it becomes harder to figure out their continuous time behavior. However, by making reasonable assumptions, we can determine approximate delays through circuits like adders. Since we are interested in the delay through adders, the ability to make simplifying assumptions and calculate reasonable delay estimates is very important.

When designing large register-transfer systems, such as data paths, we may abstract one more level to generic adders:

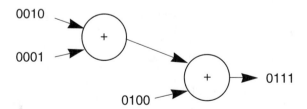

At this point, since we don't know how the adders are built, we don't have any delay information. These components are pure combinational elements—they produce an output value given an input value. The adder abstraction helps us concentrate on the proper function before we worry about the details of performance.

Figure 1-8 shows a typical design abstraction ladder for digital systems:

- **Specification**. The customer specifies what the chip should do, how fast it should run, etc. A specification is almost always incomplete—it is a set of requirements, not a design.

- **Behavior.** The behavioral description is much more precise than the specification. Specifications are usually written in English, while behavior is generally modeled as some sort of executable program.

- **Register-transfer**. The system's time behavior is fully-specified—we know the allowed input and output values on every clock cycle—but the logic isn't specified as gates. The system is specified as Boolean functions stored in abstract memory elements. Only the vaguest delay and area estimates can be made from the Boolean logic functions.

- **Logic**. The system is designed in terms of Boolean logic gates, latches, and flip-flops. We know a lot about the structure of the system but still cannot make extremely accurate delay calculations.

- **Circuit**. The system is implemented as transistors.

- **Layout**. The final design for fabrication. Parasitic resistance and capacitance can be extracted from the layout to add to the circuit description for more accurate simulation.

Design always requires working down from the top of the abstraction hierarchy and up from the least abstract description. Obviously, work must begin by adding detail to the abstraction—**top-down** design adds functional detail. But top-down design decisions are made with limited information: there may be several alternative designs at each level of abstraction; we want to choose the candidate which best fits our speed, area, and power requirements. We often cannot accurately judge those costs until we have an initial design. **Bottom-up** analysis and design percolates cost information back to higher-levels of abstraction; for instance, we may use more accurate delay information from the circuit design to redesign the logic. Experience will help you judge costs before you complete the implementation, but most designs require cycles of top-down design followed by bottom-up redesign.

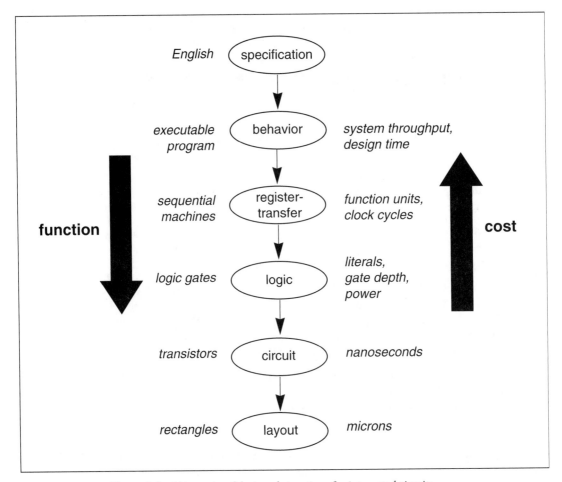

Figure 1-8: A hierarchy of design abstractions for integrated circuits.

1.4.3 Computer-Aided Design

The only realistic way to design chips given performance and design time constraints is to automate the design process, using **computer-aided design (CAD)** tools which automate parts of the design process. Using computers to automate design, when done correctly, actually helps us solve all three problems: dealing with multiple levels of abstraction is easier when you are not absorbed in the details of a particular design step; computer programs, because they are more methodical, can do a better job of analyzing cost trade-offs; and, when given a well-defined task, computers can work much more quickly than humans.

Computer-aided design tools can be categorized by the design task they handle. The simplest of CAD tool handles design entry—for example, an interactive schematic drawing package. Design entry tools capture a design in machine-readable form for use by other programs, and they often allow easier modification of a design, but they don't do any real design work.

Analysis and verification tools are more powerful. The Spice circuit simulator, for example, solves the differential equations which govern how the circuit responds to an input waveform over time. Such a program doesn't tell us how to change the circuit to make it do what we want, but many analysis tasks are too difficult to perform manually.

Synthesis tools actually create a design at a lower level of abstraction from a higher level description. Some layout synthesis programs can synthesize a layout from a circuit description like that in Figure 1-2. Using computers for design is not a panacea. Computer programs cannot now, nor are they ever likely to be able to transform marketing brochures directly into finished IC designs. Designers will always be necessary to find creative designs and to perform design tasks which are too subtle to be left to algorithms.

Both hierarchical design and design abstraction are as important to CAD tools as they are to humans—the most powerful synthesis and analysis tools operate on a very restricted design model. CAD tools can help us immensely with pieces of the design task, but algorithms that have the detailed knowledge required to solve one design problem usually do not have the broad range of data required to balance broad requirements.

CAD tools must be used judiciously by a human designer to be most effective. Nonetheless, CAD tools are an essential part of the future of IC design because they are the only way to manage the complexity of designing large integrated circuits. Manual design of a ten-million transistor chip, or even a 10,000 transistor chip, quickly overwhelms the designer with decisions. Not all decisions are equally important—some may have only a minor effect on chip size and speed while others may profoundly change the chip's costs. By concentrating on the wrong decisions, a designer may cause problems that are not easily correctable later. CAD tools, by automating parts of the design process, help the designer eliminate mundane decisions quickly and concentrate on the make-or-break problems posed by the chip.

For example, long wires can introduce excessive delay, increase power consumption, and create opportunities for crosstalk. Such problems can be found by a program that analyzes delays through the chip, but when designing a chip by hand, it may be easy to miss this single connection, and the error will not be found until the chip comes back from fabrication. CAD tools are particularly important for evaluating complex situations in which solving one problem creates other problems—for example, making one wire shorter makes other wires longer. When two constraints compete, solutions to problems may not be so easy. Making one part of the design faster may, for example, make another part of the design unacceptably large and slow. CAD tools help us solve these problems with analytical methods to evaluate the cost of decisions and synthesis methods that let us quickly construct a candidate solution to a problem. Evaluation of candidate designs is critical to designing systems to satisfy multiple costs because optimizing a complete system cannot be done simply by optimizing all the parts individually—making each part in a chip run as fast as possible in isolation by no means ensures that the entire chip will run as fast as possible. Using CAD tools to propose and analyze solutions to problems lets us examine much larger problems than is possible by hand.

1.5 A Look into the Future

Moore's Law is likely to hold for quite some time to come. In a short amount of time from this writing, we will be able to design and fabricate in large quantities circuits with several hundred million transistors. We are already in the age of deep-submicron VLSI—the typical fabrication process constructs transistors that are much smaller than one micron in size. As we move toward even smaller transistors and even more transistors per chip, several types of challenges must be faced.

The first challenge is interconnect. In the early days of the VLSI era, wires were recognized to be important because they occupied valuable chip area, but properly-designed wiring did not pose a bottleneck to performance. Today, wires cannot be ignored—the delay through a wire can easily be longer than the delay through the gate driving it. And because the parasitic components of wires are so significant, crosstalk between signals on wires can cause major problems as well. Proper design methodologies and careful analysis are keys to taming the problems introduced by interconnect.

Another challenge is power consumption. Power consumption is a concern on every large chip because of the large amount of activity generated by so many transistors. Excessive power consumption can make a chip so hot that it becomes unreliable. Careful analysis of power consumption at all stages of design is essential for keeping power consumption within acceptable limits.

And we must certainly face the challenge of design complexity as we start to be able to create complete systems-on-silicon. In about ten years, we will be able to fabricate chips with a billion transistors—a huge design task at all levels of abstraction, ranging from layout and circuit to architecture. Over the long run, VLSI designers will have to become even more skilled at programming as some fraction of the system is implemented as on-chip software. We will look at systems-on-chips in more detail in Section 8.6.

1.6 Summary

Integrated circuit manufacturing is a key technology—it makes possible a host of important, useful new devices. ICs help us make better digital systems because they are small, stingy with power, and cheap. However, the temptation to build ever more complex systems by cramming more functions onto chips leads to an enormous design problem. Integrated circuits are so complex that the only way to effectively design them is to use computers to automate parts of the design process, a situation not unlike that in Isaac Asimov's robot stories, where positronic brains are employed to design the next, more advanced generation of robot brains. But humans are not out of control of the design process—by giving up control of some details, you can obtain a clearer view of the broad horizon and avoid problems that don't lie exactly at your feet.

1.7 References

The data points in the Moore's Law chart of Figure 1-3 were taken from articles in the *IEEE Journal of Solid State Circuits* (JSSC) and from a 1967 survey article by Petritz [Pet67]. The October issue of JSSC is devoted each year to logic and memory—those articles describe state-of-the-art integrated circuits. Business magazines and newspapers, such as *The Wall Street Journal*, *Business Week*, *Fortune*, and *Forbes* provide thorough coverage of the semiconductor industry. Following business developments in the industry provides valuable insight into the economic forces which shape technical decisions.

1.8 Problems

1-1. How many CPU chips are sold each year? Given that figure, how much dynamic RAM should be sold each year?

1-2. Plot the transistor count data from Example 1-1 as a function of introduction date.

1-3. Hierarchy helps you reuse pieces of a design. The typical large CPU has about one million transistors; assume the layout for one transistor requires drawing five rectangles.

> a) How long would it take you to complete a CPU layout if you drew each rectangle separately?
>
> b) How long would it take you if half the transistors consisted of cache memory, which was designed from a six-transistor cell which could be replicated to complete the cache?
>
> c) How long would it take if the non-cache transistors were implemented in 32 identical bit slices, requiring one copy of the bit slice to be drawn, with the rest created by replication?

1-4. Draw a logic diagram for a full-adder (consult Chapter 6 if you don't know how a full adder works). Name each logic gate. Draw a four-bit adder from four full adders. Name each component in the four-bit adder and define the four-bit adder as a type. Draw the component hierarchy, showing the four-bit adder, the full adder, and the logic gates; when a component is repeated, you can draw its sub-components once and refer to them elsewhere in the diagram.

2

Transistors and Layout

Highlights:

Fabrication methods.

Transistor structures.

Characteristics of transistors and wires.

Design rules.

Layout design.

2.1 Introduction

We will start our study of VLSI design by learning about transistors and wires and how they are fabricated. The basic properties of transistors are clearly important for logic design. Going beyond a minimally-functional logic circuit to a high-performance design requires the consideration of **parasitic circuit elements**—capacitance and resistance. Those parasitics are created as necessary by-products of the fabrication process which creates the wires and transistors, which gives us a very good reason to understand the basics of how integrated circuits are fabricated. We will also study the rules which must be obeyed when designing the masks used to fabricate a chip and the basics of layout design.

Our first step is to understand the basic fabrication techniques as described in Section 2.2. This material will describe how the basic structures for transistors and wires are made. We will then study transistors and wires, both as integrated structures and as circuit elements, in Section 2.3 and Section 2.4, respectively. We will study design rules for layout in Section 2.5. Finally, we will introduce some basic concepts and tools for layout design in Section 2.6.

2.2 Fabrication Processes

We need to study fabrication processes and the design rules that govern layout. Examples are always helpful. We will use as our example the SCMOS rules, which have been defined by MOSIS, the MOS Implementation Service. (MOSIS is supported by the United States National Science Foundation. Similar services, such as Euro-Chip/EuroPractice in the European Community, VDEC in Japan, and CIC in Taiwan, serve educational VLSI needs in other countries.) SCMOS is unusual in that it is not a single fabrication process, but a collection of rules that hold for a family of processes. Using generic technology rules gives greater flexibility in choosing a manufacturer for your chips. It also means that the SCMOS technology is less aggressive than any particular fabrication process developed for some special purpose—some manufacturers may emphasize transistor switching speed, for example, while others emphasize the number of layers available for wiring. We will point out advanced technology features that are not part of the SCMOS specification but may be found in a particular fabrication process.

2.2.1 Overview

A cross-section of an integrated circuit is shown in Figure 2-1. Integrated circuits are built on a silicon **substrate** provided by the wafer. Figure 2-2 shows a technician holding 200 mm and 300 mm wafers. Wafer sizes have steadily increased over the years. Larger wafers mean more chips wafer and higher productivity.

Components are formed by a combination of processes:

- **doping** the substrate with impurities to create areas such as the n+ and p+ regions;

- adding or cutting away insulating glass (**silicon dioxide**, or SiO_2) on top of the substrate;

- adding wires made of polycrystalline silicon (**polysilicon**, also known as **poly**) or metal, insulated from the substrate by SiO_2.

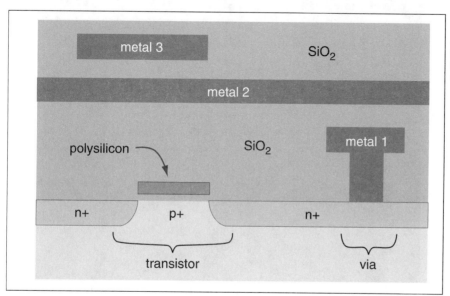

Figure 2-1: *Cross-section of an integrated circuit.*

A pure silicon substrate contains equal numbers of two types of electrical carriers: electrons and holes. While we will not go into the details of device physics here, it is important to realize that the interplay between electrons and holes is what makes transistors work. The goal of doping is to create two types of regions in the substrate: an **n-type** region which contains primarily electrons and a **p-type** region which is dominated by holes. (Heavily doped regions are referred to as n+ and p+.) Transistor action occurs at properly formed boundaries between n-type and p-type regions.

The n-type and p-type regions can be used to make wires as well as transistors, but polysilicon (which is also used to form transistor gates) and metal are the primary materials for wiring together transistors because of their superior electrical properties. There may be several levels of metal wiring to ensure that enough wires can be made to create all the necessary connections. Glass insulation lets the wires be fabricated on top of the substrate using processes like those used to form transistors. The integration of wires with components (the invention of Robert Noyce), which eliminates the

*Figure 2-2: A VLSI
manufacturing line
(courtesy **IBM**).*

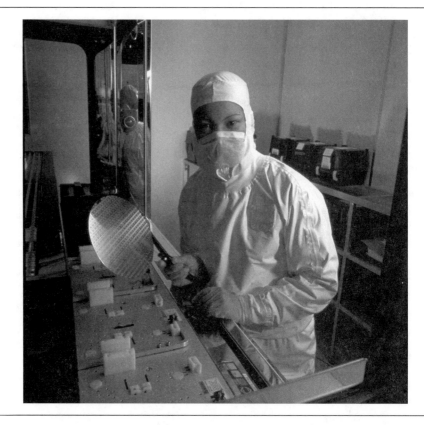

need to mechanically wire together components on the substrate, was one of the key inventions that made the integrated circuit feasible.

The key figure of merit for a fabrication process is the size—more specifically, the channel length—of the smallest transistor it can manufacture. Transistor size helps determine both circuit speed and the amount of logic that can be put on a single chip. Fabrication technologies are usually identified by their minimum transistor length, so a process which can produce a transistor with a 0.5 µm minimum channel length is called a 0.5 µm process. When we discuss design rules, we will recast the on-chip dimensions to a scalable quantity λ. Our $\lambda = 0.25$ µm CMOS process is also known as a 0.5 µm CMOS process; if λ is not referred to explicitly, the size of the process gives the minimum channel length.

2.2.2 Fabrication Steps

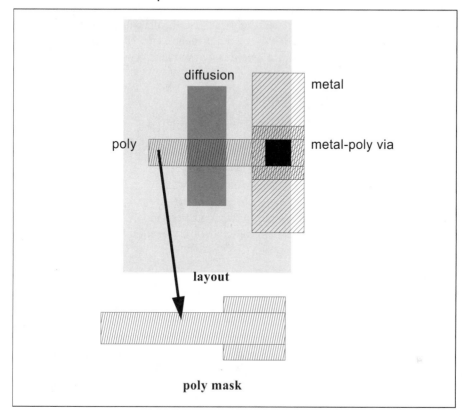

Figure 2-3: The relationship between layouts and fabrication masks.

Features are patterned on the wafer by a photolithographic process; the wafer is covered with light-sensitive material called **photoresist**, which is then exposed to light with the proper pattern. The patterns left by the photoresist after development can be used to control where SiO_2 is grown or materials are placed on the surface of the wafer.

A layout contains summary information about the patterns to be made on the wafer. Photolithographic processing steps are performed using **masks** which are created from the layout information supplied by the designer. In simple processes there is roughly one mask per layer in a layout, though in more complex processes some masks may be built from several layers while one layer in the layout may contribute to several masks. Figure 2-3 shows a simple layout and the mask used to form the polysilicon pattern.

Transistors are fabricated within regions called **tubs** or **wells**: an n-type transistor is built in a p-tub, and a p-type transistor is built in an n-tub. The wells prevent undesired conduction from the drain to the substrate. (Remember that the transistor type refers to the minority carrier which forms the inversion layer, so an n-type transistor pulls electrons out of a p-tub.) There are three ways to form tubs in a substrate:

- start with a p-doped wafer and add n-tubs;

- start with an n-doped wafer and add p-tubs;

- start with an undoped wafer and add both n- and p-tubs.

CMOS processes were originally developed from nMOS processes, which use p-type wafers into which n-type transistors are added. However, the **twin-tub process**, which uses an undoped wafer, has become the most commonly used process because it produces tubs with better electrical characteristics. We will therefore use a twin-tub process as an example.

Figure 2-4 illustrates important steps in a twin-tub process. Details can vary from process to process, but these steps are representative. The first step is to put tubs into the wafer at the appropriate places for the n-type and p-type wafers. Regions on the wafer are selectively doped by implanting ionized dopant atoms into the material, then heating the wafer to heal damage caused by ion implantation and further move the dopants by diffusion. The tub structure means that n-type and p-type wires cannot directly connect. Since the two diffusion wire types must exist in different type tubs, there is no way to build a via which can directly connect them. Connections must be made by a separate wire, usually metal, which runs over the tubs.

The next steps form an oxide covering of the wafer and the polysilicon wires. The oxide is formed in two steps: first, a thick field oxide is grown over the entire wafer. The field oxide is etched away in areas directly over transistors; a separate step grows a much thinner oxide which will form the insulator of the transistor gates. After the field and thin oxides have been grown, the polysilicon wires are formed by depositing polysilicon crystalline directly on the oxide.

Note that the polysilicon wires have been laid down before the diffusion wires were made—that order is critical to the success of MOS processing. Diffusion wires are laid down immediately after polysilicon deposition to create **self-aligned** transistors—the polysilicon masks the formation of diffusion wires in the transistor channel. For the transistor to work properly, there must be no gap between the ends of the source and drain diffusion regions and the start of the transistor gate. If the diffusion were laid down first with a hole left for the polysilicon to cover, it would be very dif-

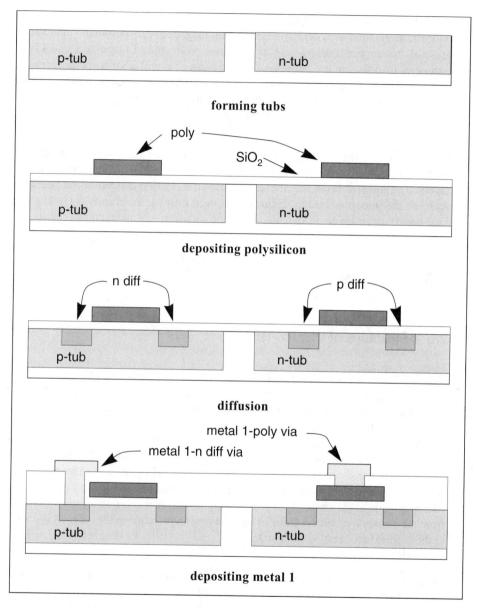

Figure 2-4: *Steps in processing a wafer.*

ficult to hit the gap with a polysilicon wire unless the transistor were made very large. Self-aligned processing allows much smaller transistors to be built.

After the diffusions are complete, another layer of oxide is deposited to insulate the polysilicon and metal wires. Aluminum has long been the dominant interconnect material, but copper has now moved into mass production. Copper is a much better conductor than aluminum, but even trace amounts of it will destroy the properties of semiconductors. Chips with copper interconnect include a special protection layer between the substrate and the first layer of copper. That layer prevents the copper from entering the substrate during processing.

Holes are cut in the field oxide where vias to the substrate are desired. The metal 1 is then deposited where desired. The metal fills the cuts to make connections between layers. The metal 2 layer requires an additional oxidation/cut/deposition sequence. After all the important circuit features have been formed, the chip is covered with a final **passivation layer** of SiO_2 to protect the chip from chemical contamination.

2.3 Transistors

2.3.1 Structure of the Transistor

Figure 2-5 shows the cross-section of an n-type MOS transistor. (The name MOS is an anachronism. The first such transistors used a metal wire for a gate, making the transistor a sandwich of metal, silicon dioxide, and the semiconductor substrate. Even though transistor gates are now made of polysilicon, the name MOS has stuck.) An n-type transistor is embedded in a p-type substrate; it is formed by the intersection of an n-type wire and a polysilicon wire. The region at the intersection, called the **channel**, is where the transistor action takes place. The channel connects to the two n-type wires which form the source and drain, but is itself doped to be p-type. The insulating silicon dioxide at the channel (called the **gate oxide**) is much thinner than it is away from the channel (called the **field oxide**); having a thin oxide at the channel is critical to the successful operation of the transistor.

Figure 2-6 shows a photomicrograph of an MOS transistor's cross-section. The photograph makes clear just how thin and sensitive the gate oxide is. The gate of this transistor is made of a sandwich of polysilicon and silicide. The sandwich's resistance is much lower than that of straight polysilicon.

The transistor works as a switch because the gate-to-source voltage modulates the amount of current that can flow between the source and drain. When the gate voltage (V_{gs}) is zero, the p-type channel is full of holes, while the n-type source and drain

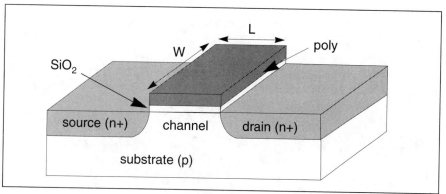

Figure 2-5:
Cross-section of
an n-type transis-
tor.

contain electrons. The p-n junction at the source terminal forms a diode, while the junction at the drain forms a second diode that conducts in the opposite direction. As a result, no current can flow from the source to the drain.

As V_{gs} rises above zero, the situation starts to change. While the channel region contains predominantly p-type carriers, it also has some n-type carriers. The positive voltage on the polysilicon which forms the gate attracts the electrons. Since they are stopped by the gate oxide, they collect at the top of the channel along the oxide boundary. At a critical voltage called the **threshold voltage** (V_t), enough electrons have collected at the channel boundary to form an **inversion layer**—a layer of electrons dense enough to conduct current between the source and the drain.

The size of the channel region is labeled relative to the direction of current flow: the channel **length** (L) is along the direction of current flow between source and drain, while the **width** (W) is perpendicular to current flow. The amount of current flow is a function of the W/L ratio, for the same reasons that bulk resistance changes with the object's width and length: widening the channel gives a larger cross-section for conduction, while lengthening the channel increases the distance current must flow through the channel. Since we can choose W and L when we draw the layout, we can very simply design the transistor current magnitude.

P-type transistors have identical structures but complementary materials: trade p's and n's in Figure 2-5 and you have a picture of a p-type transistor. The p-type transistor conducts by forming an inversion region of holes in the n-type channel; therefore, the gate-to-source voltage must be negative for the transistor to conduct current.

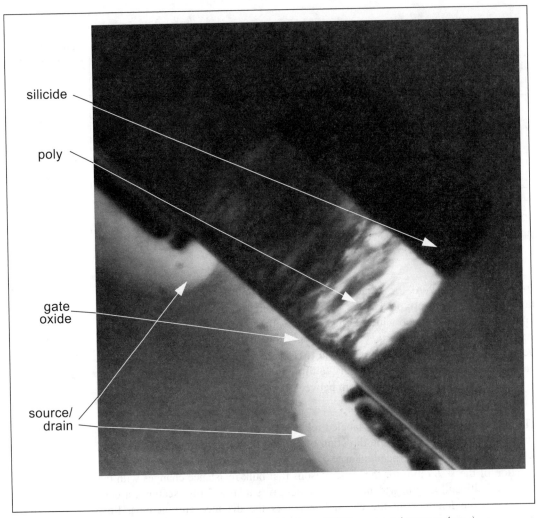

silicide

poly

gate
oxide

source/
drain

Figure 2-6: *Photomicrograph of a submicron MOS transistor (courtesy Agere).*

Example 2-1: Layout of n-type and p-type transistors

The basic layout of an n-type transistor is simple:

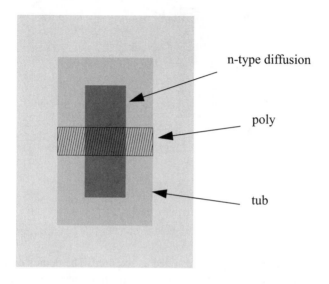

n-type diffusion

poly

tub

This layout is of a minimum-size transistor. Current flows through the channel vertically.

The layout of a p-type transistor is very similar:

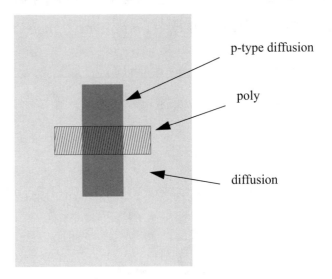

p-type diffusion

poly

diffusion

In both cases, the tub rectangles are added as required. The details of which tub must be specified vary from process to process; many designers use simple programs to generate the tubs required around rectangles.

Fabrication engineers may sometimes refer to the **drawn length** of a transistor. Photolithography steps may affect the length of the channel. As a result, the actual channel length may not be the drawn length. The drawn length is usually the parameter of interest to the digital designer, since that is the size of rectangle that must be used to get a transistor of the desired size.

We can also draw a wider n-type transistor, which delivers more current:

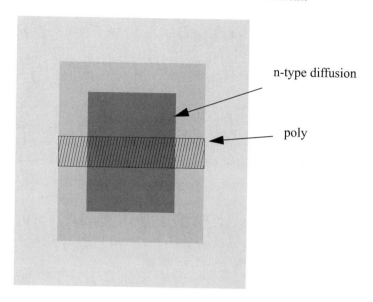

n-type diffusion

poly

2.3.2 A Simple Transistor Model

The behavior of both n-type and p-type transistors is described by two equations and two physical constants; the sign of one of the constants distinguishes the two types of transistors. The variables that describe a transistor's behavior, some of which we have already encountered, are:

- V_{gs}—the gate-to-source voltage;

- V_{ds}—the drain-to-source voltage (remember that $V_{ds} = -V_{sd}$);

- I_d—the current flowing between the drain and source.

The constants that determine the magnitude of source-to-drain current in the transistor are:

- V_t—the transistor threshold voltage, which is positive for an n-type transistor and negative for a p-type transistor;

- k'—the transistor transconductance, which is positive for both types of transistors;

- W/L—the width-to-length ratio of the transistor.

Both V_t and k' are measured, either directly or indirectly, for a fabrication process. W/L is determined by the layout of the transistor, but since it does not change during operation, it is a constant of the device equations.

Figure 2-7: *The I_d curves of an n-type transistor.*

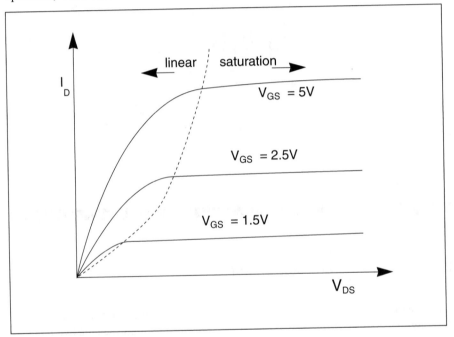

The equations that govern the transistor's behavior are traditionally written to show the drain current as a function of the other parameters. A reasonably accurate model for the transistor's behavior, written in terms of the drain current I_d, divides operation into **linear** and **saturated** [Yan78]. For an n-type transistor, we have:

- *Linear region* $V_{ds} < V_{gs}\text{-}V_t$:

$$I_d = k'\frac{W}{L}\left[(V_{gs}\text{-}V_t)\left(V_{ds}\text{-}\frac{1}{2}V_{ds}^2\right)\right]$$

(EQ 2-1)

- *Saturated region* $V_{ds} \geq V_{gs}\text{-}V_t$:

$$I_d = \frac{1}{2}k'\frac{W}{L}(V_{gs}\text{-}V_t)^2$$

(EQ 2-2)

For a p-type transistor, the drain current is negative and the device is on when V_{gs} is below the device's negative threshold voltage. Figure 2-7 plots these equations over some typical values for an n-type device. Each curve shows the transistor current as V_{gs} is held constant and V_{ds} is swept from 0 V to a large voltage.

The transistor's switch action occurs because the density of carriers in the channel depends strongly on the gate-to-substrate voltage. For $|V_{gs}| < |V_t|$, there are not enough carriers in the inversion layer to conduct an appreciable current. (To see how much current is conducted in the subthreshold region, check Section 2.3.5.) Beyond that point and until saturation, the number of carriers is directly related to V_{gs}: the greater the gate voltage applied, the more carriers are drawn to the inversion layer and the greater the transistor's conductivity.

The relationship between *W/L* and source-drain current is equally simple. As the channel width increases, more carriers are available to conduct current. As channel length increases, however, the drain-to-source voltage diminishes in effect. V_{ds} is the potential energy available to push carriers from drain to source; as the distance from drain to source increases, it takes longer to push carriers across the transistor for a fixed V_{ds}, reducing current flow.

	k'	V_t
n-type	$k'_n = 73\mu A/V^2$	$0.7V$
p-type	$k'_p = 21\mu A/V^2$	$-0.8V$

Table 2-1 *Typical transistor parameters for our 0.5 µm process.*

Table 2-1 shows typical values of k' and V_t for a 0.5 µm process. The next example calculates the current through a transistor.

Example 2-2: Current through a transistor

A minimum-size transistor in the SCMOS rules is formed by a $L = 2\,\lambda$ and $W = 3\,\lambda$. Given this size of transistor and the 0.5 μm transistor characteristics, the current through a minimum-sized n-type transistor at the boundary between the linear and saturation regions when the gate is at the low voltage $V_{gs} = 2\text{V}$ would be

$$I_d = \frac{1}{2}\left(73\frac{\mu A}{V^2}\right)\left(\frac{3\lambda}{2\lambda}\right)(2V\text{-}0.7V)^2 = 93\mu A \quad .$$

The saturation current when the transistor's gate is connected to a 5 V power supply would be

$$I_d = \frac{1}{2}\left(73\frac{\mu A}{V}\right)\left(\frac{3\lambda}{2\lambda}\right)(5V\text{-}0.7V)^2 = 1.0mA \quad .$$

2.3.3 Transistor Parasitics

Real devices have parasitic elements that are necessary artifacts of the device structure. The transistor itself introduces significant **gate capacitance**, C_g. This capacitance, which comes from the parallel plates formed by the poly gate and the substrate, forms the majority of the capacitive load in small logic circuits; $C_g = 0.9\text{fF}/\mu m^2$ for both n-type and p-type transistors in a typical 2 μm process. The total gate capacitance for a transistor is computed by measuring the area of the active region (or $W \times L$) and multiplying the area by the unit capacitance C_g.

We may, however, want to worry about the **source/drain overlap capacitances**. During fabrication, the dopants in the source/drain regions diffuse in all directions, including under the gate as shown in Figure 2-8. The source/drain overlap region tends to be a larger fraction of the channel area in deep submicron devices. The overlap region is independent of the transistor length, so it is usually given in units of Far-

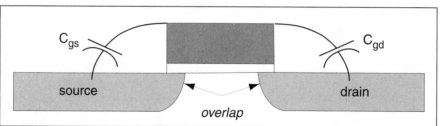

Figure 2-8:
Parasitic capaci-
tances from the
gate to the
source/drain-
overlap regions.

ads per unit gate width. Then the total source overlap capacitance for a transistor
would be

$$Cgs = C_{ol}W. \qquad \textbf{(EQ 2-3)}$$

There is also a **gate/bulk overlap capacitance** due to the overhang of the gate past
the channel and onto the bulk.

The source and drain regions also have a non-trivial capacitance to the substrate and a
very large resistance. Circuit simulation may require the specification of source/drain
capacitances and resistances. However, the techniques for measuring the source/drain
parasitics at the transistor are the same as those used for measuring the parasitics of
long diffusion wires. Therefore, we will defer the study of how to measure these parasitics to Section 2.4.1.

2.3.4 Tub Ties and Latchup

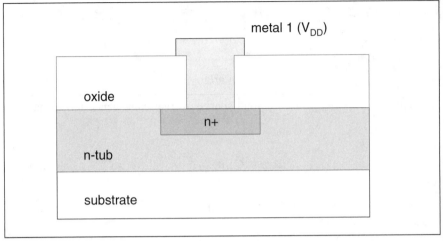

An MOS transistor is actually a four-terminal device, but we have up to now ignored the electrical connection to the substrate. The substrates underneath the transistors must be connected to a power supply: the p-tub (which contains n-type transistors) to V_{SS} and the n-tub to V_{DD}. These connections are made by special vias called **tub ties**.

Figure 2-9 shows the cross-section of a tub tie connecting to an n-tub and Figure 2-10 shows a tub tie next to a via and an n-type transistor. The tie connects a metal wire connected to the V_{DD} power supply directly to the substrate. The connection is made through a standard via cut. The substrate underneath the tub tie is heavily doped with n-type dopants (denoted as n+) to make a low-resistance connection to the tub. The SCMOS rules make the conservative suggestion that tub ties be placed every one to two transistors. Other processes may relax that rule to allow tub ties every four to five transistors. Why not place one tub tie in each tub—one tub tie for every 50 or 100 transistors? Using many tub ties in each tub makes a low-resistance connection between the tub and the power supply. If that connection has higher resistance, parasitic bipolar transistors can cause the chip to **latch-up**, inhibiting normal chip operation.

Figure 2-11 shows a chip cross-section which might be found in an inverter or other logic gate. The MOS transistor and tub structures form parasitic bipolar transistors: npn transistors are formed in the p-tub and pnp transistors in the n-tub. Since the tub regions are not physically isolated, current can flow between these parasitic transistors along the paths shown as wires. Since the tubs are not perfect conductors, some

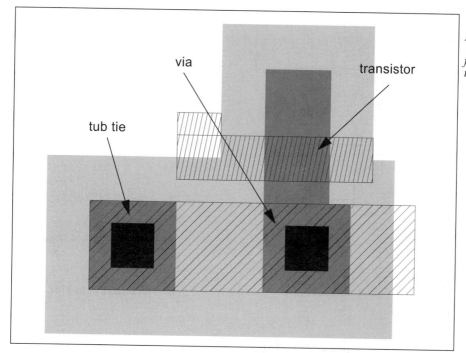

Figure 2-10: *A layout section featuring a tub tie.*

tub tie

via

transistor

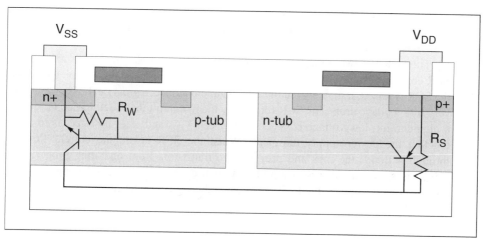

Figure 2-11: *Parasitics that cause latch-up.*

of these paths include parasitic resistors; the key resistances are those between the power supply terminals and the bases of the two bipolar transistors.

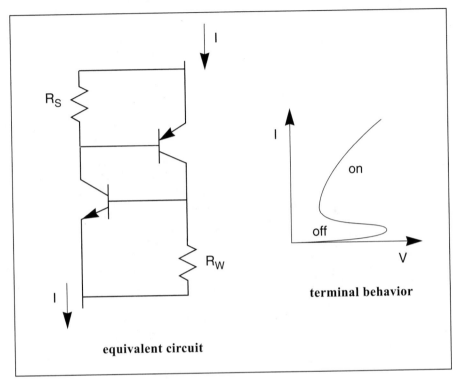

Figure 2-12:
Characteristics of
a silicon-con-
trolled rectifier.

equivalent circuit

terminal behavior

The parasitic bipolar transistors and resistors create a parasitic **silicon-controlled rectifier**, or SCR. The schematic for the SCR and its behavior are shown in Figure 2-12. The SCR has two modes of operation. When both bipolar transistors are off, the SCR conducts essentially no current between its two terminals. As the voltage across the SCR is raised, it may eventually turn on and conducts a great deal of current with very little voltage drop. The SCR formed by the n- and p-tubs, when turned on, forms a high-current, low-voltage connection between V_{DD} and V_{SS}. Its effect is to short together the power supply terminals. When the SCR is on, the current flowing through it floods the tubs and prevents the transistors from operating properly. In some cases, the chip can be restored to normal operation by disconnecting and then reconnecting the power supply; in other cases the high currents cause permanent damage to the chip.

The switching point of the SCR is controlled by the values of the two power supply resistances R_s and R_w. Each bipolar transistor in the SCR turns on when its base-to-emitter voltage reaches 0.7 V; that voltage is controlled by the voltage across the two resistors. The higher the resistance, the less stray current through the tub is required to cause a voltage drop across the parasitic resistance that can turn on the associated transistor. Adding more tub ties reduces the values of R_s and R_w. The maximum distance between tub ties is chosen to ensure that the chip will not latch-up during normal operation.

2.3.5 Advanced Transistor Characteristics

In order to better understand the transistor, we will derive the basic device characteristics that were stated in Section 2.3.2. Along the way we will be able to identify some second-order effects that can become significant when we try to optimize a circuit design.

The parallel plate capacitance of the gate determines the characteristics of the channel. We know from basic physics that the parallel-plate oxide capacitance per unit area (in units of Farads per cm^2) is

$$C_{ox} = \varepsilon_{ox}/x_{ox},$$ **(EQ 2-4)**

where ε_{ox} is the permittivity of silicon dioxide (about $3.9\varepsilon_0$, where ε_0, the permittivity of free space, is $8.854 \times 10^{-14} F/cm$) and x_{ox} is the oxide thickness in centimeters.

charge of an electron	q	$1.6 \times 10^{-19} C$
Si intrinsic carrier concentration	n_i	$1.45 \times 10^{10} C/cm^3$
permittivity of free space	ε_0	$8.854 \times 10^{-14} F/cm^2$
permittivity of Si	ε_{Si}	$11.9\varepsilon_0$
thermal voltage (300K)	kT/q	$0.026 V$

Table 2-2 *Values of some physical constants.*

The intrinsic carrier concentration of silicon is denoted as n_i. N-type doping concentrations are written as N_d (donor) while p-type doping concentrations are written as N_a (acceptor). Table 2-2 gives the values of some important physical constants.

Applying a voltage of the proper polarity between the gate and substrate pulls minority carriers to the lower plate of the capacitor, namely the channel region near the gate oxide. The threshold voltage is defined as the voltage at which the number of minority carriers (electrons in an n-type transistor) in the channel region equals the number of majority carriers in the substrate. (This actually defines the **strong threshold condition**.) So the threshold voltage may be computed from the component voltages which determine the number of carriers in the channel. The threshold voltage (assuming that the source/substrate voltage is zero) has four major components:

$$V_{t0} = V_{fb} + \phi_s + \frac{Q_b}{C_{ox}} + V_{II} .$$

(EQ 2-5)

Let us consider each of these terms.

- The first component, V_{fb}, is the **flatband voltage**, which in modern processes has two main components:

$$V_{fb} = \Phi_{gs} - (Q_f / C_{ox})$$

(EQ 2-6)

Φ_{gs} is the difference in work functions between the gate and substrate material, while Q_f is the fixed surface charge. (Trapped charge used to be a significant problem in MOS processing which increased the flatband voltage and therefore the threshold voltage. However, modern processing techniques control the amount of trapped charge.)

If the gate polysilicon is n-doped at a concentration of N_{dp}, the formula for the work function difference is

$$\Phi_{gs} = -\frac{kT}{q} \ln\left(\frac{N_a N_{dp}}{n_i^2}\right) .$$

(EQ 2-7)

If the gate is p-doped at a concentration of N_{ap}, the work function difference is

$$\Phi_{gs} = \frac{kT}{q} \ln\left(\frac{N_{ap}}{N_a}\right) .$$

(EQ 2-8)

- The second term is the surface potential. At the threshold voltage, the surface potential is twice the **Fermi potential** of the substrate:

$$\phi_s \approx 2|\phi_F| = 2\frac{kT}{q}\ln\frac{N_a}{n_i}. \tag{EQ 2-9}$$

- The third component is the voltage across the parallel plate capacitor. The value of the charge on the capacitor Q_b is

$$\sqrt{2q\varepsilon_{si}N_a\phi_s}. \tag{EQ 2-10}$$

  (We will not derive this value, but the square root comes from the value for the depth of the depletion region.)

- An additional ion implantation step is also performed to adjust the threshold voltage—the fixed charge of the ions provides a bias voltage on the gate. The voltage adjustment V_{II} has the value qD_I/C_{ox}, where D_I is the ion implantation concentration; the voltage adjustment may be positive or negative, depending on the type of ion implanted.

When the source/substrate voltage is not zero, we must add another term to the threshold voltage. Variation of threshold voltage with source/substrate voltage is called **body effect**, which can significantly affect the speed of complex logic gates. The amount by which the threshold voltage is increased is

$$\Delta V_t = \gamma_n(\sqrt{\phi_s + V_{sb}} - \sqrt{\phi_s}) \tag{EQ 2-11}$$

The term γ_n is the **body effect factor**, which depends on the gate oxide thickness and the substrate doping:

$$\gamma_n = \frac{\sqrt{2q\varepsilon_{Si}N_A}}{C_{ox}}. \tag{EQ 2-12}$$

(To compute γ_p, we substitute the n-tub doping N_D for N_A.) We will see how body effect must be taken into account when designing logic gates in Section 3.3.4.

Example 2-3: Threshold voltage of a transistor

First, we will calculate the value of the threshold voltage of an n-type transistor at zero source/substrate bias. First, some reasonable values for the parameters:

- $x_{ox} = 200\overset{\circ}{A}$;
- $\varepsilon_{ox} = 3.5 \times 10^{-13} F/cm$;
- $\phi_s = 0.6V$;
- $Q_f = q \times 10^{11} = 1.6 \times 10^{-8} C/cm^2$;
- $\varepsilon_{si} = 1.0 \times 10^{-12}$;
- $N_A = 10^{15}$ cm^{-3} ;
- $N_{ap} = 10^{19}$ cm^{-3} ;
- $N_{II} = 1 \times 10^{12}$.

Let's compute each term of V_{t0} :

- $C_{ox} = \varepsilon_{ox}/x_{ox}$

$$= 3.45 \times 10^{-13}/2 \times 10^{-6} = 1.73 \times 10^{-7} C/cm^2 .$$

- $\Phi_{gs} = -\dfrac{kT}{q} \ln\left(\dfrac{N_a N_{dp}}{n_i^2}\right)$

$$= -0.026 \ln\left(\dfrac{10^{15} 10^{19}}{(1.45 \times 10^{10})^2}\right)$$

$$= -0.82 V$$

- $V_{fb} = \Phi_{gs} + Q_f/C_{ox}$

$$= -0.82 - (1.6 \times 10^{-8}/1.73 \times 10^{-7})$$

$$= -0.91 V .$$

- $\phi_s = 2\dfrac{kT}{q} \ln\dfrac{N_a}{n_i}$

$$= 2 \times 0.026 \times \ln\left(\frac{10^{15}}{1.45 \times 10^{10}}\right)$$

$$= 0.58\,V$$

- $Q_b = \sqrt{2q\varepsilon_{si}N_a\phi_s}$

$$= \sqrt{2 \times (1.6 \times 10^{-19}) \times 1.0 \times 10^{-12} \times 10^{15} \times 0.58}$$

$$= 1.4 \times 10^{-8}$$

- $V_{II} = qD_I/C_{ox}$

$$= (1.6 \times 10^{-19}) \times (1 \times 10^{12})/(1.73 \times 10^{-7})$$

$$= 0.92\,V$$

So,

$$V_{t0} = V_{fb} + \phi_s + \frac{Q_b}{C_{ox}} + V_{II}$$

$$= -0.91\,V + 0.58\,V + \frac{1.4 \times 10^{-8}}{1.73 \times 10^{-7}} + 0.92\,V$$

$$= 0.68\,V.$$

Note that it takes a significant ion implantation to give a threshold voltage that is reasonable for digital circuit design.

What is the value of the body effect at a source/substrate voltage of 5 V? That is the voltage the source will be raised to when it is in a chain of transistors in a logic gate. First, we compute the body effect factor:

$$\gamma_n = \frac{\sqrt{2q\varepsilon_{Si}N_A}}{C_{ox}}$$

$$= \frac{\sqrt{2 \times (1.6 \times 10^{-19}) \times 1.0 \times 10^{-12} \times 10^{15}}}{1.73 \times 10^{-7}}$$

$$= 0.1 \; .$$

Then

$$\Delta V_t \; = \; \gamma_n(\sqrt{\phi_s + V_{sb}} - \sqrt{\phi_s})$$

$$= \; 0.1(\sqrt{0.58V + 5} - \sqrt{0.58V})$$

$$= \; 0.16V \; .$$

While 0.16 V may not seem like much, it is 24% of the threshold voltage, a value large enough to cause delays under some conditions.

The drain current equation of Equation 2-1 can be found by integrating the charge over the channel. The charge at a point y is given simply by the definition of a parallel plate capacitance:

$$Q(y) \; = \; C_{ox}(V_{gs} - V_t - V(y)) \quad . \tag{EQ 2-13}$$

The voltage differential over a differential distance in the channel is

$$dV \; = \; \frac{I_d dy}{\mu Q W}, \tag{EQ 2-14}$$

where μ is the (n- or p-) mobility at the surface and W is, of course, the channel width. Therefore, the total channel current is

$$I_d \; = \; \mu C_{ox} \frac{W}{L} \int_0^V (V_{gs} - V_t - V)(dV)ds \; . \tag{EQ 2-15}$$

The factor μC_{ox} is given the name k' or **process transconductance**. We sometimes call $k'W/L$ the **device transconductance** β. This integral gives us the linear-region drain current formula of Equation 2-1. At saturation, our first-order model assumes that the drain current becomes independent of the drain voltage and maintains that value as V_{ds} increases. As shown in Figure 2-13, the depth of the inversion layer varies with the voltage drop across the length of the channel and, at saturation, its height has been reduced to zero.

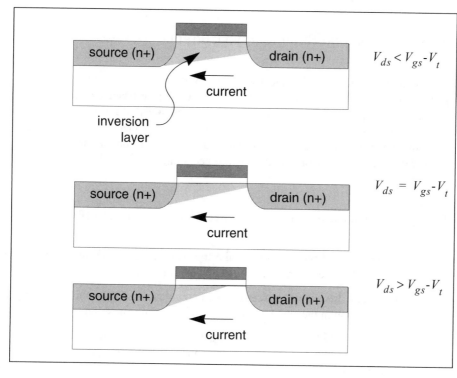

Figure 2-13: *Shape of the inversion layer as a function of gate voltage.*

But this basic drain current equation ignores the small dependence of drain current on V_{ds} in saturation. Increasing V_{ds} while in saturation causes the channel to shorten slightly, which in turn slightly increases the drain current. This phenomenon can be modeled by multiplying Equation 2-2 by a factor $(1 + \lambda V_{ds})$. (Unfortunately, the **channel length modulation parameter** λ is given the same symbol as the scaling factor λ which will be introduced in the next chapter.) The value of λ is measured empirically, not derived. This gives us the new drain current equation for the saturation region

$$I_d = \frac{1}{2}k'\frac{W}{L}(V_{gs}-V_t)^2(1 + \lambda V_{ds}) \quad . \tag{EQ 2-16}$$

Unfortunately, the λ term causes a slight discontinuity between the drain current equations in the linear and saturation regions—at the transition point, the $\lambda V ds$ term introduces a small jump in I_d. A discontinuity in drain current is clearly not physically possible, but the discontinuity is small and usually can be ignored during manual analysis of the transistor's behavior. Circuit simulation, however, may require using a slightly different formulation that keeps drain current continuous.

2.3.6 Leakage and Subthreshold Currents

The drain current through the transistor does not drop to zero once the gate voltage goes below the threshold voltage. A variety of **leakage currents** continue to flow through various parts of the transistor, including a **subthreshold current** through the channel. Those currents are small, but they are becoming increasingly important in low-power applications. Not only do many circuits need to operate under very low current drains, but subthreshold currents are becoming relatively larger as transistor sizes shrink.

Leakage currents come from a variety of effects within the transistor [Roy00]:

- Reverse-biased pn junctions in the transistor, such as the one between the drain and its well, carry small reverse bias currents.

- The **weak inversion current** (also known as the subthreshold current) is carried through the channel when the gate is below threshold.

- **Drain-induced barrier lowering** is an interaction between the drain's depletion region and the source that causes the source's potential barrier to be lowered.

- **Gate-induced drain leakage** current happens around the high electric field under the gate/drain overlap.

- **Punchthrough currents** flow when the source and drain depletion regions connect within the channel.

- Gate oxide tunneling currents are caused by high electric fields in the gate.

- Hot carriers can be injected into the channel.

Different mechanisms dominate at different drain voltages, with weak inversion dominating at low drain voltages.

The subthreshold current can be written as [Roy00]:

$$I_{sub} = ke^{\left(\frac{V_{gs} - V_t}{S/\ln 10}\right)}[1 - e^{-qV_{ds}/kT}].$$ **(EQ 2-17)**

The **subthreshold slope** S characterizes the magnitude of the weak inversion current in the transistor. The subthreshold slope is determined by a plot of log I_d vs. V_{gs}. An S value of 100 mV/decade indicates a very leaky transistor, with lower values indicating lower leakage currents.

The subthreshold current is a function of the threshold voltage V_t. The threshold voltage is primarily determined by the process. However, since the threshold voltage is measured relative to the substrate, we can adjust V_t by changing the substrate bias. We will take advantage of this effect in Section 3.6.

2.3.7 Advanced Transistor Structures

The modern MOS transistor is more complex than the basic transistor shown in Figure 2-5. A number of improvements to the basic MOS structure have been introduced over time to increase performance and permit the construction of efficient, short-channel devices [Bre90]. These new structures increase process complexity. The region between the source and drain is more heavily doped than the region underneath, allowing the source and drain to be put closer together. An epitaxial layer—a crystalline layer grown on top of the wafer during processing—allows fine control of doping in large regions; the lightly-doped region created underneath the source and drain by epitaxial growth reduces the junction capacitance between the two. The contacts to the source and drain are also designed to reduce source/drain capacitance by reducing the contact area. The diffusion is made thin near the channel area to reduce the penetration of the source/drain electric fields into the channel area. Many transistors also use lightly-doped drains to reduce the generation of **hot electrons**—high-energy electrons which can physically damage the drain region. We saw a silicided poly gate in Figure 2-6; silicides can also be applied to diffusions to reduce their resistance.

2.3.8 Spice Models

A circuit simulator, of which Spice [Nag75] is the prototypical example, provides the most accurate description of system behavior by solving for voltages and currents over time. The basis for circuit simulation is Kirchoff's laws, which describe the relationship between voltages and currents. Linear elements, like resistors and capacitors, have constant values in Kirchoff's laws, so the equations can be solved by standard linear algebra techniques. However, transistors are non-linear, greatly complicating the solution of the circuit equations. The circuit simulator uses a model—an equivalent circuit whose parameters may vary with the values of other circuits voltages and currents—to represent a transistor. Unlike linear circuits, which can be solved analytically, numerical solution techniques must be used to solve non-linear circuits. The solution is generated as a sequence of points in time. Given the circuit solution at time t, the simulator chooses a new time $t+\delta$ and solves for the new voltages and currents. The difficulty of finding the $t+\delta$ solution increases when the circuit's voltages and currents are changing very rapidly, so the simulator chooses the time step δ based on the derivatives of the Is and Vs. The resulting values can be plotted in a variety of ways using interactive tools.

A circuit simulation is only as accurate as the model for the transistor. Most versions of Spice offer three MOS transistor models, called naturally enough level 1, level 2, and level 3 [Gei90, Rab96]. The level 1Spice model is roughly the device equations of Section 2.3. The level 2Spice model provides more accurate determination of effective channel length and the transition between the linear and saturation regions, but is used less frequently today. The level 3 Spice model uses empirical parameters to provide a better fit to the measured device characteristics. Each model requires a number of parameters which should be supplied by your fabrication vendor. The level 4 model, also known as the BSIM model, uses some extracted parameters, but is smaller and more efficient. Even more recent models, including level 28 (BSIM2) and level 47 (BSIM3) have been developed recently to more accurately model deep submicron transistors.

Table 2-3 gives the Spice names for some common parameters of Spice models and their correspondence to names used in the literature. Process vendors typically supply customers with Spice model parameters directly. You should use these values rather than try to derive them from some other parameters.

2.4 Wires and Vias

Figure 2-14 illustrates the cross-section of a nest of wires and vias. N-diffusion and p-diffusion wires are created by doping regions of the substrate. Polysilicon and metal wires are laid over the substrate, with silicon dioxide to insulate them from the substrate and each other. Wires are added in layers to the chip, alternating with SiO_2: a layer of wires is added on top of the existing silicon dioxide, then the assembly is covered with an additional layer of SiO_2 to insulate the new wires from the next layer. Vias are simply cuts in the insulating SiO_2; the metal flows through the cut to make

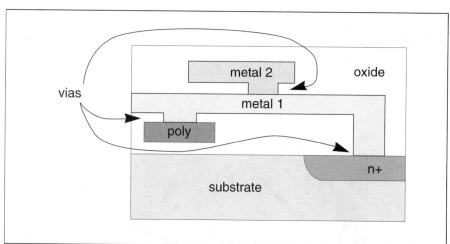

Figure 2-14:
A cross-section of a chip showing wires and vias.

the connection on the desired layer below. Figure 2-15 shows a photomicrograph of a multi-level interconnect structure, including four levels of metal and one level of polysilicon.

Figure 2-15: *Cross-section of 4-level metal/1-level poly interconnect (courtesy UMC).*

As mentioned in Section 2.2, copper interconnect can now be produced in volume thanks to a special protection layer that keeps the copper from poisoning the semiconductors in the substrate. Figure 2-16 shows a cross section of a chip with six levels of copper interconnect. The fabrication methods, and therefore the design rules, for copper interconnect are similar to those used for aluminum wires. However, as we will see in Chapter 3, the circuit characteristics of copper differ radically from those of aluminum.

Figure 2-16:
Cross-section of six layers of copper interconnect (courtesy IBM).

In addition to carrying signals, metal lines are used to supply power throughout the chip. On-chip metal wires have limited current-carrying capacity, as does any other wire. (Poly and diffusion wires also have current limitations, but since they are not used for power distribution those limitations do not affect design.) Electrons drifting through the voltage gradient on a metal line collide with the metal grains which form the wire. A sufficiently high-energy collision can appreciably move the metal grain. Under high currents, electron collisions with metal grains cause the metal to move; this process is called **metal migration** (also known as **electromigration**) [Mur93].

The **mean time to failure (MTF)** for metal wires—the time it takes for 50% of testing sites to fail—is a function of current density:

$$\text{MTF} \propto j^{-n} e^{Q/kT}, \tag{EQ 2-18}$$

where j is the current density, n is a constant between 1 and 3, and Q is the diffusion activation energy. This equation is derived from the drift velocity relationship.

Metal wires can handle 1.5 mA of current per micron of wire width under the SCMOS rules. (Width is measured perpendicular to current flow.) A 3 μm wire can handle 4.5 mA of current. We will see in Chapter 3 that 4.5 mA is enough current to supply a large number of logic gates, so in small designs, metal migration limits are not a major problem. In larger designs, however, sizing power supply lines is critical to ensuring that the chip does not fail once it is installed in the field.

2.4.1 Wire Parasitics

Wires, vias and transistors all introduce parasitic elements into our circuits. Inductance is not a significant problem in current generations of integrated circuit technology (though, as we will see in Chapter 7, it is an important concern in IC packaging), so the parasitics of interest are resistance and capacitance. It is important to understand the structural properties of our components that introduce parasitic elements, and how to measure parasitic element values from layouts.

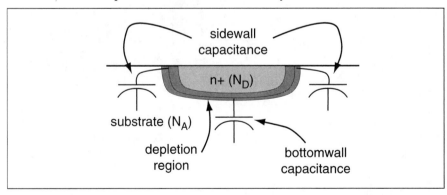

Figure 2-17: Sidewall and bottomwall capacitances of a diffusion region.

Diffusion wire capacitance is introduced by the p-n junctions at the boundaries between the diffusion and underlying tub or substrate. While these capacitances change with the voltage across the junction, which varies during circuit operation, we

Table 2-3
*Names of some
Spice parame-
ters.*

parameter	symbol	Spice name
channel drawn length	L	L
channel width	W	W
source, drain areas		AS, AD
source, drain perimeters		PS, PD
source/drain resistances	R_s, R_d	RS, RD
source/drain sheet resistance		RSH
zero-bias bulk junction capacitance	C_{j0}	CJ
bulk junction grading coefficient	m	MJ
zero-bias sidewall capacitance	C_{jsw0}	CJSW
sidewall grading coefficient	m_{sw}	MJSW
gate-bulk/source/drain overlap capacitances	$C_{gb0}/C_{gs0}/$ C_{gd0}	CGBO, CGSO, CGDO
bulk junction leakage current	I_s	IS
bulk junction leakage current density	J_s	JS
bulk junction potential	ϕ_0	PB
zero-bias threshold voltage	V_{t0}	VT0
transconductance	k'	KP
body bias factor	γ	GAMMA
channel modulation	λ	LAMBDA
oxide thickness	t_{ox}	TOX
lateral diffusion	x_d	LD
metallurgical junction depth	x_j	XJ
surface inversion potential	$2\lvert\phi_F\rvert$	PHI
substrate doping	N_A, N_D	NSUB
surface state density	Q_{ss}/q	NSS
surface mobility	μ_0	U0
maximum drift velocity	v_{max}	VMAX
mobility critical field	E_{crit}	UCRIT
critical field exponent in mobility degradation		UEXP
type of gate material		TPG

generally assume worst-case values. An accurate measurement of diffusion wire capacitance requires separate calculations for the bottom and sides of the wire—the doping density, and therefore the junction properties, vary with depth. To measure total capacitance, we measure the diffusion area, called **bottomwall** capacitance, and perimeter, called **sidewall** capacitance, as shown in Figure 2-17, and sum the contributions of each.

The **depletion region capacitance** value is given by

$$C_{j0} = \frac{\varepsilon_{si}}{x_d}.$$

(EQ 2-19)

This is the **zero-bias depletion capacitance**, assuming zero voltage and an abrupt change in doping density from N_a to N_d. The depletion region width x_{d0} is shown in Figure 2-17 as the dark region; the depletion region is split between the n+ and p+ sides of the junction. Its value is given by

$$x_{d0} = \sqrt{\left(\frac{1}{N_A} + \frac{1}{N_D}\right)\frac{2\varepsilon_{si}V_{bi}}{q}},$$

(EQ 2-20)

where the built-in voltage V_{bi} is given by

$$V_{bi} = \frac{kT}{q}\ln\left(\frac{N_A N_D}{n_i^2}\right).$$

(EQ 2-21)

The junction capacitance is a function of the voltage across the junction V_r:

$$C_j(V_r) = \frac{C_{j0}}{\sqrt{1 + \frac{V_r}{V_{bi}}}}.$$

(EQ 2-22)

So the junction capacitance decreases as the reverse bias voltage increases.

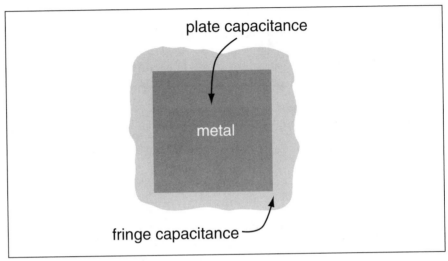

Figure 2-18:
Plate and fringe capacitances of a parallel-plate capacitor.

The capacitance mechanism for poly and metal wires is, in contrast, the parallel plate capacitor from freshman physics. We must also measure area and perimeter on these layers to estimate capacitance, but for different reasons. The **plate capacitance** per unit area assumes infinite parallel plates. We take into account the changes in the electrical fields at the edges of the plate by adding in a **fringe capacitance** per unit perimeter. These two capacitances are illustrated in Figure 2-18. Capacitances can form between signal wires. In conservative technologies, the dominant parasitic capacitance is between the wire and the substrate, with the silicon dioxide layer forming the insulator between the two parallel plates.

However, as the number of metal levels increases and the substrate capacitance decreases, wire-to-wire parasitics are becoming more important. Both capacitance between two different layers and between two wires on the same layer are basic parallel plate capacitances. The parasitic capacitance between two wires on different layers, such as C_{m1m2} in Figure 2-19, depends on the area of overlap between the two wires. In a typical 0.5 µm process, the plate capacitance between metal 1 and metal 2 is 0.3 fF/cm^2 and the metal 1-metal3 plate capacitance is 0.1 fF/cm^2. When two wires run together for a long distance, with one staying over the other, the layer-to-layer capacitance can be very large. The capacitance between two wires on the same layer, C_{w1w2} in the figure, is formed by the vertical sides of the metal wires. Metal wires can be very tall in relation to their width, so the vertical wall coupling is non-negligible. However, this capacitance depends on the distance between two wires. The values given in process specifications are for minimum-separation wires, and the capacitance decreases by a factor of $1/x$ as distance increases. When two wires on the same layer run in parallel for a long distance, the coupling capacitance can become very large.

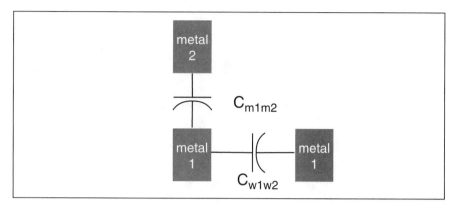

The following example illustrates how to measure parasitic capacitance from a layout.

Example 2-4: Parasitic capacitance measurement

N-diffusion wires in our typical 0.5 µm process have a bottomwall capacitance of 0.6 fF/µm^2 and a sidewall capacitance of 0.2 fF/µm. P-diffusion wires have bottomwall and sidewall capacitances of 0.9 fF/µm^2 and 0.3 fF/µm, respectively. The sidewall capacitance of a diffusion wire is typically as large or larger its bottomwall capacitances because the well/substrate doping is highest near the surface. Typical metal 1 capacitances in a process are 0.04 fF/µm^2 for plate and 0.09 fF/µm for fringe; typical poly values are 0.09 fF/µm^2 plate and 0.04 fF/µm fringe. The fact that diffusion capacitance is an order of magnitude larger than metal or poly capacitance suggests that we should avoid using large amounts of diffusion.

Here is our example wire, made of n-diffusion and metal connected by a via:

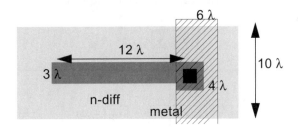

To measure wire capacitance of a wire, simply measure the area and perimeter on each layer, compute the bottomwall and sidewall capacitances, and add them together. The only potential pitfall is that our layout measurements are probably, as in this example, in λ units, while unit capacitances are measured in units of μm. The n-diffusion section of the wire occupies $3 \mu m \times 0.75 \mu m + 1 \mu m \times 1 \mu m = 3.25 \mu m^2$, for a bottomwall capacitance of $2 fF$. In this case, we count the n-diffusion which underlies the via, since it contributes capacitance to the substrate. The n-diffusion's perimeter is, moving counterclockwise from the upper left-hand corner, $0.75 \mu m + 3 \mu m + 0.25 \mu m + 1 \mu m + 1 \mu m + 4 \mu m = 10 \mu m$, giving a total sidewall capacitance of $2 fF$. Because the sidewall and bottomwall capacitances are in parallel, we add them to get the n-diffusion's contribution of 4 fF.

The metal 1 section has a total area of $2.5 \mu m \times 1.5 \mu m = 3.75 \mu m^2$, giving a plate capacitance of 0.15 fF. The metal's perimeter is $2.5 \mu m \times 2 + 1.5 \mu m \times 2 = 8 \mu m$ for a fringe capacitance of 0.72 fF and a total metal contribution of 0.87 fF. A slightly more accurate measurement would count the metal area overlying the n-diffusion differently—strictly speaking, the metal forms a capacitance to the n-diffusion, not the substrate, since the diffusion is the closer material. However, since the via area is relatively small, approximating the metal 1-n-diffusion capacitance by a metal 1-substrate capacitance doesn't significantly change the result.

The total wire capacitance is the sum of the layer capacitances, since the layer capacitors are connected in parallel. The total wire capacitance is 4.9 fF; the n-diffusion capacitance dominates the wire capacitance, even though the metal 1 section of the wire is larger.

Wire resistance is also computed by measuring the size of the wire in the layout, but the unit of resistivity is **ohms per square** ($\Omega / \square$), not ohms per square micron. The resistance of a square unit of material is the same for a square of any size; to under-

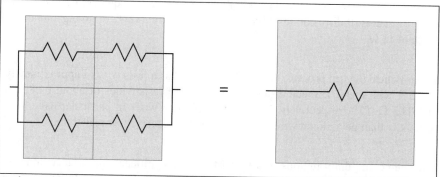

stand, consider Figure 2-20. Assume that a unit square of material has a resistance of
$1\,\Omega$. Two squares of material connected in parallel have a total resistance of $1/2\,\Omega$.
Connecting two such rectangles in series creates a 2×2 square with a resistance of 1
Ω. We can therefore measure the resistance of a wire by measuring its aspect ratio.

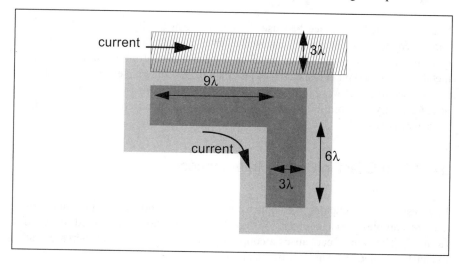

Figure 2-21: *An
example of resis-
tance calculation.*

Figure 2-21 shows two example wires. The upper wire is made of polysilicon, which
has a resistivity of 4 $\Omega/\square$ in our 0.5 μm process. Current flows in the direction
shown; wire length is along the direction of current flow, while wire width is perpendicular to the current. The wire is composed of 18/3 squares connected in series, giving a total resistance of 24 Ω.

The second wire is more interesting because it is bent. A 90° bend in a wire offers
less resistance because electrons nearer the corner travel a shorter distance. A simple
and common approximation is to count each square corner rectangle as 1/2 squares of

resistance. The wire can be broken into three pieces: 9/3 = 3 squares, 1/2 squares, and 6/3 = 2 squares. P-diffusion resistivity is approximately 2 $\Omega/\square$, giving a total resistance of 11 Ω

In our typical 0.5 µm process, an n-diffusion wire has a resistivity of approximately 2 $\Omega/\square$, with metal 1, metal 2, and metal 3 having resistivities of about 0.08, 0.07, and 0.03 $\Omega/\square$, respectively. Note that p-diffusion wires in particular have higher resistivity than polysilicon wires, and that metal wires have low resistivities.

The source and drain regions of a transistor have significant capacitance and resistance. These parasitics are, for example, entered into a Spice simulation as device characteristics rather than as separate wire models. However, we measure the parasitics in the same way we would measure the parasitics on an isolated wire, measuring area and perimeter up to the gate-source/drain boundary.

Vias have added resistance because the cut between the layers is smaller than the wires it connects and because the materials interface introduces resistance. The resistance of the via is usually determined by the resistance of the materials: a metal 1-metal 2 via has a typical resistance of less than 0.5 Ω while a metal1-poly contact has a resistance of 2.5 Ω We rarely worry about the exact via resistance in layout design; instead, we try to avoid introducing unnecessary vias in current paths for which low resistance is critical.

2.4.2 Skin Effect in Copper Interconnect

Low-resistance conductors like copper not only exhibit inductance, they also display a more complex resistance relationship due to a phenomenon called **skin effect** [Ram65]. The skin effect causes a copper conductor's resistance to increase (and its inductance to decrease) at high frequencies.

An ideal conductor would conduct currents only on its surface. The current at the surface is a boundary effect—any current within the conductor would set up an electromagnetic force that would induce an opposing and cancelling current. The copper wiring used on ICs is a non-ideal conductor; at low frequencies, the electromagnetic force is low enough and resistance is high enough resistance that current is conducted throughout the wire's cross section. However, as the signal's frequency increases, the electromagnetic forces increase. As illustrated in Figure 2-22, the current through an isolated conductor migrates toward the edges as frequency increases; when the conductor is close to a ground, the current in both move toward each other.

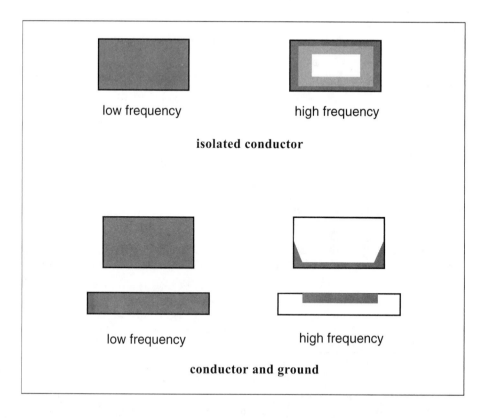

Figure 2-22:
How current
changes with fre-
quency due to
skin effect.

Skin effect causes the conductor's resistance to increase with frequency. The **skin depth** δ is the depth at which the conductor's current is reduced to $1/e = 37\%$ of its surface value [Che00]:

$$\delta = \frac{1}{\sqrt{\pi f \mu \sigma}},\qquad \text{(EQ 2-23)}$$

where f is the signal frequency, μ is the magnetic permeability, and σ is the wire's conductivity. The skin depth goes down as the square root of frequency.

Cheng et al [Che00] provide an estimation of the delay per unit length of a wire suffering from skin effect. Two values, R_{dc} and R_{hf}, estimate the resistance at low and high frequencies:

$$\left(R_{dc} = \frac{1}{\sigma w t}\right), R_{hf} = \frac{1}{2\sigma\delta(w+t)}, \qquad \text{(EQ 2-24)}$$

where w and t are the width and height of the conductor, respectively. The skin depth δ ensures that R_{hf} depends on frequency. The resistance per unit length can be estimated as

$$R_{ac} = \sqrt{R_{dc}^2 + (\kappa R_{hf})^2}, \qquad \text{(EQ 2-25)}$$

where κ is a weighting factor typically valued at 1.2

Skin effect typically becomes important at gigahertz frequencies in ICs. Some microprocessors already run at those frequencies and more chips will do so in the near future.

2.5 Design Rules

Layouts are built from three basic component types: transistors, wires, and vias. We have seen the structures of these components created during fabrication. Now we will consider the design of the layouts which determine the circuit that is fabricated. **Design rules** govern the layout of individual components and the interactions—spacings and electrical connections—between those components. Design rules determine the low-level properties of chip designs: how small individual logic gates can be made; how small the wires connecting gates can be made, and therefore, the parasitic resistance and capacitance which determine delay.

Design rules are determined by the conflicting demands of component packing and chip yield. On the one hand, we want to make the components as small as possible, to put as many functions as possible on-chip. On the other hand, since the individual transistors and wires are about as small as the smallest feature that our manufacturing process can produce, errors during fabrication are inevitable: wires may short together or never connect, transistors may be faulty, etc. One common model for yield of a single type of structure is a Gamma distribution [Mur93]:

$$Y_i = \left(\frac{1}{1 + A\beta_i}\right)^{\alpha_i}.$$
                                                                                      (EQ 2-26)

The total yield for the process is then the product of all the yield components:

$$Y = \prod_{i=1}^{n} Y_i.$$
                                                                                      (EQ 2-27)

This formula suggests that low yield for even one of the process steps can cause serious final yield problems. But being too conservative about design rules leads to chips
that are too large (which itself reduces yield) and too slow as well. We try to balance
chip functionality and manufacturing yield by following rules for layout design
which tell us what layout constructs are likely to cause the greatest problems during

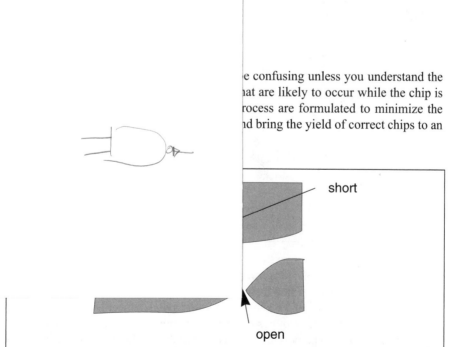

e confusing unless you understand the
nat are likely to occur while the chip is
rocess are formulated to minimize the
nd bring the yield of correct chips to an

Figure 2-23:
Problems when
wires are too wide
or narrow.

The most obvious type of fabrication problem is a wire or other feature being made
too wide or too narrow. This problem can occur for a variety of reasons: photolitho-

graphic errors may leave an erroneous pattern for later steps; local materials variations may cause different rates of diffusion or deposition; processing steps at a nearby feature may cause harmful interactions. One important problem in fabrication is **planarization** [Gha94]—poly and metal wires leave hills in the oxide. The bumps in the oxide can be smoothed by several different chemical or mechanical methods; failure to do so causes **step coverage** problems which may lead to breaks in subsequent metallization layers. In any case, the result is a wire that is too narrow or too wide. As shown in Figure 2-23, a wire that is too narrow may never conduct current, or may burn out after some use. A too-wide wire may unintentionally short itself to another wire or, as in the case of a poly wire overlapping a parallel diffusion wire, cut into another element.

The simplest remedy for these problems is the introduction of **spacing** and **minimum-width** rules, which take a variety of forms in our design rules. Minimum width rules give a minimum size for a layout element; they help ensure that even with minor variations in the position of the lines that form the element, the element will be of an acceptable size. Spacing rules give a minimum distance between the edges of layout elements, so that minor processing variations will not cause the element to overlap nearby layout elements.

Figure 2-24:
Potential problems in transistor fabrication.

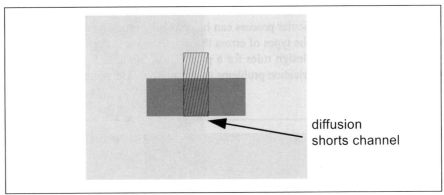

diffusion
shorts channel

We also have a number of **composition** rules to ensure that components are well-formed. Consider the transistor layout in Figure 2-24—the transistor action itself takes place in the channel, at the intersection of the polysilicon and diffusion regions, but a valid transistor layout requires extensions of both the poly and diffusion regions beyond the boundary. The poly extensions ensure that no strand of diffusion shorts together the source and drain. The diffusion extensions ensure that adequate contact can be made to the source and drain.

Figure 2-25:
Potential prob-
lems in via fabri-
cation.

Vias have construction rules as well: the material on both layers to be connected must extend beyond the SiO_2 cut itself; and the cut must be of a fixed size. As shown in Figure 2-25, the overlap requirement simply ensures that the cut will completely connect the desired layout elements and not mistakenly connect to the substrate or another wire. The key problem in via fabrication, however, is making the cuts. A large chip may contain millions of vias, all of which must be opened properly for the chip to work. The acid etching process which creates cuts must be very uniform—cuts may be neither too small and shallow nor too large. It isn't hard to mount a bookcase in a wall with an electric drill—it is easy to accurately size and position each hole required. Now imagine making those holes by covering the wall with acid at selected points, then wiping the wall clean after a few minutes, and you should empathize with the problems of manufacturing vias on ICs. The cut must also be filled with material without breaking as the material flows over the edge of the cut. The size, shape, and spacing of via cuts are all strictly regulated by modern fabrication processes to give maximum via yield.

2.5.2 Scalable Design Rules

Manufacturing processes are constantly being improved. The ability to make ever-smaller devices is the driving force behind Moore's Law. But many characteristics of the fabrication process do not change as devices shrink—layouts do not have to be completely redesigned, simply shrunk in size. We can take best advantage of process scaling by formulating our design rules to be explicitly scalable.

We will scale our design rules by expressing them not in absolute physical distances, but in terms of λ, the size of the smallest feature in a layout. All features can be measured in integral multiples of λ. By choosing a value for λ we set all the dimensions in a scalable layout.

Scaling layouts makes sense because chips actually get faster as layouts shrink. As a result, we don't have to redesign our circuits for each new process to ensure that speed doesn't go down as packing density goes up. If circuits became slower with smaller transistors, then circuits and layouts would have to be redesigned for each process.

Digital circuit designs scale because the capacitive loads that must be driven by logic gates shrink faster than the currents supplied by the transistors in the circuit [Den74]. To understand why, assume that all the basic physical parameters of the chip are shrunk by a factor $1/x$:

- lengths and widths: $W \rightarrow W/x$, $L \rightarrow L/x$;

- vertical dimensions such as oxide thicknesses: $t_{ox} \rightarrow t_{ox}/x$;

- doping concentrations: $N_d \rightarrow N_d/x$;

- supply voltages: $V_{DD} - V_{SS} \rightarrow (V_{DD} - V_{SS})/x$.

We now want to compute the values of scaled physical parameters, which we will denote by variables with hat symbols. One result is that the transistor transconductance scales: since $k' = (\mu_{eff}\varepsilon_{ox})/t_{ox}$ [Mul77], $\hat{k}'/k' = x$. (μ_{eff} is the carrier mobility and ε_{ox} is the dielectric constant.) The threshold voltage scales with oxide thickness, so $\hat{V}_t = V_t/x$. Now compute the scaling of the saturation drain current W/L:

$$
\begin{aligned}
\frac{\hat{I}_d}{I_d} &= \left(\frac{\hat{k}'}{k'}\right)\left(\frac{\hat{W}/\hat{L}}{W/L}\right)\left[\frac{(\hat{V}_{gs}-\hat{V}_t)^2}{(V_{gs}-V_t)^2}\right] \\
&= x\left(\frac{1/x}{1/x}\right)\left(\frac{1}{x}\right)^2 \\
&= \frac{1}{x}
\end{aligned}
$$

(EQ 2-28)

The scaling of the gate capacitance is simple to compute: $C_g = \varepsilon_{ox}WL/t_{ox}$, so $\hat{C}_g/C_g = 1/x$. The total delay of the logic circuit depends on the capacitance to be charged, the current available, and the voltage through which the capacitor must be charged; we will use CV/I as a measure of the speed of a circuit over scaling. The voltage through which the logic circuit swings is determined by the power supply, so the voltage scales as $1/x$. When we plug in all our values,

$$\frac{\hat{C}\hat{V}/\hat{I}}{CV/I} = \frac{1}{x}.$$
(EQ 2-29)

So, as the layout is scaled from λ to $\hat{\lambda} = \lambda/x$, the circuit is actually speeded up by a factor x.

In practice, few processes are perfectly λ-scalable. As process designers learn more, they inevitably improve some step in the process in a way that does not scale. High-performance designs generally require some modification when migrated to a smaller process as detailed timing properties change. However, the scalability of VLSI systems helps contain the required changes.

2.5.3 SCMOS Design Rules

Finally, we reach the SCMOS design rules themselves[1]. We will cast these rules in terms of λ. For the SCMOS rules, a 0.5 μm process, the nominal value for λ is 0.25 μm. However, fabrication lines often make minor adjustments to the masks (bloating or shrinking) to adjust the final physical dimensions of the features. We will concentrate on the rules for the basic two-level metal process, with some discussion of the rules for metal 3, since the rules for the higher levels of metal are not as well standardized as for the two-level metal process. At this writing, MOSIS provides up to four metal layers in some processes.

Design rules are generally specified as pictures illustrating basic situations, with notes to explain features not easily described graphically. While this presentation may be difficult to relate to a real layout, practice will teach you to identify potential design rule violations in a layout from the prototype situations in the rules. Many layout editor programs, such as Magic [Ost84], have built-in design-rule checkers which will identify design-rule violations on the screen for you. Using such a program is a big help in learning the process design rules.

Figure 2-26 summarizes the basic spacing and minimum size design rules. Classifying the situations described in these pictures as separation, minimum size, or composition will help you distinguish and learn the rules. Many of these rules hold for any

---

1. We will describe the most recent set of rules available to us. Since these rules change over time, it is always best to consult MOSIS (http://www.mosis.edu) before starting a design that will be fabricated using the SCMOS rules.

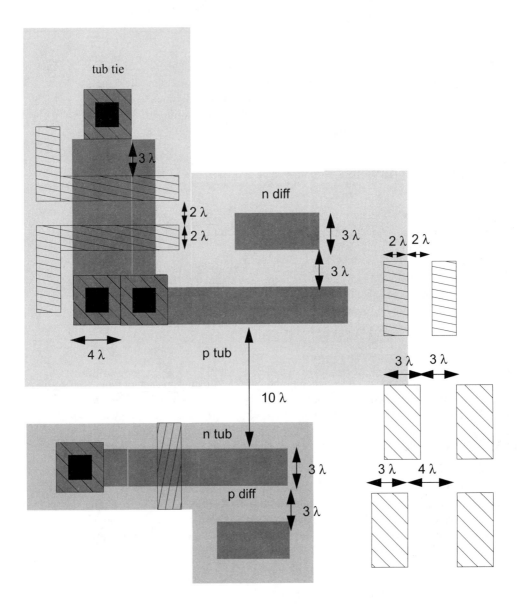

Figure 2-26: *A summary of the SCMOS design rules for two-level metal processes.*

tub structure: n-tub, p-tub, or twin-tub. The rules regarding tubs and tub ties necessarily depend on the tub structure, however.

The basic separation and minimum size rules are:

- **metal 1** Minimum width is 3 λ, minimum separation is 3 λ.

- **metal 2** Minimum width is 3 λ, minimum separation is 4 λ.

- **polysilicon** Minimum width is 2 λ, minimum poly–poly separation is 2 λ.

- **p-, n-diffusion** Minimum width is 3 λ, minimum separation between same-type diffusions is 3 λ, minimum p-diff–n-diff separation is 10 λ.

- **tubs** Tubs must be at least 10 λ wide. The minimum distance from the tub edge to source/drain active area is 5 λ.

The basic construction rules are:

- **transistors** The smallest transistor is of width 3 λ and length 2 λ; poly extends 2 λ beyond the active region and diffusion extends 3 λ. The active region must be at least 1 λ from a poly-metal via, 2 λ from another transistor, and 3 λ from a tub tie.

- **vias** Cuts are 2 $\lambda \times$ 2 λ; the material on both layers to be connected extends 1 λ in all directions from the cut, making the total via size 4 $\lambda \times$ 4 λ. (MOSIS also suggests another via construction with 1.5 λ of material around the cut. This construction is safer but the fractional design rule may cause problems with some design tools.) Available via types are:

  - n/p-diffusion–poly;

  - poly–metal 1;

  - n/p-diffusion–metal 1;

  - metal 1–metal 2;

If several vias are placed in a row, successive cuts must be at least 2 λ apart. Spacing to a via refers to the complete 4 $\lambda \times$ 4 λ object, while spacing to a via cut refers to the 2×2 λ cut.

- **tub ties** A p-tub tie is made of a 2 $\lambda \times$ 2 λ cut, a 4 $\lambda \times$ 4 λ metal element, and a 4 $\lambda \times$ 4 λ p^+ diffusion. An n-tub tie is made with an n^+ diffusion replacing the p^+ diffusion. A tub tie must be at least 2 λ from a diffusion contact.

It is important to remember that different rules have different dependencies on electrical connectivity. Spacing rules for wires, for example, depend on whether the wires are on the same electrical node. Two wire segments on the same electrical node may touch. However, two via cuts must be at least 2 λ apart even if they are on the same electrical net. Similarly, two active regions must always be 2 λ apart, even if they are parallel transistors.

The rules for metal 3 are:

- Minimum metal 3 width is 6 λ, minimum separation is 4 λ.

- Available via from metal 3 is to metal 2. Connections from metal 3 to other layers must be made by first connecting to metal 2.

There are some specialized rules that do not fit into the separation/minimum size/composition categorization.

- A cut to polysilicon must be at least 3 λ from other polysilicon.

- Polysilicon cuts and diffusion cuts must be at least 2 λ apart.

- A cut must be at least 2 λ from a transistor active region.

- A diffusion contact must be at least 4 λ away from other diffusion.

- A metal 2 via must not be directly over polysilicon.

One final special rule is to avoid generating small **negative features**. Consider the layout of Figure 2-27: the two edges of the notch are 1 λ apart, but both sides of the notch are on the same electrical node. The two edges are not in danger of causing an inadvertent short due to a fabrication error, but the notch itself can cause processing errors. Some processing steps are, for convenience, done on the negative of the mask given, as shown in the figure. The notch in the positive mask forms a 1λ wide protrusion on the negative mask. Such a small feature in the photoresist, called a **negative mask feature**, can break off during processing, float around the chip, and land elsewhere, causing an unwanted piece of material. We can minimize the chances of stray photoresist causing problems by requiring all negative features to be at least 2 λ in size.

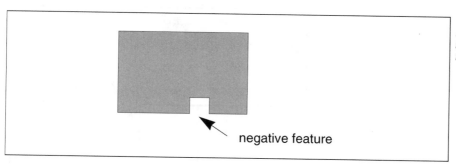

negative feature

Figure 2-27: *A negative mask feature.*

2.5.4 Typical Process Parameters

Typical values of process parameters for a 0.5 μm fabrication process are given in Table 2-4. We use the term typical loosely here; these are estimated values which do not reflect a particular manufacturing process and the actual parameter values can vary widely. You should always request process parameters from your vendor when designing a circuit that you intend to fabricate.

2.6 Layout Design and Tools

2.6.1 Layouts for Circuits

We ultimately want to design layouts for circuits. Layout design requires not only a knowledge of the components and rules of layout, but also strategies for designing layouts which fit together with other circuits and which have good electrical properties.

Since layouts have more physical structure than schematics, we need to augment our terminology. Chapter 1 introduced the term *net* to describe a set of electrical connections; a net corresponds to a variable in the voltage equations, but since it may connect many pins, it is hard to draw. A **wire** is a set of point-to-point connections; as shown in Figure 2-28, a wire may contain many branches. The straight sections are called **wire segments**.

The starting point for layout is a **circuit schematic**. The schematic symbols for n- and p-type transistors are shown in Figure 2-29. The schematic shows all electrical

Figure 2-28:
Wires and wire segments.

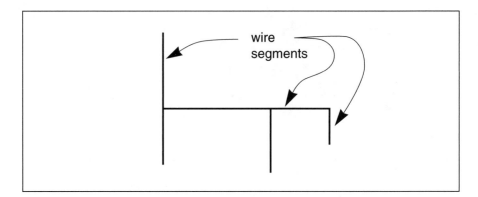

Figure 2-29:
Schematic symbols for transistors.

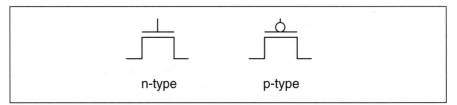

connections between transistors (except for tub ties, which are often omitted to simplify the diagram); it must also be annotated with the W/L of each transistor. We will discuss the design of logic circuits from transistors in detail in Chapter 3. At this point, we will treat the circuit schematic as a specification for which we must implement the transistors and connections in layout. (Most professional layout designers, in fact, have no training in electrical engineering and treat layout design strictly as an artwork design problem.) The next example walks through the design of an inverter's layout.

n-type transconductance	k'_n	$73\mu A/V^2$
p-type transconductance	k'_p	$-21\mu A/V^2$
n-type threshold voltage	V_{tn}	$0.7V$
p-type threshold voltage	V_{tp}	$-0.8V$
gate capacitance	C_g	$0.9fF/\mu m^2$
n-diffusion bottomwall capacitance	$C_{ndiff,bot}$	$0.6fF/\mu m^2$
n-diffusion sidewall capacitance	$C_{ndiff,side}$	$0.2fF/\mu m$
p-diffusion bottomwall capacitance	$C_{pdiff,bot}$	$0.9fF/\mu m^2$
p-diffusion sidewall capacitance	$C_{pdiff,side}$	$0.3fF/\mu m$
n-type source/drain resistivity	R_{ndiff}	$2\Omega/\square$
p-type source/drain resistivity	R_{pdiff}	$2\Omega/\square$
poly-substrate plate capacitance	$C_{poly,plate}$	$0.09fF/\mu m^2$
poly-substrate fringe capacitance	$C_{poly,fringe}$	$0.04fF/\mu m$
poly resistivity	R_{poly}	$4\Omega/\square$
metal 1-substrate plate capacitance	$C_{metal1,plate}$	$0.04fF/\mu m^2$
metal 1-substrate fringe capacitance	$C_{metal1,fringe}$	$0.09fF/\mu m$
metal 2-substrate capacitance	$C_{metal2,plate}$	$0.02fF/\mu m^2$
metal 2-substrate fringe capacitance	$C_{metal2,fringe}$	$0.06fF/\mu m$
metal 3-substrate capacitance	$C_{metal3,plate}$	$0.009fF/\mu m^2$
metal 3-substrate fringe capacitance	$C_{metal3,fringe}$	$0.02fF/\mu m$
metal 1 resistivity	R_{metal1}	$0.08\Omega/\square$
metal 2 resistivity	R_{metal2}	$0.07\Omega/\square$
metal 3 resistivity	R_{metal3}	$0.03\Omega/\square$
metal current limit	$I_{m,max}$	$1.5mA/\mu m$

Table 2-4 *Typical parameters for our 0.5 μm process.*

Example 2-5: Design of an inverter layout

The inverter circuit is simple (+ is V_{DD} and the triangle is V_{SS}):

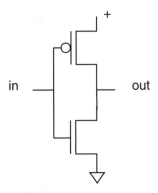

In thinking about how the layout will look, a few problems become clear. First, we cannot directly connect the p-type and n-type transistors with pdiff and ndiff wires. We must use vias to go from ndiff to metal and then to pdiff. Second, the *in* signal is naturally in polysilicon, but the *out* signal is naturally in metal, since we must use a metal strap to connect the transistors' source and drain. Third, we must use metal for the power and ground connections. We probably want to place several layouts side-by-side, so we will run the power/ground signals from left to right across the layout.

Assuming that both transistors are minimum size, here is one layout for the inverter:

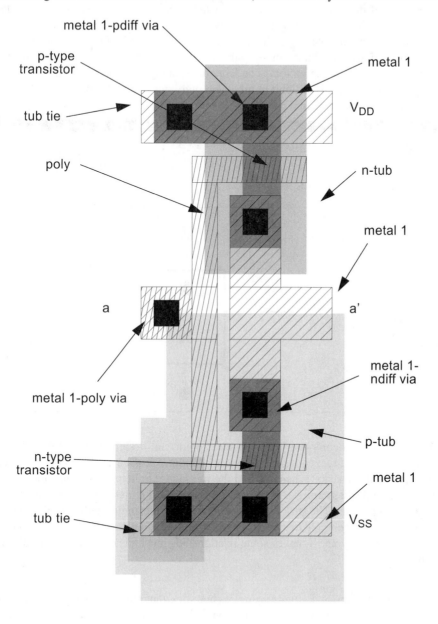

We chose to put a metal-poly via at the inverter's input so the signal would be on the same layer at input and output; we might want to connect the output of one inverter directly to the input of another. We ran power and ground along the top and bottom of the cell, respectively, placing the p-type transistor in the top half and the n-type in the bottom half. Larger layouts with many transistors follow this basic convention: p-type on the top, n-type on the bottom. The large tub spacing required between p-type and n-type devices makes it difficult to mix them more imaginatively. We also included a tub tie for both the n-tub and p-tub.

2.6.2 Stick Diagrams

Figure 2-30: A stick diagram for an inverter.

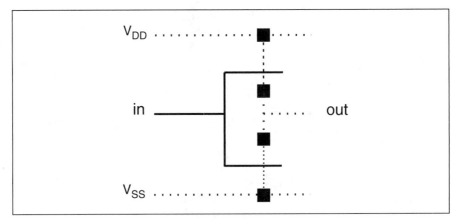

We must design a complete layout at some point, but designing a complex system directly in terms of rectangles can be overwhelming. We need an abstraction between the traditional transistor schematic and the full layout to help us organize the layout design. A **stick diagram** is a cartoon of a chip layout. Figure 2-30 shows a stick diagram for an inverter. The stick diagram represents the rectangles with lines which represent wires and component symbols. While the stick diagram does not represent all the details of a layout, it makes some relationships much clearer and it is simpler to draw.

Layouts are constructed from rectangles, but stick diagrams are built from cartoon symbols for components and wires. The symbols for wires used on various layers are shown in Figure 2-31. You probably want to draw your own stick diagrams in color: red for poly, green for n-diffusion, yellow for p-diffusion, and shades of blue for metal are typical colors. A few simple rules for constructing wires from straight-line segments ensure that the stick diagram corresponds to a feasible layout. First, wires cannot be drawn at arbitrary angles—only horizontal and vertical wire segments are allowed. Second, two wire segments on the same layer which cross are electrically connected. Vias to connect wires that do not normally interact are drawn as black dots. Figure 2-32 shows the stick figures for transistors—each type of transistor is represented as poly and diffusion crossings, much as in the layout.

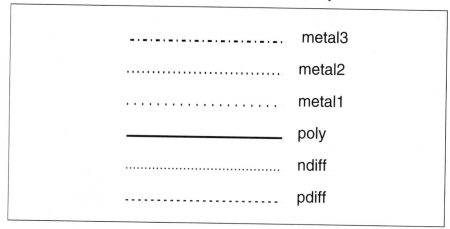

Figure 2-31: Stick diagram symbols for wires.

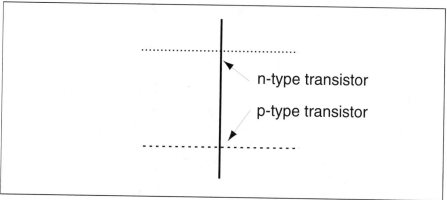

Figure 2-32: Symbols for components in stick diagrams.

The complete rules which govern how wires on different layers interact are shown in Table 2-5; they tell whether two wires on given layers are allowed to cross and, if so, the electrical properties of the new construct. This table is derived from the manufacturing design rules.

Stick diagrams are not exact models of layouts. Most of the differences are caused by the use of zero-width lines and zero-area transistors in stick diagrams. When you draw a layout using a stick diagram as a guide, you may find it necessary to move

metal3	metal2	metal1	poly	ndiff	pdiff	
short	open	open	open	open	open	metal3
	short	open	open	open	open	metal2
		short	open	open	open	metal1
			short	n-type	p-type	poly
				short	illegal	ndiff
					short	pdiff

Table 2-5 *Rules for possible interactions between layers.*

transistors and vias and to reroute wires. Area and aspect ratio are also difficult to estimate from stick diagrams. But a stick diagram can be drawn much faster than a full-fledged layout and lets you evaluate a candidate design with relatively little effort. Stick diagrams are especially important tools for layouts built from large cells and for testing the connections between cells—tangled wiring within and between cells quickly becomes apparent when you sketch the stick diagram of a cell.

2.6.3 Hierarchical Stick Diagrams

Drawing a large chip as a single stick diagram—covering a huge sheet of butcher paper with arcane symbols—usually leads to a spaghetti layout. We can make use of hierarchy to organize stick diagrams and layouts just as with schematics. Components in a layout or hierarchical stick diagram are traditionally called **cells**. In schematics, we either invent a symbol for a type (*e.g.*, logic gate symbols) or we use a box; however, the shape of the component symbol has no physical significance. Layouts and stick diagrams have physical extent. The simplest representation for a cell is its **bounding box**: a rectangle which just encloses all the elements of the cell. Bounding boxes are easy to generate; some layout tools require that cells be represented by rectangular bounding boxes. However, in some cases, we use non-rectangular **cell boundaries** to represent cells with very non-rectangular shapes.

Figure 2-33 shows a hierarchical stick diagram built from two copies of an inverter cell. The top-level cell in the hierarchy, *pair1*, includes some wires used to connect the cells together and to make external connections. Note that *pair1*'s wiring implies that the *inv1* stick diagram has been redesigned so that, unlike the stick diagram of Figure 2-30, its input and output are both on the polysilicon layer. We sometimes

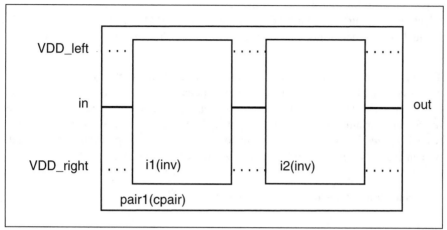

Figure 2-33: A hierarchical stick diagram.

want to show sticks cells in their entirety, and sometimes as outlines—some relationships between cells are apparent only when detail within a cell is suppressed. Hierarchical design is particularly useful in layout and sticks design because we can reuse sections of layout. Many circuits are designed by repeating the same elements over and over. Repeating cells saves work and makes it easier to correct mistakes in the design of cells.

Example 2-6: Sticks design of a multiplexer

A more interesting example of a stick diagram which takes advantage of hierarchy is a multiplexer (also known as a mux):

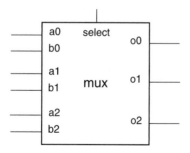

A two-input, n-bit multiplexer (in this case, $n = 3$) has two n-bit data inputs and a select input, along with an n-bit data output. When select = 0, the data output's value is equal to the a data input's value; if select = 1, the data output's value is equal to b.

The multiplexer can be designed as a one-bit slice which is replicated to create an n-bit system. The Boolean logic formula which determines the output value of one bit is $o_i = (a_i\ select) + (b_i\ select')$; the value of o_i depends only on a_i, b_i, and $select$. We can rewrite this formula in terms of two-input NAND gates: $o_i = NAND(NAND(a_i, select), NAND(b_i, select'))$. Since we know how to design the stick diagram for a NAND gate, we can easily design the one-bit multiplexer out of NAND cells.

Here is the transistor schematic for a two-input NAND gate:

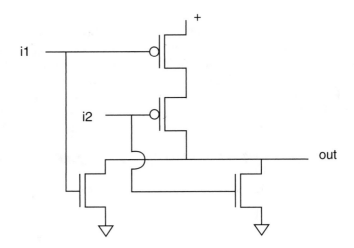

And here is a stick diagram for the two-input NAND:

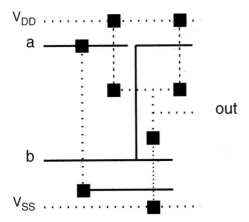

We can use the NAND cell to build a one-bit multiplexer cell:

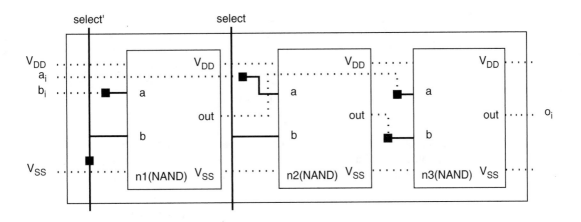

In this case we've drawn the hierarchical stick diagram using bounding boxes; to design the complete layout we would have to look into the cells. The connections designed between NAND cells were designed to avoid creating unwanted shorts with wires inside the NANDs; to be completely sure the intercell wires do not create problems, you must expand the view of the bit slice to include the internals of the NAND cells. However, making an initial wiring design using the NANDs as boxes, remembering the details of their internals as you work, makes it easier to see the relationships between wires that go between the cells.

We can build a three-bit multiplexer from our bit slice by stacking three instances of the slice cell along with a few wires:

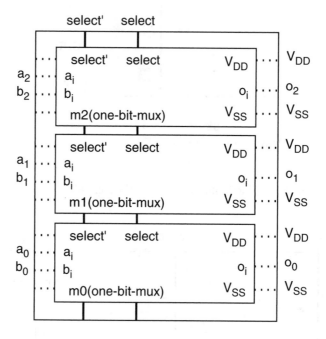

The **select** signal was designed to run vertically through the cell so vertical connections could easily be made between stacked cells. The multiplexer inputs arrive at the left edge of the stack, while the multiplexer's outputs leave at the right edge.

Constructing this three-bit multiplexer required very little labor—given a NAND cell, we were able to construct the bit slice with only a few extra wires; and given the bit slice building the complete multiplexer was almost trivial. Changing n, the width of the data word, is very simple. And last but not least, building large stick diagrams out of previously-designed smaller cells means the complete design is more likely to be correct: cells we have used before are likely to have been previously checked, and repeating cells gives us fewer opportunities to make simple mistakes while copying simple constructs.

2.6.4 Layout Design and Analysis Tools

A variety of CAD tools help us design and verify layouts. The most important tools are **layout editors**, **design rule checkers**, and **circuit extractors**.

A layout editor is an interactive graphic program that lets you create and delete layout elements. Most layout editors work on hierarchical layouts, organizing the layout into cells which may include both primitive layout elements and other cells. Some layout editing programs, such as Magic, work on **symbolic layouts**, which include somewhat more detail than do stick diagrams but are still more abstract than pure layouts. A via, for example, may be represented as a single rectangle while you edit the symbolic layout; when a final physical layout is requested, the symbolic via is fleshed out into all the rectangles required for your process. Symbolic layout has several advantages: the layout is easier to specify because it is composed of fewer elements; the layout editor ensures that the layouts for the symbolic elements are properly constructed; and the same symbolic layout can be used to generate several variations, such as n-tub, p-tub, and twin-tub versions of a symbolic design.

A design rule checker (often called a **DRC** program), as the name implies, looks for design rule violations in the layout. It checks for minimum spacing and minimum size and ensures that combinations of layers form legal components. The results of the DRC are usually shown as highlights. on top of the layout. Some layout editors, including Magic, provide on-line design rule checking. Design rule checking algorithms are briefly described in Chapter 10.

Circuit extraction is an extension of design rule checking and uses similar algorithms. A design rule checker must identify transistors and vias to ensure proper checks—otherwise, it might highlight a transistor as a poly-diffusion spacing error. A circuit extractor performs a complete job of component and wire extraction. It produces a net list which lists the transistors in the layout and the electrical nets which connect their terminals. Vias do not appear in the net list—a via simply merges two nets into a single larger net. The circuit extractor usually measures parasitic resistance and capacitance on the wires and annotates the net list with those parasitic values. The next example describes how we can extract a circuit from a layout.

Example 2-7: Circuit extraction

We will extract the circuit by successively identifying, then deleting components. After all component types have been extracted, only the wires will remain.

Identifying components from the layout requires manipulating masks singly and in combination. **Grow** and **shrink** are two important operations:

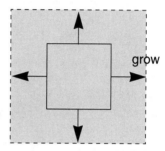

The grow operation increases the extent of each polygon in the mask by a fixed amount in every direction; the shrink operation does the converse. We will also need to form Boolean combinations of masks: the NOT of a mask covers all the area not covered by the mask itself; the AND of two masks covers only the area under both masks; and the OR includes the area covered by either mask. Boolean and grow/shrink operations generate new masks.

When we extract the circuit, we will assume the layout has no design-rule errors; we can always DRC the layout before extraction. We can identify all the transistors in the layout very easily: the n-type transistors' active areas are exactly the AND of the poly and the n-diff masks, with analogous definition for the p-type transistors. After identifying the transistors, we can remove them from the layout of the active-area mask, which leaves the gate, source, and drain connections hanging. We will mark and remember the locations of the transistors' terminals for the final step of extraction.

Identifying vias requires a little more effort. To identify poly-metal1 vias, we first grow the cut mask by 2λ, then we form the AND of the grown-cut, metal, and poly masks. The result is one 4λ-by-4λ square for each poly-metal1 via. After identifying all the vias, we remove them while marking their place. We can identify tub ties, but we won't need them for the later stages of analysis, since they don't make new electrical connections.

At this point, only the wires are left in the layout. A polygon on one layer forms an electrically connected region. However, we're not quite done, because connections may have been made by vias or by wires through transistors. To take account of all connections, we must first identify where each wire touches a connection point to a via or transistor. We then form the transitive closure of all the connection points: if one wire connects points A and B, and another wire connects B and C, then A, B, and C are all electrically connected.

Once we have traced through all the connections, we have a basic circuit description. We have not yet taken parasitics into account. To do so, we must count parasitics for each wire, via, and transistor, then mark each electrical node appropriately. However, for simple functional analysis, extracting parasitics may not be necessary. Here is a fragment of an extracted circuit written in Magic's *ext* format:

```
node "6_38_29#" 122 55 19 -14 green 0 0 0 0 54 34 0 0 92 62 0 0 0 0 0 0
node "6_50_15#" 120 10 25 -7 green 0 0 0 0 12 16 0 0 0 0 0 0 0 0 0 0
node "6_50_7#" 521 92 25 -3 green 0 0 60 44 30 22 0 0 80 64 0 0 0 0 0 0
node "6_36_19#" 825 12 18 -9 p 110 114 0 0 0 0 0 0 0 0 0 0 0 0 0 0
node "6_36_11#" 690 9 18 -5 p 92 96 0 0 0 0 0 0 0 0 0 0 0 0 0 0
node "6_40_40#" 559 83 20 20 brown 0 0 80 54 0 0 0 0 68 58 0 0 0 0 0 0
cap "6_36_19#" "6_50_7#" 1
fet nfet 25 -9 26 -8 12 16 "GND!" "6_36_19#" 4 0 "6_38_29#" 6 0 "6_50_15#" 6 0
fet nfet 25 -5 26 -4 12 16 "GND!" "6_36_11#" 4 0 "6_50_15#" 6 0 "6_50_7#" 6 0
fet pfet 39 17 40 18 12 16 "Vdd!" "6_36_19#" 4 0 "6_50_7#" 6 0 "6_40_40#" 6 0
fet pfet 25 17 26 18 12 16 "Vdd!" "6_36_11#" 4 0 "6_50_7#" 6 0 "6_40_40#" 6 0
```

The exact format of this file isn't important, but a few details should help make this information less forbidding. A node record defines an electrical node in the circuit—explicit declaration of the nodes simplifies the program which reads the file. The record gives total resistance and capacitance for the node, an *x, y* position which can be used to identify the node in the layout, and area and perimeter information for resistance extraction. A cap record gives two nodes and the capacitance between them. A fet record describes the type of transistor, the corners of its channel, and the electrical nodes to which the source, drain, and gate are connected.

The simplest extraction algorithm works on a layout without cells—this is often called flat circuit extraction because the component hierarchy is flattened to a single level before extraction. However, a flattened layout is very large: a layout built of one 100-rectangle cell repeated 100 times will have 100 rectangles plus 100 (small) cell records; the same layout flattened to a single cell will have 10,000 rectangles. That added size claims penalties in disk storage, main memory, and CPU time.

Hierarchical circuit extraction extracts circuits directly on the hierarchical layout description. Dealing with cell hierarchies requires more sophisticated algorithms which are beyond our scope. Hierarchical extraction may also require design restrictions, such as eliminating overlaps between cells. However, one problem which must be solved illustrates the kinds of problems introduced by component hierarchies.

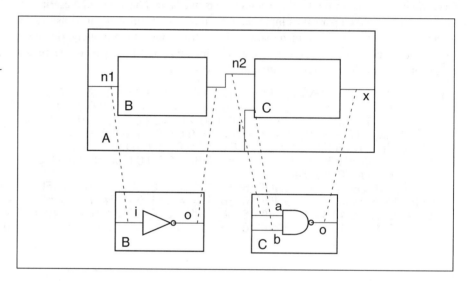

Figure 2-34:
Tracing nets for
hierarchical cir-
cuit extraction.

Consider the example of Figure 2-34. Each cell has its own net list. The net lists of leaf cells make sense on their own, but A's net list is written in terms of its components. We often want to generate a flattened net list—flattening the net list after extraction makes sense because the net list is much smaller than the layout. To create the flattened net list, we must make correspondences between nets in the cells and nets in the top-level component. Once again, we use transitive closure: if net o in cell B is connected to n2 in A, which in turn is connected to net a in C, then B.o, A.n2, and C.a are all connected. Flattening algorithms can be very annoying if they choose the wrong names for combined elements. In this case, n2, the top-level component's name for the net, is probably the name most recognizable to the designer.

An extracted layout has two important uses. First, the extracted circuit can be simulated and the results compared to the specified circuit design. Serious layout errors, such as a missing transistor or wire, should show up as a difference in the specified and extracted circuits. Second, extracted parasitics can be used to calculate actual delays. Circuit performance may have been estimated using standard parasitic values or parasitics may have been ignored entirely, but long wires can slow down logic gates. Comparing the actual performance of the extracted layout to the predicted performance tells you whether the logic and circuits need to be modified and, if so, where critical delay problems exist.

2.6.5 Automatic Layout

Hierarchical stick diagrams are a good way to design large custom cells. But you will probably design large cells from scratch infrequently. You are much more likely to use layouts generated by one of two automated methods: **cell generators** (also known **macrocell generators**), which create optimized layouts for specialized functions such as ALUs; or **standard cell placement and routing**, which use algorithms to build layouts from gate-level cells.

A cell generator is a parameterized layout—it is a program written by a person to generate the layout for a particular cell or a families of cells. The generator program is usually written textually, though some graphical layout editors provide commands to create parameterized layouts. If the generator creates only one layout, it may as well have been created with a graphical layout editor. But designers often want to create variations on a basic cell: changing the sizes of transistors, choosing the number of busses which run through a cell, perhaps adding simple logic functions. Specialized functions like ALUs, register files, and RAMs often require careful layout and circuit design to operate at high speed. Generator languages let skilled designers create parameterized layouts for such cells which can be used by chip designers whose expertise is in system design, not circuit and layout design.

Figure 2-35 shows a fragment of a layout generator written in the CLL language [Sax83]. The wire command works like a pen—it lays down material on a layer as it moves from point to point. Movement may be specified by absolute coordinate or by relative movement; a layer switch instantiates a via at the point of the switch. This small example doesn't show how to build layout procedures, which can take parameters much like procedures in programming languages. Parameters passed in at run time can control whether layout elements are placed at all and where they are placed.

Place-and-route programs take a very different approach to layout synthesis: they break the problem into placing components on the plane, then routing wires to make the necessary connections. Placement and routing algorithms may not be able to match the quality of hand-designed layouts for some specialized functions, but they

Figure 2-35: A layout generation program.

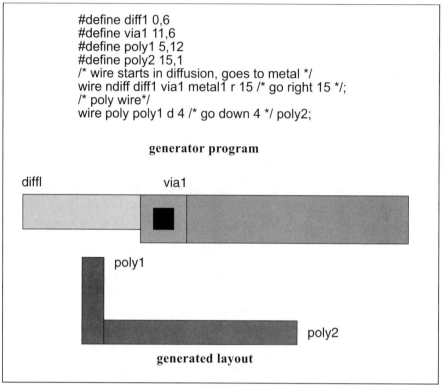

often do better than people on large random logic blocks because they have greater patience to search through large, unstructured problems to find good solutions.

The most common placement-and-routing systems use **standard cells**, which are logic gates, latches, flip-flops, or occasionally slightly larger functions like full adders. Figure 2-36 shows the architecture of a standard cell layout: the component cells, which are of standard height but of varying width, are arranged in rows; wires are run in **routing channels** between the cell rows, along the sides, and occasionally through **feedthroughs** (spaces left open for wires in the component cells). The layout is designed in two stages: components are placed using approximations to estimate the amount of wire required to make the connections; then the wires are routed. We will discuss standard cell layout synthesis in more detail in Chapter 6. Figure 2-37 shows a small standard cell layout generated by the *wolfe* program [San84, Sec85].

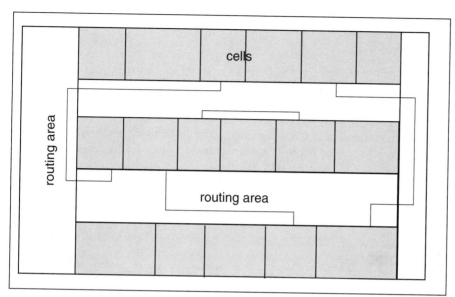

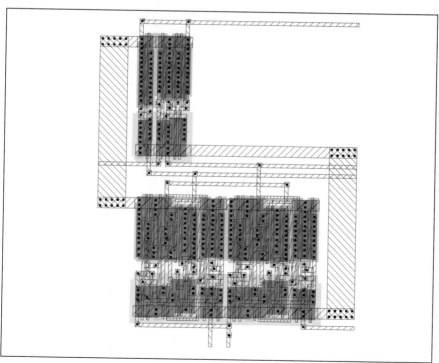

Figure 2-36:
Architecture of a standard cell layout.

Figure 2-37: *An example of standard cell layout.*

2.7 References

As mentioned in the last chapter, Dennard *et al.* [Den74] first explained why shrinking IC feature sizes led to higher performance as well as smaller chips. That observation led to the development of scalable design rules, which were first introduced by Mead and Conway [Mea80]. The specifications for the MOSIS SCMOS process were derived from MOSIS data. Complete documentation on the SCMOS rules is available on the World Wide Web at http://www.mosis.edu or by sending electronic mail to mosis@mosis.edu. The MOSIS SCMOS rules do occasionally change, so it is always best to consult MOSIS for the latest design rules before starting a design. Cheng et al. [Che00] survey modeling techniques for interconnect.

2.8 Problems

Use process parameters from Table 2-4 as required.

2-1. Draw the cross-section of the inverter shown below along a cut through the middle of the p-type and n-type transistors.

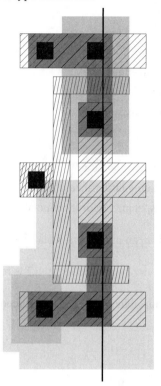

cut

2-2. Assuming that $V_{gs} = 3.3V$, compute the drain current through n-type transistors of these sizes at V_{ds} values of 1V, 2V, 3.3V, and 5V:

    a) W/L = 5/2.

    b) W/L = 8/2.

    c) W/L =12/2.

    d) W/L = 25/2.

2-3. Using the parameter values given in Example 2-3:

    a) How much would you have to change the gate oxide thickness to reduce the threshold voltage by 0.1 V?

    b) How much would you have to change the doping N_A to reduce the threshold voltage by 0.1 V?

2-4. Plot I_d vs. V_{gs} at $V_{ds} = 5V$ for two values of the channel length modulation parameter: $\lambda = 0$ and $\lambda = 0.03$.

2-5. Justify each of these design rules:

    a) 2λ poly-poly spacing;

    b) no required poly-metal spacing;

    c) 1λ of diffusion and metal surrounding cut;

    d) 2λ overhang of poly at transistor gate.

2-6. Explain:

    a) Why is ndiff-to-pdiff spacing so large?

    b) Why is metal-metal spacing larger than poly-poly spacing?

    c) Why is metal2-metal2 spacing larger than metal1-metal1 spacing?

2-7. What distinguishes a tub tie from an ndiff-metal1 via?

2-8. Design the layout for an ndiff wire connected to a pdiff wire.

2-9. Compute the resistance and capacitance for each polysilicon wire below:

    a)

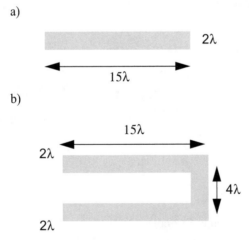

    b)

2-10. Compute the parasitic resistance and capacitance of the source/drain region of transistors of the types and sizes given below, assuming that these source drain-regions have the required 3 λ overhang past the gate:

    a) p-type; W/L = 3/2;

     b) n-type, W/L = 4/2;

     c) p-type; W/L = 6/2;

     d) n-type; W/L = 12/2.

2-11. Recompute the parasitic resistances and capacitances of the transistors of Question 2-10. Assume in this case that the source/drain regions have the required overhang but also include as many diffusion/metal 1 vias as will fit in the given gate width. Calculate the total resistance including the via resistance.

2-12. Calculate the parasitic resistance and capacitance on the input to the inverter shown on the next page. The transistors have W/L = 3 λ / 2 λ and the V_{DD} and V_{SS} lines are each 4 λ wide. Assume that the resistance of a via is 4Ω. Give the total resistance and capacitance of all the material connected to the inverter's input, showing how you broke the layout elements into sections for measurement.

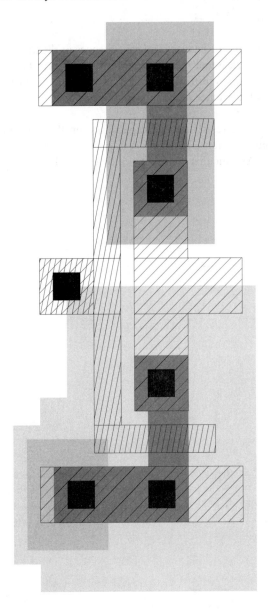

2-13. Reverse-engineer the layout shown on the next page. Draw:

a) a stick diagram corresponding to the layout;

b) a transistor schematic.

Label inputs and outputs in your drawings in accordance with the labels in the layout.

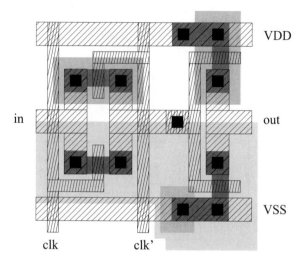

2-14. Design a layout for the circuit fragment below in two stages:

a) a stick diagram;

b) the complete layout.

2-15. Design two different stick diagrams for the inverter circuit of Example 2-5. Your only constraint is that both V_{DD} and V_{SS} must be available on two opposite sides of the cell.

2-16. Expand the stick diagram for the one-bit-mux cell of Example 2-6 to include the wires in the NAND cells.

2-17. Redesign the three-bit multiplexer of Example 2-6 so that adjacent one-bit-mux cells share V_{DD} and V_{SS} lines. You can overlap cells to force sharing of wires.

2-18. Design two different stick diagrams for this circuit, such that both V_{DD} and V_{SS} are available on two opposite sides of the cell:

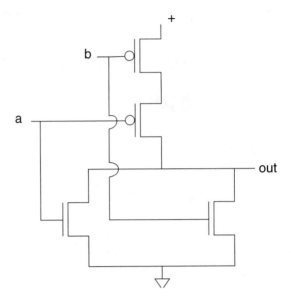

2-19. Design a layout for the circuit shown below in two stages: first draw a stick diagram, then design the complete layout. Assume that all transistors are 3 λ / 2 λ. The power supply lines should run through the cell in metal 1, available on opposite sides of the cell.

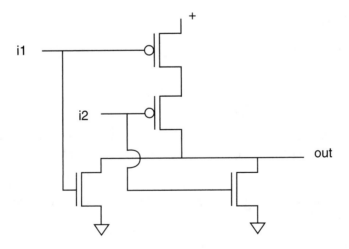

2-20. Design a stick diagram for the circuit shown below, assuming that transistor sizes have W/L = 3 λ / 2 λ except as indicated. The cell should be designed to be area-efficient and to be horizontally tilable—you should be able to place two of your cells side-by-side and make all electrical connections without adding any wires.

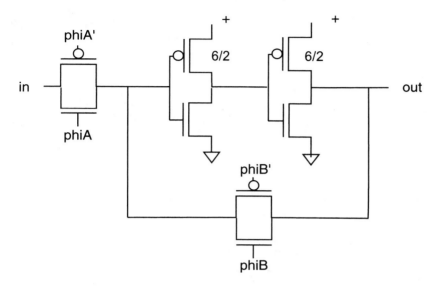

2-21. Design a layout for the stick diagram you designed in Question 2-20. Show that the cell tiles horizontally.

3

Logic Gates

Highlights:

Combinational logic.

Static logic gates.

Delay and power.

Alternate gate structures: switch, domino, etc.

Wire delay models.

3.1 Introduction

This chapter concentrates on the design of combinational logic functions. The knowledge gained in the last chapter on fabrication is important for combinational logic design—technology-dependent parameters for minimum size, spacing, and parasitic values largely determine how big a gate circuit must be and how fast it can run. We will start by reviewing some important facts about combinational logic functions. The first family of logic gate circuits we will consider are **static, fully complementary** gates, which are the mainstay of CMOS design. We will analyze the properties of these gates in detail: speed, power consumption, layout design, testability. We will also study some more advanced circuit families—pseudo-nMOS, DCVS, domino, and low-power gates—that are important in special design situations. We will also study the delays through wires, which can be much longer than the delays through the gates.

3.2 Combinational Logic Functions

First, it is important to distinguish between *combinational logic expressions* and *logic gate networks*. A combinational logic expression is a mathematical formula which is to be interpreted using the laws of Boolean algebra: given the expression $a + b$, for example, we can compute its truth value for any given values of a and b; we can also evaluate relationships such as $a + b = c$. A logic gate computes a specific Boolean function, such as $(a + b)'$. The goal of logic design or optimization is to find a network of logic gates which together compute the combinational logic function we want. Logic optimization is interesting and difficult for two reasons:

- We may not have a logic gate for every possible function, or even for every function of n inputs. It therefore may be a challenge to rewrite our combinational logic expression so that each term represents a gate.

- Not all gate networks that compute a given function are alike—networks may differ greatly in their area and speed. We want to find a network that satisfies our area and speed requirements, which may require drastic restructuring of our original logic expression.

Figure 3-1: Two logic gate implementations of a Boolean function.

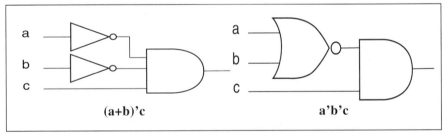

(a+b)'c   a'b'c

Figure 3-1 illustrates the relationship between logic expressions and gate networks. The two expressions are logically equivalent: $(a + b)'c = a'b'c$. We have shown a logic gate network for each expression which directly implements each function—each term in the expression becomes a gate in the network. The two logic networks have very different structures. Which is best depends on the requirements—the relative importance of area and delay—and the characteristics of the technology. But we must work with both logic expressions and gate networks to find the best implementation of a function, keeping in mind the relationships:

- combinational logic expressions are the specification;
- logic gate networks are the implementation;
- area, delay, and power are the costs.

We will use fairly standard notation for logic expressions: if a and b are variables, then a' (or $\bar{a}$) is the complement of a, $a \cdot b$ (or ab) is the AND of the variables, and $a + b$ is the OR of the variables. In addition, for the NAND function (ab)' we will use the | symbol[1], for the NOR function $(a + b)$' we will use a NOR b, and for exclusive-or (a XOR $b = ab' + a'b$) we will use the $\oplus$ symbol. (Students of algebra know that XOR and AND form a ring.) We use the term **literal** for either the true form (a) or complemented form (a') of a variable. Understanding the relationship between logical expressions and gates lets us study problems in the model that is simplest for that problem, then transfer the results. Two problems that are of importance to logic design but easiest to understand in terms of logical expressions are **completeness** and **irredundancy**.

A set of logical functions is complete if we can generate every possible Boolean expression using that set of functions—that is, if for every possible function built from arbitrary combinations of +, ·, and ', an equivalent formula exists written in terms of the functions we are trying to test. We generally test whether a set of functions is complete by inductively testing whether those functions can be used to generate all logic formulas. It is easy to show that the NAND function is complete, starting with the most basic formulas:

- 1: $a|(a|a) = a|a$'$= 1$.
- 0: $\{a|(a|a)\}|\{a|(a|a)\} = 1|1 = 0$.
- a': $a|a = a$'.
- ab: $(a|b)|(a|b) = ab$.
- $a + b$:$(a|a)|(b|b) = a$'$|b$' $= a + b$.

---

1. The Scheffer stroke is a dot with a negation line through it. C programmers should note that this character is used as OR in the C language.

From these basic formulas we can generate all the formulas. So the set of functions {|} can be used to generate any logic function. Similarly, any formula can be written solely in terms of NORs.

The combination of AND and OR functions, however, is not complete. That is fairly easy to show: there is no way to generate either 1 or 0 directly from any combination of AND and OR. If NOT is added to the set, then we can once again generate all the formulas: $a + a' = 1$, etc. In fact, both {', ·} and {',+} are complete sets.

Any circuit technology we choose to implement our logic functions must be able to implement a complete set of functions. Static, complementary circuits naturally implement NAND or NOR functions, but some other circuit families do not implement a complete set of functions. Incomplete logic families place extra burdens on the logic designer to ensure that the logic function is specified in the correct form.

A logic expression is irredundant if no literal can be removed from the expression without changing its truth value. For example, $ab + ab'$ is redundant, because it can be reduced to a. An irredundant formula and its associated logic network have some important properties: the formula is smaller than a logically equivalent redundant formula; and the logic network is guaranteed to be testable for certain kinds of manufacturing defects. However, irredundancy is not a panacea. Irredundancy is not the same as **minimality**—there are many irredundant forms of an expression, some of which may be smaller than others, so finding one irredundant expression may not guarantee you will get the smallest design. Irredundancy often introduces added delay, which may be difficult to remove without making the logic network redundant. However, simplifying logic expressions before designing the gate network is important for both area and delay. Some obvious simplifications can be done by hand; CAD tools can perform more difficult simplifications on larger expressions.

3.3 Static Complementary Gates

This section concentrates on one family of logic gate circuits: the static complementary gate. These gates are static because they do not depend on stored charge for their operation. They are complementary because they are built from complementary (dual) networks of p-type and n-type transistors. The important characteristics of a logic gate circuit are its layout area, delay, and power consumption. We will concentrate our analysis on the inverter because it is the simplest gate to analyze and its analysis extends straightforwardly to more complex gates.

3.3.1 Gate Structures

A static complementary gate is divided into a **pullup network** made of p-type transistors and a **pulldown network** made of n-type transistors. The gate's output can be connected to V_{DD} by the pullup network or V_{SS} by the pulldown network. The two networks are complementary to ensure that the output is always connected to exactly one of the two power supply terminals at any time: connecting the output to neither would cause an indeterminate logic value at the output, while connecting it to both would cause not only an indeterminate output value, but also a low-resistance path from V_{DD} to V_{SS}. The structures of an inverter, a two-input NAND gate, and a two-input NOR gate are shown in Figure 3-2, Figure 3-3, and Figure 3-4, respectively; + stands for V_{DD} and the triangle stands for V_{SS}. Inspection shows that they satisfy the complementarity requirement: for any combination of input values, the output value is connected to exactly one of V_{DD} or V_{SS}.

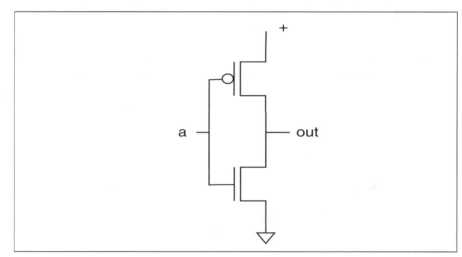

Figure 3-2: *Transistor schematic of a static complementary inverter.*

Gates can be designed for functions other than NAND and NOR by designing the proper pullup and pulldown networks. Networks that are series-parallel combinations of transistors can be designed directly from the logic expression the gate is to implement. In the pulldown network, series-connected transistors or subnetworks implement AND functions in the expression and parallel transistors or subnetworks implement OR functions. The converse is true in the pullup network because p-type transistors are off when their gates are high. Consider the design of a two-input NAND gate as an example. To design the pulldown network, write the gate's logic expression to have negation at the outermost level: $(ab)'$ in the case of the NAND. This expression specifies a series-connected pair of n-type transistors. To design the pullup network, rewrite the expression to have the inversion pushed down to the

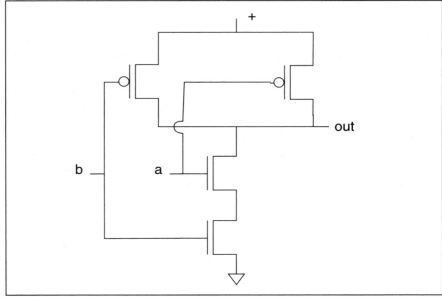

Figure 3-3: A
static complemen-
tary NAND gate.

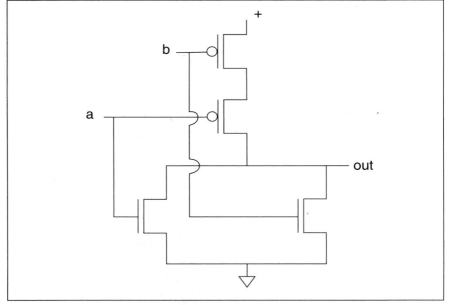

Figure 3-4: A
static complemen-
tary NOR gate.

innermost literals: $a' + b'$ for the NAND. This expression specifies a parallel pair of
p-type transistors, completing the NAND gate design of Figure 3-3. Figure 3-5 shows
the topology of a gate which computes $[a(b+c)]'$: the pulldown network is given by

the expression, while the rewritten expression $a' + (b'c')$ determines the pullup network.

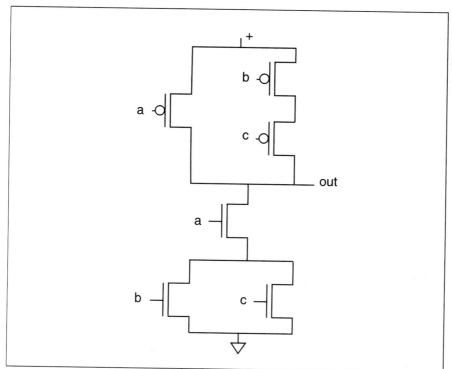

Figure 3-5: *A static complementary gate that computes [a(b+c)]'.*

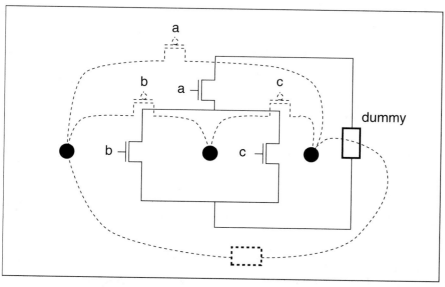

Figure 3-6: *Constructing the pullup network from the pulldown network.*

You can also construct the pullup network of an arbitrary logic gate from its pulldown network, or vice versa, because they are **duals**. Figure 3-6 illustrates the dual construction process using the pulldown network of Figure 3-5. First, add a dummy component between the output and the V_{SS} (or V_{DD}) terminals. Assign a node in the dual network for each region, including the area not enclosed by wires, in the non-dual graph. Finally, for each component in the non-dual network, draw a dual component which is connected to the nodes in the regions separated by the non-dual component. The dual component of an n-type transistor is a p-type, and the dual of the dummy is the dummy. You can check your work by noting that the dual of the dual of a network is the original network.

Common forms of complex logic gates are **and-or-invert** (AOI) and **or-and-invert** (OAI) gates, both of which implement sum-of-products/product-of-sums expressions. The function computed by an AOI gate is best illustrated by its logic symbol, shown in Figure 3-7: groups of inputs are ANDed together, then all products are ORed together and inverted for output. An AOI-21 gate, like that shown in the figure, has two inputs to its first product and one input (effectively eliminating the AND gate) to its second product; an AOI-121 gate would have two one-input products and one two-input product.

It is possible to construct large libraries of complex gates with different input combinations. An OAI gate computes an expression in product-of-sums form: it generates sums in the first stage which are then ANDed together and inverted. An AOI or OAI function can compute a sum-of-products or product-of-sums expression faster and using less area than an equivalent network of NAND and NOR gates. Human designers rarely make extensive use of AOI and OAI gates, however, because people have difficulty juggling a large number of gate types in their heads. Logic optimization programs, however, can make very efficient use of AOI, OAI, and other complex gates to produce very efficient layouts.

3.3.2 Basic Gate Layouts

Figure 3-8 shows a layout of an inverter, Figure 3-10 shows a layout of a static NAND gate, and Figure 3-11 shows a layout of a static NOR gate. Transistors in a gate can be densely packed—the NAND gate is not much larger than the inverter. Layouts can vary greatly, depending on the requirements of the cell: transistor sizes, positions of terminals, layers used to route signals. CMOS technology allows few major variations of the basic cell organization: V_{DD} and V_{SS} lines run in metal along the cell, with n-type transistors along the V_{SS} rail and p-types along the V_{DD} rail. The input and output signals of the NAND are presented at the cell's edge on different layers: the inputs are in poly while the output is in metal 1. If we want to cascade two cells, with the

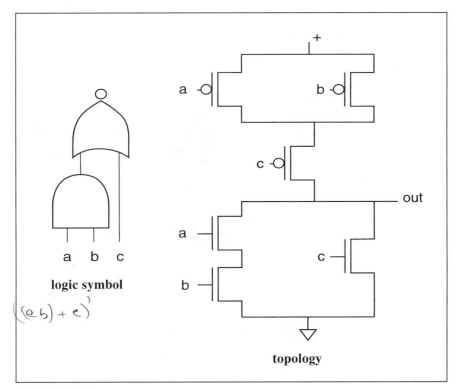

Figure 3-7: An
and-or-invert-21
(AOI-21) gate.

output of one feeding an input of another, we will have to add a via to switch layers; we will also have to add the space between the cells required for the via and make sure that the gaps in the V_{DD} and V_{SS} caused by the gap are bridged. The p-type transistors in the NAND and NOR gate were made wide to compensate for their lower current capability; in practice, the inverter layout would probably have a wider pullup as well. We routed both input wires of the NAND to the transistor gates entirely in poly, while we used a metal 1 jumper in one of the NOR inputs.

If you are truly concerned with cell size, many variations are possible. Figure 3-9 shows a very wide transistor. A very wide transistor can create too much white space in the layout, especially if the nearby transistors are smaller. We have split this transistor into two pieces, each half as wide, and turned one piece 180 degrees, so that the outer two sections of diffusions are used as drains and the inner sections become sources.

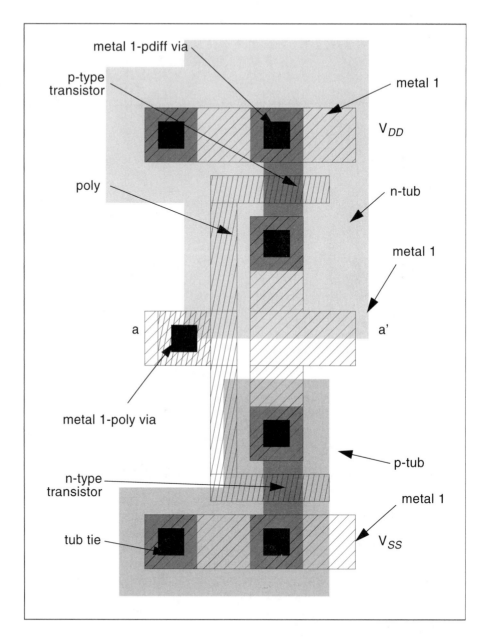

Figure 3-8: A layout of an inverter.

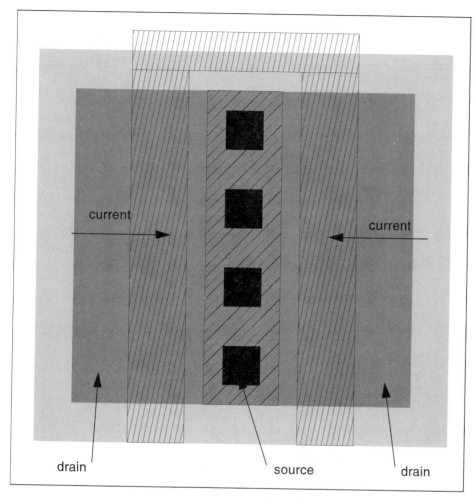

Figure 3-9: A
wide transistor
split into two
sections.

current →

← current

drain

source

drain

3.3.3 Logic Levels

Since we must use voltages to represent logic values, we must define the relationship between the two. As Figure 3-12 shows, a range of voltages near V_{DD} corresponds to logic 1 and a band around V_{SS} corresponds to logic 0. The range in between is X, the unknown value. Although signals must swing through the X region while the chip is operating, no node should ever achieve X as its final value.

We want to calculate the upper boundary of the logic 0 region and the lower boundary of the logic 1 region. In fact, the situation is slightly more complex, as shown in

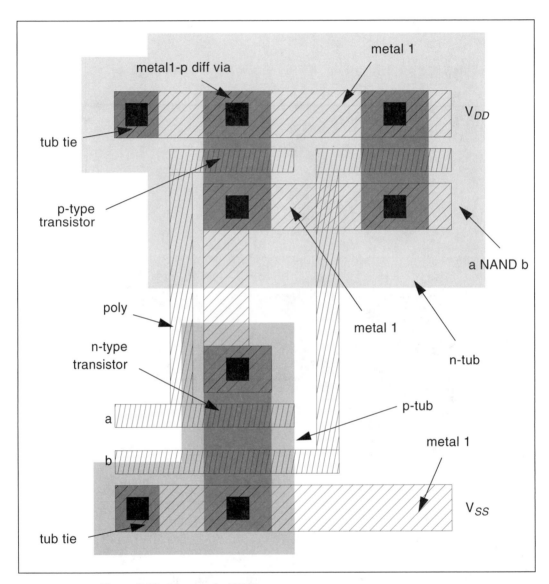

Figure 3-10: *A layout of a NAND gate.*

Figure 3-13, because we must consider the logic levels produced at outputs and required at inputs. Given our logic gate design and process parameters, we can guarantee that the maximum voltage produced for a logic 0 will be some value V_{OL} and that the minimum voltage produced for a logic 0 will be V_{OH}. These same constraints place limitations on the input voltages which will be interpreted as a logic 0 (V_{IL}) and

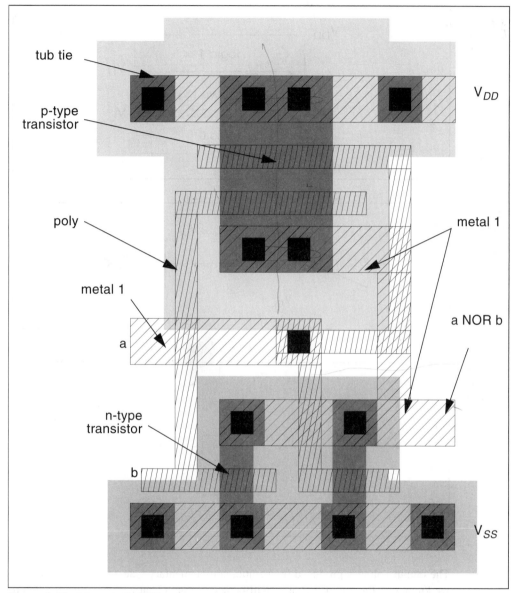

Figure 3-11: *A layout of a NOR gate.*

logic 1 (V_{IH}). If the gates are to work together, we must ensure that $V_{OL} < V_{IL}$ and $V_{OH} > V_{IH}$.

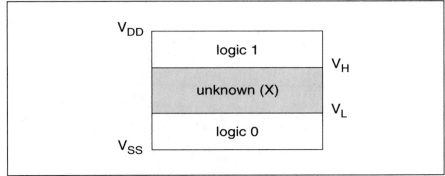

Figure 3-12:
How voltages correspond to logic levels.

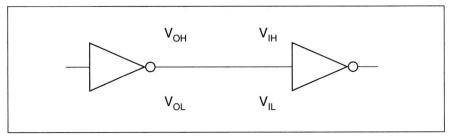

Figure 3-13: *Logic levels on cascaded gates.*

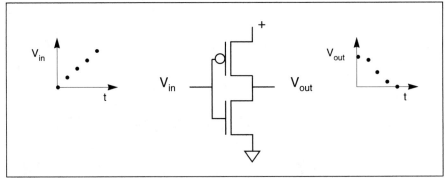

Figure 3-14: *The inverter circuit used to measure transfer characteristics.*

The output voltages produced by a static, complementary gate are V_{DD} and V_{SS}, so we know that the output voltages will be acceptable. (That isn't true of all gate circuits; the pseudo-nMOS circuit of Section 3.5.1 produces a logic 0 level well above V_{SS}.) We need to compute the values of V_{IL} and V_{IH} and to do the computation, we need to define those values. A standard definition is based on the transfer characteristic of the inverter—its output voltage as a function of its input voltage, assuming that the input voltage and all internal voltages and currents are at equilibrium. Figure 3-14 shows the circuit we will use to measure an inverter's transfer characteristic. We

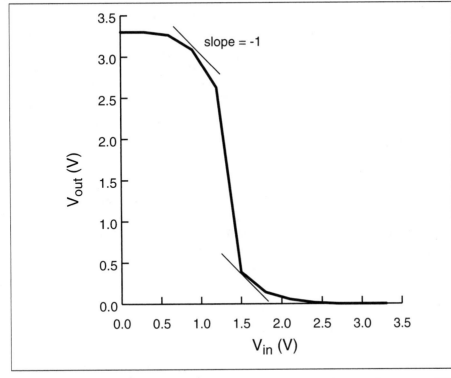

Figure 3-15: Voltage transfer curve of an inverter with minimum-size transistors.

apply a sequence of voltages to the input and measure the voltage at the output. (We can also sweep the input voltage if we do at a much slower rate than the circuit's transients.) Alternatively, we can solve the circuit's voltage and current equations to find V_{out} as a function of V_{in}: we equate the drain currents of the two transistors and set their gate voltages to be complements of each other (since the n-type's gate voltage is measured relative to V_{SS} and the p-type's to V_{DD}).

Figure 3-15 shows a **transfer characteristic** (simulated using Spice level 3 models) of an inverter with minimum-size transistors for both pullup and pulldown. We define V_{IL} and V_{IH} as the points at which the curve's tangent has a slope of -1. Between these two points, the inverter has high gain—a small change in the input voltage causes a large change in the output voltage. Outside that range, the inverter has a gain less than 1, so that even a large change at the input causes only a small change at the output, attenuating the noise at the gate's input. The curve is not symmetric because the pullup supplies less current than the pulldown when both are minimum size. In particular, the valid logic 1 range is smaller than the valid logic 0 range because the pullup's resistance is too high relative to the pulldown's.

The difference between V_{OL} and V_{IL} (or between V_{OH} and V_{IH}) is called the **noise margin**—the size of the safety zone that prevents production of an illegal X output value. Since real circuits function under less-than-ideal conditions, adequate noise margins are essential for ensuring that the chip operates reliably. Noise may be introduced by a number of factors: it may be introduced by off-chip connections; it may be generated by capacitive coupling to other electrical nodes; or it may come from variations in the power supply voltage.

3.3.4 Delay and Transition Time

Delay is one of the most important properties of a logic gate—the majority of chip designs are limited more by speed than by area. An analysis of logic gate delay not only tells us how to compute the speed of a gate, it also points to parasitics that must be controlled during layout design to minimize delay. Later, in Section 3.3.7, we will apply what we have learned from delay analysis to the design of logic gate layouts.

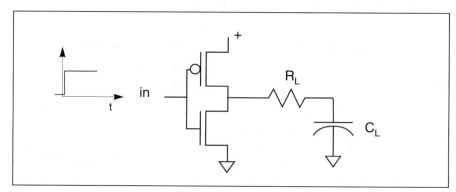

Figure 3-16: The inverter circuit used for delay analysis.

There are two interesting but different measures of combinational logic effort:

- **Delay** is generally used to mean the time it takes for a gate's output to arrive at 50% of its final value.

- **Transition time** is generally used to mean the time it takes for a gate to arrive at 10% (for a logic 0) or 90% (for a logic 1) of its final value; both **fall time** t_f and **rise time** t_r are transition times.

We will analyze delay and transition time on the simple inverter circuit shown in Figure 3-16; our analysis easily extends to more complex gates as well as more complex loads. We will assume that the inverter's input changes voltage instantaneously; since the input signal to a logic gate is always supplied by another gate, that assumption is optimistic, but it simplifies analysis without completely misleading us.

It is important to recognize that we are analyzing not just the gate delay but delay of the combination of the gate and the load it drives. CMOS gates have low enough gain to be quite sensitive to their load, which makes it necessary to take the load into account in even the simplest delay analysis. The load on the inverter is a single resistor-capacitor (RC) circuit; the resistance and capacitance come from the logic gate connected to the inverter's output and the wire connecting the two. We will see in Section 4.5.1 that other models of the wire's load are possible. There are two cases to analyze: the output voltage V_{out} is pulled down (due to a logic 1 input to the inverter); and V_{out} is pulled up. Once we have analyzed the $1 \rightarrow 0$ output case, modifying the result for the $0 \rightarrow 1$ case is easy.

While the circuit of Figure 3-16 has only a few components, a detailed analysis of it is difficult due to the complexity of the transistor's behavior. We need to further simplify the circuit. A detailed circuit analysis would require us to consider the effects of both pullup and pulldown transistors. However, our assumption that the inverter's input changes instantaneously between the lowest and highest possible values lets us assume that one of the transistors turns off instantaneously. Thus, when V_{out} is pulled low, the p-type transistor is off and out of the circuit; when V_{out} is pulled high, the n-type transistor can be ignored.

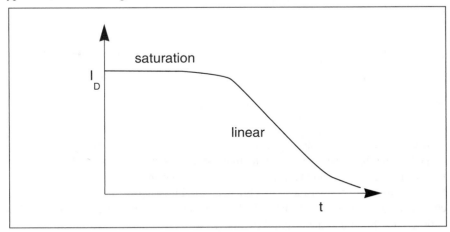

Figure 3-17:
Current through the pulldown during a $1 \rightarrow 0$ transition.

There are several different models that people use to compute delay and transition time. The first is the τ **model**, which was introduced by Mead and Conway [Mea80] as a simple model for basic analysis of digital circuits. This model reduces the delay of the gate to an RC time constant which is given the name τ. As the sizes of the transistors in the gate are increased, the delay scales as well.

At the heart of the τ model is the assumption that the pullup or pulldown transistor can be modeled as a resistor. The transistor does not obey Ohm's law as it drives the gate's output, of course. As Figure 3-17 shows, the pulldown spends the first part of the $1 \rightarrow 0$ transition in the saturation region, then moves into the linear region. But the resistive model will give sufficiently accurate results to both estimate gate delay and to understand the sources of delay in a logic circuit.

Figure 3-18:
How to approxi-
mate a transistor
with a resistor.

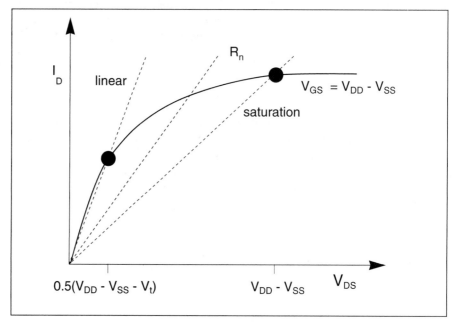

How do we choose a resistor value to represent the transistor over its entire operating range? A standard resistive approximation for a transistor is to measure the transistor's resistance at two points in its operation and take the average of the two values [Hod83]. We find the resistance by choosing a point along the transistor's I_d vs. V_{ds} curve and computing the ratio V/I, which is equivalent to measuring the slope of a line between that point and the origin. Figure 3-18 shows the approximation points for an n-type transistor: the inverter's maximum output voltage, $V_{DS} = V_{DD} - V_{SS}$, where the transistor is in the saturation region; and the middle of the linear region, $V_{DS} = (V_{DD} - V_{SS} - V_t)/2$. We will call the first value $R_{sat} = V_{sat}/I_{sat}$ and the second value $R_{lin} = V_{lin}/I_{lin}$. This gives the basic formula

$$R_n = \left(\frac{V_{sat}}{I_{sat}} + \frac{V_{lin}}{I_{lin}}\right)/2 \qquad \text{(EQ 2-1)}$$

for which we must find the Vs and Is.

The current through the transistor at the saturation-region measurement point is

$$I_{sat} = \frac{1}{2}k'\frac{W}{L}(V_{DD}\text{-}V_{SS}\text{-}V_t)^2. \qquad \text{(EQ 3-2)}$$

The voltage across the transistor at that point is

$$V_{sat} = V_{DD} - V_{SS}. \qquad \text{(EQ 3-3)}$$

At the linear region point,

$$V_{lin} = (V_{DD}\text{-}V_{SS}\text{-}V_t)/2, \qquad \text{(EQ 3-4)}$$

so the drain current is

$$\begin{aligned} I_{lin} &= k'\frac{W}{L}\left[\frac{1}{2}(V_{DD}\text{-}V_{SS}\text{-}V_t)^2 - \frac{1}{2}\left(\frac{V_{DD}\text{-}V_{SS}\text{-}V_t}{2}\right)^2\right] \\ &= \frac{3}{8}k'\frac{W}{L}(V_{DD}\text{-}V_{SS}\text{-}V_t)^2 \end{aligned} \qquad \text{(EQ 3-5)}$$

We can compute the effective resistances of transistors in the 0.5 μm process by plugging in the technology values of Table 2-4. The resistance values for minimum-size n-type and p-type transistors are shown in Table 3-1 for two power supply voltages: 5V and 3.3V. The effective resistance of a transistor is scaled by L/W. The p-type transistor has about three-and-a-half times the effective resistance of an n-type transistor for this set of process parameters, which is what we expect from the ratio k'_n/k'_p. and the variation in threshold voltages between the two types of transistors. Note that the effective resistance of the transistors increases as the power supply voltage goes down.

Table 3-1
*Effective
resistance values
for minimum-size
transistors in our
0.5 μm process.*

type	V_{DD}-V_{SS} = 5V	V_{DD}-V_{SS} = 3.3V
R_n	3.9 kΩ	6.8 kΩ
R_p	14 kΩ	25 kΩ

Given these resistance values, we can then analyze the delay and transition time of the gate.

Figure 3-19:
*The circuit model
for τ model
delay.*

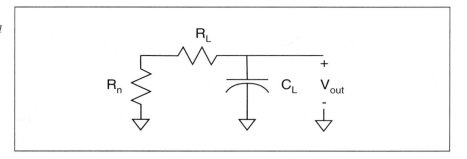

We can now develop the τ model that helps us compute delay and transition time. Figure 3-19 shows the circuit model we use: R_n is the transistor's effective resistance while R_L and C_L are the load. The capacitor has an initial voltage of V_{DD}. The transistor discharges the load capacitor from V_{DD} to V_{SS}; the output voltage as a function of time is

$$V_{out}(t) = V_{DD}e^{-t/[(R_n + R_L)C_L]}.$$ (EQ 3-6)

We typically use R_L to represent the resistance of the wire which connects the inverter to the next gate; in this case, we'll assume that $R_L = 0$, simplifying the total resistance to $R = R_n$.

To measure delay, we must calculate the time required to reach the 50% point. Then

$$0.5 = e^{-t_d/[(R_n + R_L)C_L]}, \qquad \text{(EQ 3-7)}$$

$$t_d = -(R_n + R_L)C_L \ln 0.5 = 0.69(R_n + R_L)C_L. \qquad \text{(EQ 3-8)}$$

We generally measure transition time as the interval between the time at which $V_{out} = 0.9V_{DD}$ and $V_{out} = 0.1V_{DD}$; let's call these times t_1 and t_2. Then

$$t_f = t_2 - t_1 = -(R_n + R_L)C_L \ln\frac{0.1}{0.9} = 2.2(R_n + R_L)C_L. \qquad \text{(EQ 3-9)}$$

The next example illustrates how to compute delay and transition time using the τ model.

Example 3-1: *Inverter delay and transition time using the τ model*

Once the effective resistance of a transistor is known, delay calculation is easy. What is a minimum inverter delay and fall time with our 0.5 μm process parameters? Assume a minimum-size pulldown, no wire resistance, and a capacitive load equal to two minimum-size transistors' gate capacitance. First, the τ model parameters:

$$R_n = 3.9k\Omega$$

$$C_L = 0.9\frac{fF}{\mu m^2} \times \left(3\lambda \times 2\lambda \times \frac{0.0625\mu m^2}{\lambda^2}\right) \times 2$$

$$= 0.68fF$$

Then delay is

$$t_d = 0.69 \cdot 3.9k\Omega \cdot 0.68\times 10^{-15} = 1.8ps$$

and fall time is

$$t_f = 2.2 \cdot 3.9k\Omega \cdot 0.68 \times 10^{-15} = 5.8ps$$

If the transistors are not minimum size, their effective resistance is scaled by L/W. To compute the delay through a more complex gate, such as a NAND or an AOI, compute the effective resistance of the pullup/pulldown network using the standard Ohm's law simplifications, then plug the effective R into the delay formula.

If we decrease the supply voltage to 3.3 V, the load capacitance does not change but the effective resistance of the transistor does:

$$R_n = 6.8k\Omega,$$

$$t_d = 0.69 \cdot 6.8k\Omega \cdot 0.68 \times 10^{-15} = 3.1ps,$$

$$t_f = 2.2 \cdot 6.8k\Omega \cdot 0.68 \times 10^{-15} = 10ps$$

This simple RC analysis tells us two important facts about gate delay. First, if the pullup and pulldown transistor sizes are equal, the $0 \rightarrow 1$ transition will be about one-half to one-third the speed of the $1 \rightarrow 0$ transition. That observation follows directly from the ratio of the n-type and p-type effective resistances. Put another way, to make the high-going and low-going transition times equal, the pullup transistor must be twice to three times as wide as the pulldown. Second, complex gates like NANDs and NORs require wider transistors where those transistors are connected in series. A NAND's pulldowns are in series, giving an effective pulldown resistance of $2R_n$. To give the same delay as an inverter, the NAND's pulldowns must be twice as wide as the inverter's pulldown. The NOR gate has two p-type transistors in series for the pullup network. Since a p-type transistor must be two to three times wider than an n-type transistor to provide equivalent resistance, the pullup network of a NOR can take up quite a bit of area.

A second model is the **current source model,** which is sometimes used in power/delay studies because of its tractability. If we assume that the transistor acts as a current source whose V_{gs} is always at the maximum value, then the delay can be approximated as

$$t_f = \frac{C_L(V_{DD}\text{-}V_{SS})}{I_d} = \frac{C_L(V_{DD}\text{-}V_{SS})}{0.5k'(W/L)(V_{DD}\text{-}V_{SS}\text{-}V_t)^2}. \qquad \text{(EQ 3-10)}$$

A third type of model is the **fitted model.** This approach measures circuit characteristics and fits the observed characteristics to the parameters in a delay formula. This technique is not well-suited to hand analysis but it is easily used by programs that analyze large numbers of gates.

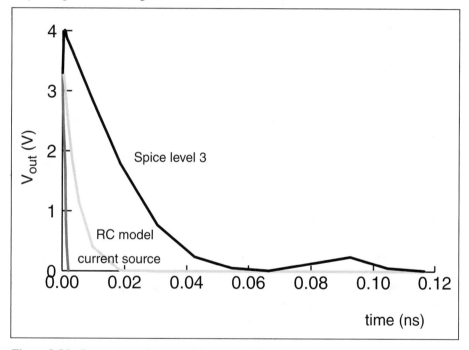

Figure 3-20: *Comparison of inverter delay to the RC and current source approximations.*

How accurate are the RC and current source approximations? Figure 3-20 shows the results of Spice simulation of three circuits: a resistance discharging a capacitance equal to the gate capacitance of an inverter with minimum-size transistors, a full inverter discharging the same capacitance, and the current source approximation. The results show that, for these process parameters, the resistance value calculated by the two-step method is somewhat optimistic and the current source approximation is even more so. We can use Spice simulation to generate a more accurate model of an inverter for a particular process, but you should always remember that the RC delay model is meant as only a rough approximation. RC and current source delay are best

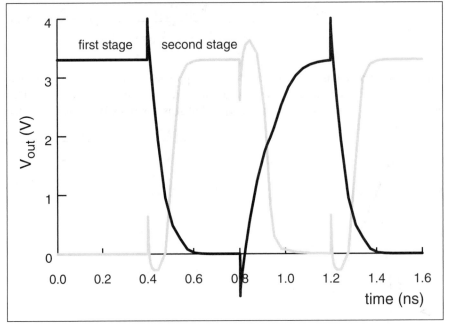

Figure 3-21:
Circuit simulation
of a pair of invert-
ers.

used as relative measures of delay, not absolute measures.

The RC model assumes that the gate's input is a step, but the input in fact comes from another gate which may generate a relatively slow signal. Figure 3-21 shows the results of Spice simulation of one inverter driving another; the first is driven by a square wave, but the second is driven by the output of the first inverter. The second inverter's output response is somewhat slower than the first's.

The fundamental reason for developing an RC model of delay is that we often can't afford to use anything more complex. Full circuit simulation of even a modest-size chip is infeasible: we can't afford to simulate even one waveform, and even if we could, we would have to simulate all possible inputs to be sure we found the worst-case delay. The RC model lets us identify sections of the circuit which probably limit circuit performance; we can then, if necessary, use more accurate tools to more closely analyze the delay problems of that section.

Body effect, as we saw in Section 2.3.5, is the modulation of threshold voltage by a difference between the voltage of the transistor's source and the substrate—as the source's voltage rises, the threshold voltage also rises. This effect can be modeled by a capacitor from the source to the substrate's ground as shown in Figure 3-22. To eliminate body effect, we want to drive that capacitor to 0 voltage as soon as possible. If there is one transistor between the gate's output and the power supply, body effect is not a problem, but series transistors in a gate pose a challenge. Not all of the gate's input signals may reach their values at the same time—some signals may arrive earlier than others. If we connect early-arriving signals to the transistors nearest the power supply and late-arriving signals to transistors nearest the gate output, the early-arriving signals will discharge the body effect capacitance of the signals closer to the output. This simple optimization can have a significant effect on gate delay [Hil89].

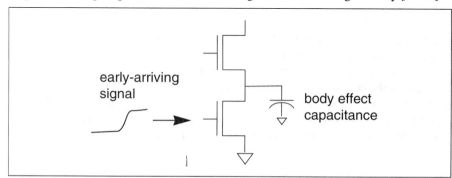

Figure 3-22:
Body effect and
signal ordering.

3.3.5 Power Consumption

Analyzing the power consumption of an inverter provides an alternate window into the cost and performance of a logic gate. Circuits can be made to go faster—up to a point—by causing them to burn more power. Power consumption always comes at the cost of heat which must be dissipated out of the chip. Static, complementary CMOS gates are remarkably efficient in their use of power to perform computation.

Once again we will analyze an inverter with a capacitor connected to its output. However, to analyze power consumption we must consider both the pullup and pulldown phases of operation. The model circuit is shown in Figure 3-23. The first thing to note about the circuit is that it has almost no steady-state power consumption. After the output capacitance has been fully charged or discharged, only one of the pullup and pulldown transistors is on. The following analysis ignores the leakage current; we will look at techniques to combat leakage current in Section 3.6.

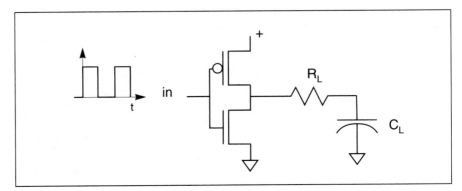

Figure 3-23:
*Circuit used for
power consump-
tion analysis.*

Power is consumed when gates drive their outputs to new values. Surprisingly, the power consumed by the inverter is independent of the sizes/resistances of its pullup and pulldown transistors—power consumption depends only on the size of the capacitive load at the output and the rate at which the inverter's output switches. To understand why, consider the energy required to drive the inverter's output high calculated two ways: by the current through the load capacitor C_L and by the current through the pullup transistor, represented by its effective resistance R_p.

The current through the capacitor and the voltage across it are:

$$i_{CL}(t) = \frac{V_{DD}-V_{SS}}{R_p}e^{-(t/R_pC_L)} \quad ,$$ (EQ 3-11)

$$v_{CL}(t) = (V_{DD}-V_{SS})[1-e^{-(t/R_pC_L)}] \quad .$$ (EQ 3-12)

So, the energy required to charge the capacitor is:

$$E_C = \int_0^\infty i_{C_L}(t)v_{C_L}(t))dt$$ (EQ 3-13)

$$= C_L(V_{DD}-V_{SS})^2\left(e^{-t/R_pC_L}-\frac{1}{2}e^{-2t/R_pC_L}\right)\Big|_0^\infty$$

$$= \frac{1}{2}C_L(V_{DD}-V_{SS})^2$$

This formula depends on the size of the load capacitance but not the resistance of the pullup transistor. The current through and voltage across the pullup are:

$$i_p(t) = i_{CL}(t),$$

(EQ 3-14)

$$v_p(t) = Ve^{-(t/R_pC_L)}.$$

(EQ 3-15)

The energy required to charge the capacitor, as computed from the resistor's point of view, is

$$
\begin{aligned}
E_R &= \int_0^\infty i_p(t)v_p(t)dt \\
&= C_L(V_{DD}\text{-}V_{SS})^2 (e^{-2t/R_pC_L})\Big|_0^\infty \\
&= \frac{1}{2}C_L(V_{DD}\text{-}V_{SS})^2
\end{aligned}
$$

(EQ 3-16)

Once again, even though the circuit's energy consumption is computed through the pullup, the value of the pullup resistance drops from the energy formula. (That holds true even if the pullup is a nonlinear resistor.) The two energies have the same value because the currents through the resistor and capacitor are equal.

The energy consumed in discharging the capacitor can be calculated the same way. The discharging energy consumption is equal to the charging power consumption: $1/2 \, C_L(V_{DD}\text{-}V_{SS})^2$. A single cycle requires the capacitor to both charge and discharge, so the total energy consumption is $C_L(V_{DD}\text{-}V_{SS})^2$.

Power is energy per unit time, so the power consumed by the circuit depends on how frequently the inverter's output changes. The worst case is that the inverter alternately charges and discharges its output capacitance. This sequence takes ~~two~~ ᴼᴺᴱ clock cycles. The clock frequency is $f = 1/t$. The total power consumption is

$$fC_L(V_{DD}\text{-}V_{SS})^2.$$

(EQ 3-17)

Power consumption in CMOS circuits depends on the frequency at which they operate, which is very different from nMOS or bipolar logic circuits. Power consumption depends on clock frequency because most power is consumed while the outputs are changing; most other circuit technologies burn most of their power while the circuit is idle. Power consumption depends on the sizes of the transistors in the circuit only in that the transistors largely determine C_L. The current through the transistors, which is determined by the transistor W/Ls, doesn't determine power consumption, though the available transistor current does determine the maximum speed at which the circuit can run, which indirectly determines power consumption.

Does it make sense that CMOS power consumption should be independent of the effective resistances of the transistors? It does, when you remember that CMOS circuits consume only dynamic power. Most power calculations are made on static circuits—the capacitors in the circuit have been fully charged or discharged, and power consumption is determined by the current flowing through resistive paths between V_{DD} and V_{SS} in steady state. Dynamic power calculations, like those for our CMOS circuit, depend on the current flowing through capacitors; the resistors determine only maximum operating speed, not power consumption.

Static complementary gates can operate over a wide range of voltages, allowing us to trade delay for power consumption. To see how performance and power consumption are related, let's consider changing the power supply voltage from its original value V to a new V'. It follows directly from Equation 3-17 that the ratio of power consumptions P'/P is proportional to V^2/V'^2. When we compute the ratio of rise times t'_r/t_r the only factor to change with voltage is the transistor's equivalent resistance R, so the change in delay depends only on R'/R. If we use the technique of Section 3.3.4 to compute the new effective resistance, we find that $t'_r/t_r \propto V/V'$. So as we reduce power supply voltage, power consumption goes down faster than does delay.

3.3.6 The Speed-Power Product

The **speed-power product**, also known as the **power-delay product**, is an important measure of the quality of a logic circuit family. Since delay can in general be reduced by increasing power consumption, looking at either power or delay in isolation gives an incomplete picture.

The speed-power product for static CMOS is easy to calculate. If we ignore leakage current and consider the speed and power for a single inverter transition, then we find that the speed-power product SP is

$$SP = \frac{1}{f}P = CV^2.$$

(EQ 3-18)

The speed-power product for static CMOS is independent of the operating frequency of the circuit. It is, however, a quadratic function of the power supply voltage. This result suggests an important method for power consumption reduction known as **voltage scaling**: we can often reduce power consumption by reducing the power supply voltage and adding parallel logic gates to make up for the lower performance. Since the power consumption shrinks more quickly than the circuit delay when the voltage is scaled, voltage scaling is a powerful technique. We will study techniques for low-power gate design in Section 3.6.

3.3.7 Layout and Parasitics

How do parasitics affect the performance of a single gate? Answering this question tells us how to design the layout of a gate to maximize performance and minimize area.

Example 3-2: *Parasitics and performance*

To answer the question, we will consider the effects of adding resistance and capacitance to each of the labeled points of this layout:

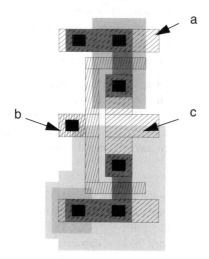

- *a*

Adding capacitance to point a (or its conjugate point on the V_{SS} wire) adds capacitance to the power supply wiring. Capacitance on this node doesn't slow down the gate's output.

Resistance at a can cause problems. Resistance in the V_{SS} line can be modeled by this equivalent circuit:

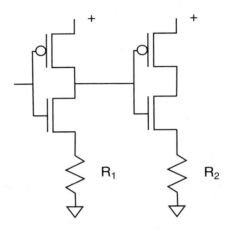

The power supply resistance is in series with the pulldown. That differential isn't a serious problem in static, complementary gates. The resistance slows down the gate, but since both the transistor gates of the pullup and pulldown are connected to the same electrical node, we can be sure that only one of them will be on in steady state. However, the dynamic logic circuits we will discuss in Section 3.5 may not work if the series power supply resistance is too high, because the voltages supplied by the gate with resistance may not properly turn on succeeding transistor gates.

The layout around point a should be designed to minimize resistance. A small length of diffusion is required to connect the transistors to the power lines, but power lines should be kept in metal as long as possible. If the diffusion wire is wider than a via (to connect to a wide transistor), several parallel vias should be used to connect the metal and diffusion lines.

- b

Capacitance at b adds to the load of the gate driving this node. However, the transistor gate capacitances are much larger than the capacitance added by the short wire feeding the transistor gates. Resistance at b actually helps isolate the previous gate from the load capacitance, as we will see when we discuss the π model in Section 4.5.1 Gate layouts should avoid making big mistakes by using large sections of diffusion wire or a single via to connect high-current wires.

- c

    Capacitance and resistance at c are companions to parasitics at b—they form part of the load that this gate must drive, along with the parasitics of the b zone of the next gate. But if we consider a more accurate model of the parasitics, we will see that not all positions for parasitic R and C are equally bad.

Up to now we have modeled the resistance and capacitance of a wire as single components. But now consider the inverter's load as two RC sections:

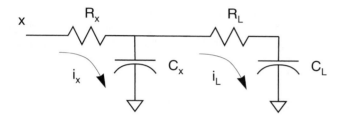

One RC section is contributed by the wires at point c, near the output; the RC section comes from the long wire connecting this gate to the next one. How does the voltage at point x—the input to the next gate—depend on the relative values of the R's? The simplified circuit shows how a large value for R_x, which is supplied by the parasitics at point c, steals current from $R_L C_L$. As R_x grows relative to R_L, the voltage drop across R_x increases, increasing the current through R_x while decreasing the current through R_L. As a result, more of the current supplied by the gate will go through C_x; only after it is fully charged will C_L get the full current supplied by the gate. Since C_L is almost certainly significantly larger than C_x, since it includes both the transistor gate capacitances and the long-wire capacitance, it is more important to charge C_L to switch the next gate as quickly as possible. But charging/discharging of C_L has been delayed while R_x diverts current into C_x.

The moral is that resistance close to the gate output is worse than resistance farther away—close-in resistance must charge more capacitors, slowing down the signal swing at the far end of the wire. Therefore, the layout around c should be designed to minimize resistance. That requires:

- using as little diffusion as possible—diffusion should be connected to metal (or perhaps poly) as close to the channel as possible;

• using parallel vias at the diffusion/metal interface to minimize resistance.

3.3.8 Driving Large Loads

Logic delay increases as the capacitance attached to the logic's output becomes larger. In many cases, one small logic gate is driving an equally small logic gate, roughly matching drive capability to load. However, there are several situations in which the capacitive load can be much larger than that presented by a typical gate:

• driving a signal connected off-chip;

• driving a long signal wire;

• driving a clock wire which goes to many points on the chip.

The obvious answer to driving large capacitive loads is to increase current by making wider transistors. However, this solution begs the question—those large transistors simply present a large capacitive load to the gate which drives them, pushing the problem back one level of logic. It is inevitable that we must eventually use large transistors to drive the load, but we can minimize delay along the path by using a sequence of successively larger drivers.

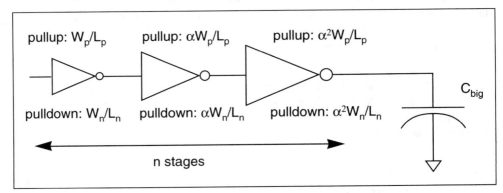

Figure 3-24: Cascaded inverters driving a large capacitive load.

The driver chain with the smallest delay to drive a given load is exponentially tapered—each stage supplies e times more current than the last [Jae75]. In the chain of inverters of Figure 3-24, each inverter can produce α times more current than the previous stage (implying that its pullup and pulldown are each α times larger). If C_g is the minimum-size load capacitance, the number of stages n is related to α by the formula $\alpha = (C_{big}/C_g)^{1/n}$. The time to drive a minimum-size load is t_{min}. We want to minimize the total delay through the driver chain:

$$t_{tot} = n\left(\frac{C_{big}}{C_g}\right)^{1/n} t_{min} \, . \qquad \textbf{(EQ 3-19)}$$

To find the minimum, we set $\dfrac{dt_{tot}}{dn} = 0$, which gives

$$n_{opt} = ln\left(\frac{C_{big}}{C_g}\right) . \qquad \textbf{(EQ 3-20)}$$

When we substitute the optimal number of stages back into the definition of α, we find that the optimum value is at $\alpha = e$. Of course, n must be an integer, so we will not in practice be able to implement the exact optimal circuit. However, delay changes slowly with n near the optimal value, so rounding n to the floor of n_{opt} gives reasonable results.

3.4 Switch Logic

How do we build switches from MOS transistors? One way is the **transmission gate** shown in Figure 3-25, built from parallel n-type and p-type transistors. This switch is built from both types of transistors so that it transmits logic 0 and 1 from drain to source equally well: when you put a V_{DD} or V_{SS} at the drain, you get V_{DD} or V_{SS} at the source. But it requires two transistors and their associated tubs; equally damning, it requires both true and complement forms of the gate signal.

An alternative is the **n-type switch**—a solitary n-type transistor. It requires only one transistor and one gate signal, but it is not as forgiving electrically: it transmits a logic 0 well, but when V_{DD} is applied to the drain, the voltage at the source is $V_{DD} - V_{tn}$. When switch logic drives gate logic, n-type switches can cause electrical problems. An n-type switch driving a complementary gate causes the complementary gate to run

*Figure 3-25: A
complementary
transmission gate.*

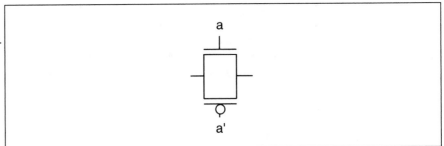

slower when the switch input is 1: since the n-type pulldown current is weaker when a lower gate voltage is applied, the complementary gate's pulldown will not suck current off the output capacitance as fast. When the n-type switch drives a pseudo-nMOS gate, disaster may occur. A pseudo-nMOS gate's ratioed transistors depend on logic 0 and 1 inputs to occur within a prescribed voltage range. If the n-type switch doesn't turn on the pseudo-nMOS pulldown strongly enough, the pulldown may not divert enough current from the pullup to force the output to a logic 0, even if we wait forever. Ratioed logic driven by n-type switches must be designed to produce valid outputs for both polarities of input.

Both types of switch logic are sensitive to noise—pulling the source beyond the power supply (above V_{DD} or below V_{SS}) causes the transistor to start conducting. We will see in Section 4.7 that logic networks made of switch logic are prone to errors introduced by parasitic capacitance.

3.5 Alternative Gate Circuits

The static complementary gate has several advantages: it is reliable, easy to use in large combinational logic networks, and does not require any separate precharging steps. It is not, however, the only way to design a logic gate with p-type and n-type transistors. Other circuit topologies have been created that are smaller or faster (or both) than static complementary gates. Still others use less power.

In this section we will review the design of several important alternative CMOS gate topologies. Each has important uses in chip design. But it is important to remember that they all have their limitations and caveats. Specialized logic gate designs often require more attention to the details of circuit design—while the details of circuit and layout design affect only the speed at which a static CMOS gate runs, circuit and layout problems can cause a fancier gate design to fail to function correctly. Particular care must be taken when mixing logic gates designed with different circuit topologies to ensure that one's output meets the requirements of the next's inputs. A good, conservative chip design strategy is to start out using only static complementary gates, then to use specialized gate designs in critical sections of the chip to meet the project's speed or area requirements.

3.5.1 Pseudo-nMOS Logic

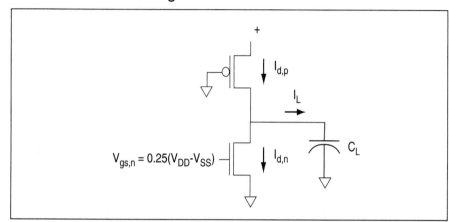

Figure 3-26: A *pseudo-nMOS NOR gate.*

The simplest non-standard gate topology is **pseudo-nMOS**, so called because it mimics the design of an nMOS logic gate. Figure 3-26 shows a pseudo-nMOS NOR gate. The pulldown network of the gate is the same as for a fully complementary gate. The pullup network is replaced by a single p-type transistor whose gate is connected to V_{SS}, leaving the transistor permanently on. The p-type transistor is used as a resistor: when the gate's inputs are **ab** = 00, both n-type transistors are off and the p-type transistor pulls the gate's output up to V_{DD}. When either **a** or **b** is 1, both the p-type and n-type transistor are on and both are fighting to determine the gate's output voltage.

We need to determine the relationship between the *W/L* ratios of the pullup and the pulldowns which provide reasonable output voltages for the gate. For simplicity, assume that only one of the pulldown transistors is on; then the gate circuit's output voltage depends on the ratio of the effective resistances of the pullup and the operating pulldown. The high output voltage of the gate is V_{DD}, but the output low voltage V_{OL} will be some voltage above V_{SS}. The chosen V_{OL} must be low enough to activate the next logic gate in the chain. For pseudo-nMOS gates which feed static or pseudo-nMOS gates, a value of $V_{OL} = 0.15(V_{DD}\text{-}V_{SS})$ is a reasonable value, though others could be chosen. To find the transistor sizes which give reasonable output voltages, we must consider the simultaneous operation of the pullup and pulldown. When the gate's output has just switched to a logic 0, the n-type pulldown is in saturation with $V_{gs,n} = V_{in}$. The p-type pullup is in its linear region: its $V_{gs,p} = V_{DD} - V_{SS}$ and its $V_{ds,p} = V_{out} - (V_{DD} - V_{SS})$. We need to find V_{out} in terms of the W/Ls of the pullup and pulldown. To solve this problem, we set the currents through the saturated pulldown and the linear pullup to be equal:

$$\frac{1}{2}k'_n\frac{W_n}{L_n}(V_{gs,n}-V_{tn})^2 = \frac{1}{2}k'_p[2(V_{gs,p}-V_{tp})V_{ds,p}-V_{ds,p}^2] \qquad . \qquad \textbf{(EQ 3-21)}$$

The simplest way to solve this equation is to substitute the technology and circuit values. Using the 0.5 μm values and assuming a 3.3V power supply and a full-swing input $(V_{gs,n} = V_{DD}-V_{SS})$, we find that

$$\frac{W_p/L_p}{W_n/L_n} \approx 3.9 \quad . \qquad \textbf{(EQ 3-22)}$$

The pulldown network must exhibit this effective resistance in the worst case combination of inputs. Therefore, if the network contains series pulldowns, they must be made larger to provide the required effective resistance.

Figure 3-27:
Currents in a
pseudo-nMOS
gate during
low-to-high
transition.

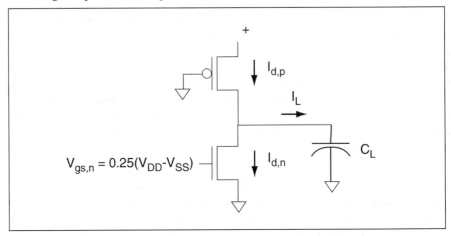

The pseudo-nMOS gate consumes static power, unlike the fully complementary gate. When both the pullup and pulldown are on, the gate forms a conducting path from V_{DD} to V_{SS}, which must be kept on to maintain the gate's logic output value. The choice of V_{OL} determines whether the gate consumes may consume static power when its output is logic 1. If pseudo-nMOS feeds pseudo-nMOS and V_{OL} is chosen to be greater than $V_{t,n}$, then the pulldown will remain on. Whether the pulldown is in the linear or saturation region depends on the exact transistor characteristics, but in either case, its drain current will be low since $V_{gs,n}$ is low. As shown in Figure 3-27, so long as the pulldown drain current is significantly less than the pullup drain current, there will be enough current to charge the output capacitance and bring the gate output to the desired level.

The ratio of the pullup and pulldown sizes also ensures that the times for $0 \rightarrow 1$ and $1 \rightarrow 0$ transitions are asymmetric. Since the pullup transistor has about three times the effective resistance of the pulldown, the $0 \rightarrow 1$ transition occurs much more slowly than the $1 \rightarrow 0$ transition and dominates the gate's delay. The long pullup time makes the pseudo-nMOS gate slower than the static complementary gate.

Why use a pseudo-nMOS gate? The main advantage of the pseudo-nMOS gate is the small size of the pullup network, both in terms of number of devices and wiring complexity. The pullup network of a static complementary gate can be large for a complex function. Furthermore, the input signals do not have to be routed to the pullup, as in a static complementary gate. The pseudo-nMOS gate is used for circuits where the size and wiring complexity of the pullup network are major concerns but speed and power are less important. We will see two examples of uses of pseudo-nMOS circuits in Chapter 6: busses and PLAs. In both cases, we are building **distributed NOR gates**—we use pulldowns spread over a large physical area to compute the output, and we do not want to have to run the signals which control the pulldowns around this large area. Pseudo-nMOS circuits allow us to concentrate the logic gate's functionality in the pulldown network.

3.5.2 DCVS Logic

Differential cascode voltage switch logic (DCVSL) is a static logic family that, like pseudo-nMOS logic, does not have a complementary pullup network, but it has a very different structure. It uses a latch structure for the pullup which both eliminates static power consumption and provides true and complement outputs.

The structure of a generic DCVSL gate is shown in Figure 3-28. There are two pulldown networks which are the duals of each other, one for each true/complement output. Each pulldown network has a single p-type pullup, but the pullups are cross-coupled. Exactly one of the pulldown networks will create a path to ground when the gate's inputs change, causing the output nodes to switch to the required values. The cross-coupling of the pullups helps speed up the transition—if, for example, the com-

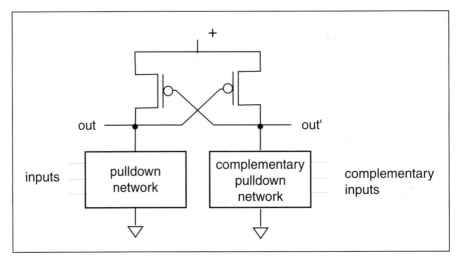

Figure 3-28:
Structure of a
DCVSL gate.

plementary network forms a path to ground, the complementary output goes toward V_{SS}, which turns on the true output's pullup, raising the true output, which in turn lowers the gate voltage on the complementary output's pullup. This gate consumes no DC power (except due to leakage current), since neither side of the gate will ever have both its pullup and pulldown network on at once.

Figure 3-29: *An*
example DCVSL
gate circuit.

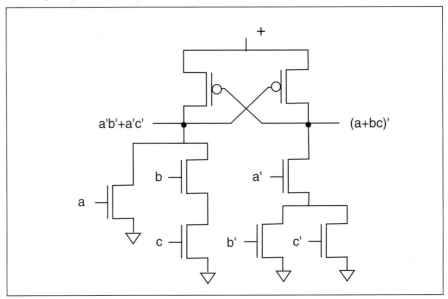

Figure 3-29 shows the circuit for a particular DCVSL gate. This gate computes $a+bc$ on one output and $(a+bc)' = a'b'+a'c'$ on its other output.

3.5.3 Domino Logic

Precharged circuits offer both low area and higher speed than static complementary gates. Precharged gates introduce functional complexity because they must be operated in two distinct phases, requiring introduction of a clock signal. They are also more sensitive to noise; their clocking signals also consume power and are difficult to turn off to save power.

The canonical precharged logic gate circuit is the **domino** circuit [Kra82]. A domino gate is shown in Figure 3-30, along with a sketch of its operation over one cycle. The gate works in two phases, first to precharge the storage node, then to selectively discharge it. The phases are controlled by the clock signal ϕ:

- **Precharge.** When ϕ goes low, the p-type transistor starts charging the precharge capacitance. The pulldown transistors controlled by the clock keep that precharge node from being drained. The length of the $\phi = 0$ phase is adjusted to ensure that the storage node is charged to a solid logic 1.

- **Evaluate.** When ϕ goes high, precharging stops (the p-type pullup turns off) and the evaluation phase begins (the n-type pulldowns at the bottom of the circuit turn on). The logic inputs a and b can now assume their desired value of 0 or 1. The input signals must monotonically rise—if an input goes from 0 to 1 and back to 0, it will inadvertently discharge the precharge capacitance. If the inputs create a conducting path through the pulldown network, the precharge capacitance is discharged, forcing its value to 0 and the gate's output (through the inverter) to 1. If neither a nor b is 1, then the storage node would be left charged at logic 1 and the gate's output would be 0.

The gate's logic value is valid at the end of the evaluation phase, after enough time has been allowed for the pulldown transistors to fully discharge the storage node. If the gate is to be used to compute another value, it must go through the precharge-evaluate cycle again.

Figure 3-31 illustrates the phenomenon which gave the domino gate its name. Since each gate is precharged to a low output level before evaluation, the changes at the primary inputs ripple through the domino network from one end to another. Signals at the far end of the network change last, with each change to a gate output causing a

Figure 3-30: A
*domino OR gate
and its opera-
tion.*

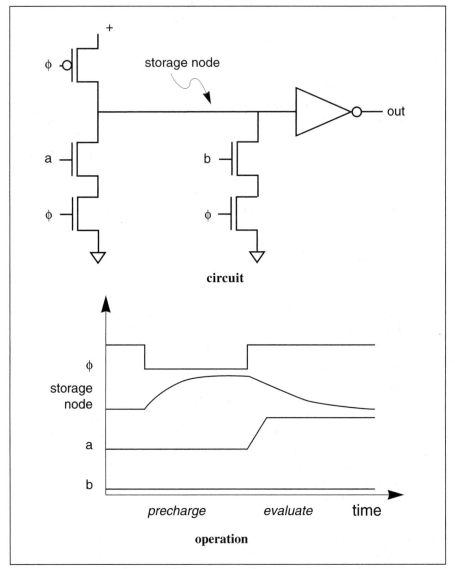

change to the next output. This sequential evaluation resembles a string of falling dominos.

Why is there an inverter at the output of the domino gate? There are two reasons: logical operation and circuit behavior. To understand the logical need for an output inverter, consider the circuit of Figure 3-32, in which the output of one domino gate is

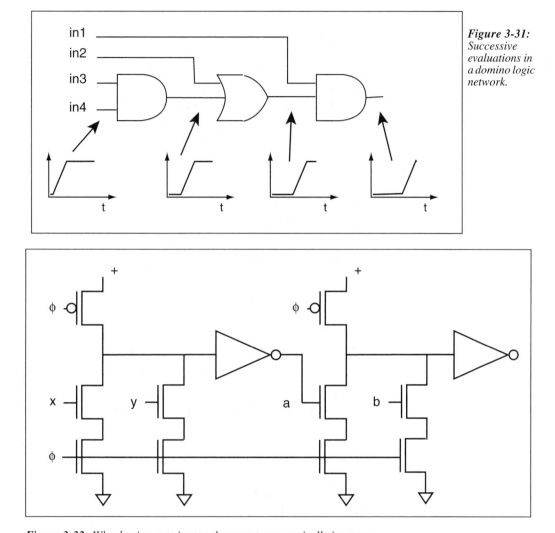

Figure 3-31: *Successive evaluations in a domino logic network.*

Figure 3-32: *Why domino gate input values must monotonically increase.*

fed into an input of another domino gate. During the precharge phase, if the inverter were not present, the intermediate signal would rise to 1, violating the requirement that all inputs to the second gate be 0 during precharging.

However, the more compelling reason for the output inverter is to increase the reliability of the gate. Figure 3-33 shows two circuit variations: one with the output inverter and one without. In both cases, the storage node is coupled to the output of the following gate by the gate-to-source/drain capacitances of the transistors in that

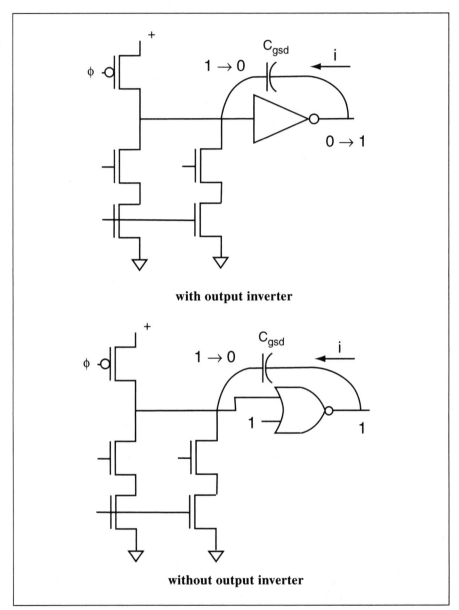

Figure 3-33:
Capacitive coupling in domino gates.

gate. This coupling can cause current to flow into the storage node, disturbing its value. Since the coupling capacitance is across the transistor, the Miller effect magnifies its value. When the storage node is connected to the output inverter, the inverter's output is at least correlated to the voltage on the storage node and we can design the circuit to withstand the effects of the coupling capacitance. However, when the stor-

age node is connected to an arbitrary gate, that gate's output is not necessarily correlated to the storage node's behavior, making it more difficult to ensure that the storage
node is not corrupted. The fact that the wire connecting the domino gate's pulldown
network to the next gate (and the bulk of the storage node capacitance) may be long
and subject to crosstalk generated by wire-to-wire coupling capacitances only makes
this circuit less attractive.

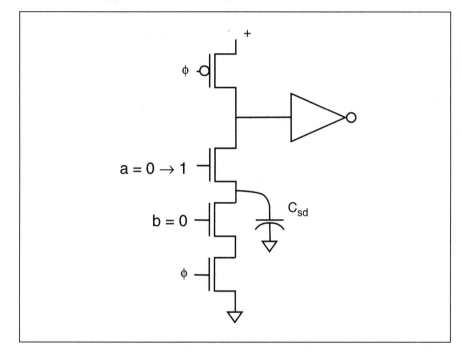

Figure 3-34:
Charge sharing
in a domino cir-
cuit.

Domino gates are also vulnerable to errors caused by **charge sharing**. Charge sharing is a problem in any network of switches, and we will cover it in more detail in
Section 4.7. However, we need to understand the phenomenon in the relatively simple
form in which it occurs in domino gates. Consider the example of Figure 3-34. C_{sd},
the stray capacitance on the source and drain of the two pulldown transistors, can
store enough charge to cause problems. In the case when the a input is 1 and the b
input is 0, the precharge node should not be discharged. However, since a is one, the
pulldown connected to the storage node is turned on, draining charge from the storage
node into the parasitic capacitance between the two pulldowns. In a static gate, charge
stored in the intermediate pulldown capacitances does not matter because the power
supply drives the output, but in the case of a dynamic gate that charge is lost to the
storage node. If the gate has several pulldown transistors, the charge loss is that much
more severe. The problem can be averted by precharging the internal pulldown net-

work nodes along with the precharge node itself, although at the cost of area and complexity.

Because dynamic gates rely on stored charge, they are vulnerable to charge leakage through the substrate. The primary threat comes from designs which do not evaluate some dynamic gates on every clock cycle; in these cases, the designer must verify that the gates are always re-evaluated frequently enough to ensure that the charge stored in the gates has not leaked away in sufficient quantities to destroy the gate's value.

Domino gates cannot invert, and so this logic family does not form a complete logic, as defined in Section 3.2. A domino logic network consists only of AND, OR, and complex AND/OR gates. However, any such function can be rewritten using De Morgan's laws to push all the inverters to the forward outputs or backward to the inputs; the bulk of the function can be implemented in domino gates with the inverters implemented as standard static gates. However, pushing back the inversions to the primary inputs may greatly increase the number of gates in the network.

3.6 Low-Power Gates

There are several different strategies for building low-power gates. Which one is appropriate for a given design depends on the required performance and power as well as the fabrication technology. In very deep submicron technologies leakage current has become a major consumer of power.

Of course, the simplest way to reduce the operating voltage of a gate is to connect it to a lower power supply. We saw the relationship between power supply voltage and power consumption in Section 3.3.5:

- For large V_t, Equation 3-10 tells us that delay changes linearly with power supply voltage.

- Equation 3-17 tells us that power consumption varies quadratically with power supply voltage.

This simple analysis tells us that reducing the power supply saves us much more in power consumption than it costs us in gate delay. Of course, the performance penalty incurred by reducing the power supply voltage must be taken care of somewhere in the system. One possible solution is architecture-driven voltage scaling, which we will study in Section 8.5, which replicates logic to make up for slower operating speeds.

It is also possible to operate different gates in the circuit at different voltages: gates on the critical delay path can be run at higher voltages while gates that are not delay-critical can be run at lower voltages. However, such circuits must be designed very carefully since passing logic values between gates running at different voltages may run into noise limits.

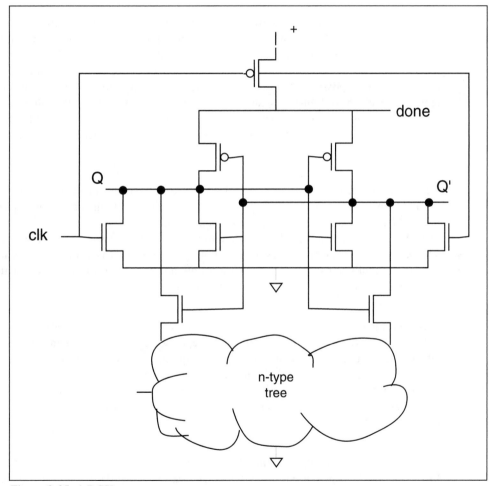

Figure 3-35: *A DCSL gate.*

After changing power supply voltages, the next step is to use different logic gate topologies. An example of this strategy is the differential current switch logic (DCSL) gate [Roy00] shown in Figure 3-35 is related to the DCVS gate of Section 3.5.2. Both use nMOS pulldown networks for both logic 0 and logic 1. However, the DCSL gate disconnects the n-type networks to reduce their power consumption. This gate is precharged with Q and Q' low. When the clock goes high, one of Q or Q' will be pulled low by the n-type evaluation tree and that value will be latched by the cross-coupled inverters.

After these techniques have been tried, two techniques can be used: reducing leakage current and turning off gates when they are not in use. Leakage current is becoming increasingly important in very deep submicron technologies. We studied leakage currents in Section 2.3.6. One simple approach to reducing leakage currents in gates is to choose, whenever possible, don't-care conditions on the inputs to reduce leakage currents. Series chains of transistors pass much lower leakage currents when both are off than when one is off and the other is on. If don't-care conditions can be used to turn off series combinations of transistors in a gate, the gate's leakage current can be greatly reduced.

The key to low leakage current is low threshold voltage. Unfortunately, there is an essential tension between low leakage and high performance. Remember from Equation 2-17 that leakage current is an exponential function of $V_{gs} - V_t$. As a result, increasing Vt decreases the subthreshold current when the transistor is off. However, a high threshold voltage increases the gate's delay since the transistor turns on later in the input signal's transition. One solution to this dilemma is to use transistors with different thresholds at different points in the circuit.

Turning off gates when they are not used saves even more power, particularly in technologies that exhibit significant leakage currents. Care must be used in choosing which gates to turn off, since it often takes 100 µs for the power supply to stabilize after it is turned on. We will discuss the implications of power-down modes in Section 8.5. However, turning off gates is a very useful technique that becomes increasingly important in very deep submicron technologies with high leakage currents.

The leakage current through a chain of transistors in a pulldown or pullup network is lower than the leakage current through a single transistor [De01]. It also depends on whether some transistors in the stack are also on. Consider the pulldown network of a NAND gate shown in Figure 3-36. If both the a and b inputs are 0, then both transistors are off. Because a small leakage current flows through transistor M_a, the parasitic capacitance between the two transistors is charged, which in turns holds the voltage at

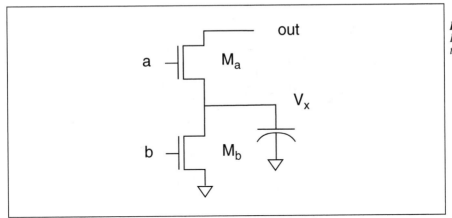

Figure 3-36:
Leakage through transistor stacks.

that node above ground. This means that V_{gs} for is M_a is negative, thus reducing the total leakage current. The leakage current is found by simultaneously solving for the currents through the two transistors. The leakage current through the chain can be an order of magnitude lower than the leakage current through a single transistor. But the total leakage current clearly depends on the gate voltages of the transistors in the chain; if some of the gate's inputs are logic 1, then there may not be chains of transistors that are turned off and thus have reduced input voltages. Algorithms can be used to find the lowest-leakage input values for a set of gates; latches can be used to hold the gates' inputs at those values in standby mode to reduce leakage.

Figure 3-37 shows a **multiple-threshold logic (MTCMOS)** [Mut98] gate that can be powered down. This circuit family uses low-leakage transistors to turn off gates when they are not in use. A **sleep transistor** is used to control the gate's access to the power supply; the gated power supply is known as a **virtual V_{DD}**. The gate uses low-threshold transistors to increase the gate's delay time. However, lowering the threshold voltage also increases the transistors' leakage current, which causes us to introduce the sleep transistor. The sleep transistor has a high threshold to minimize its leakage. The fabrication process must be able to build transistors with low and high threshold voltages.

The layout of this gate must include both V_{DD} and virtual V_{DD}: virtual V_{DD} is used to power the gate but V_{DD} connects to the pullup's substrate. The layout must include The sleep transistor must be properly sized. If the sleep transistor is too small, its impedance would cause virtual V_{DD} to bounce. If the sleep transistor is too large, the sleep transistor would occupy too much area and it would use more energy when switched.

Figure 3-37: A
*multiple-thresh-
old (MTCMOS)
inverter.*

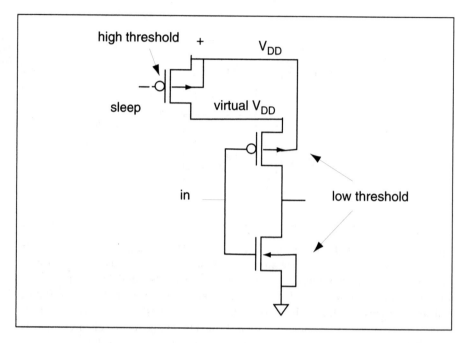

Figure 3-38: A
*variable-threshold
CMOS (VTCMOS)
gate.*

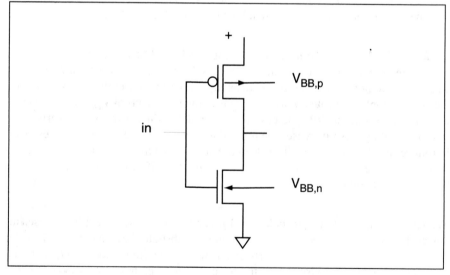

It is important to remember that some other logic must be used to determine when a
gate is not used and control the gate's power supply. This logic must be watch the

state of the chip's inputs and memory elements to know when logic can safely be turned off. It may also take more than one cycle to safely turn on a block of logic.

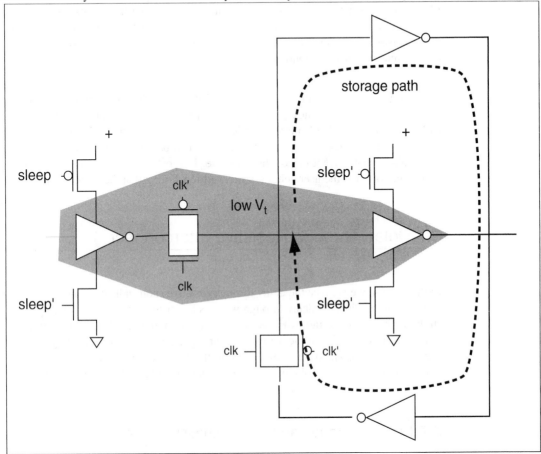

Figure 3-39: *An MTCMOS flip-flop.*

Figure 3-39 shows an MTCMOS flip-flop. The storage path is made of high V_t transistors and is always on. The signal is propagated from input to output through low V_t transistors. The sleep control transistors on the second inverter in the forward path to prevent a short-circuit path between V_{DD} and virtual V_{DD} that could flow through the storage inverter's pullup and the forward chain inverter's pullup.

A more aggressive method is **variable threshold CMOS (VTCMOS)** [Kur96], which actually can be implemented in several ways. Rather than fabricating fixed-threshold voltage transistors, the threshold voltages of the transistors in the gate are controlled by changing the voltages on the substrates. Figure 3-38 shows the structure of a VTCMOS gate. The substrates for the p- and n-type transistors are each connected to their own threshold supply voltages, $V_{BB,p}$ and $V_{BB,n}$. V_{BB} is raised to put the transistor in standby mode and lowered to put it into active mode. Rather sophisticated circuitry is used to control the substrate voltages.

VTCMOS logic comes alive faster than it falls asleep. The transition time to sleep mode depends on how quickly current can be pulled out of the substrate, which typically tens to hundreds of microseconds. Returning the gate to active mode requires injecting current back into the substrate, which can be done 100 to 1000 times faster than pulling that current out of the substrate. In most applications, a short wake-up time is important—the user generally gives little warning that the system is needed.

3.7 Delay Through Resistive Interconnect

In this section we analyze the delay through resistive (non-inductive) interconnect. In many modern chips, the delay through wires is larger than the delay through gates, so studying the delay through wires is as important as studying delay through gates. We will build a suite of analytical models, starting from the relatively straightforward Elmore model for an RC transmission line through more complex wire shapes. We will also consider the problem of where to insert buffers along wires to minimize delay.

3.7.1 Delay Through an RC Transmission Line

An **RC transmission line** models a wire as infinitesimal RC sections, each representing a differential resistance and capacitance. Since we are primarily concerned with RC transmission lines, we can use the transmission line model to compute the delay through very long wires. We can model the transmission line as having unit resistance r and unit capacitance c. The standard schematic for the RC transmission line is shown in Figure 3-40. The transmission line's voltage response is modeled by a differential equation:

$$\frac{1}{r}\frac{d^2V}{dx^2} = c\frac{dV}{dt}.$$

(EQ 3-23)

This model gives the voltage as a function of both x position along the wire and time.

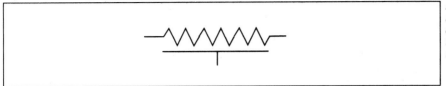

Figure 3-40:
Symbol for a
distributed RC
transmission
line.

The raw differential equation, however, is unwieldy for many circuit design tasks. **Elmore delay** [Elm48] is the most widely used metric for RC wire delay and has been shown to sufficiently accurately model the results of simulating RC wires on integrated circuits [Boe93]. Elmore defined the delay through a linear network as the first moment of the impulse response of the network:

$$\delta_E = \int_0^\infty t V_{out}(t)dt \; .$$

(EQ 3-24)

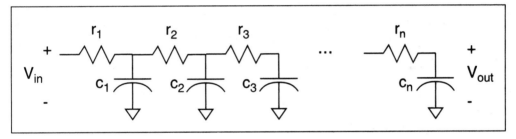

Figure 3-41: An RC transmission line for Elmore delay calculations.

It is because only the first moment is used that Elmore delay is not sufficiently accurate for inductive interconnect. However, in overdamped RC networks, the first moment is sufficiently accurate.

Elmore modeled the transmission line as a sequence of n sections of RC, as shown in Figure 3-41. In the case of a general RC network, the Elmore delay can be computed by taking the sum of RC products, where each resistance R is multiplied by the sum of all the downstream capacitors (a special case of the RC tree formulas we will introduce in Section 3.7.2). Since all the transmission line section resistances and capacitances in an n-section are identical, this reduces to

$$\delta_E = \sum_{i=1}^{n} r(\text{n-i})c = \frac{1}{2}rc \times n(\text{n-i}). \qquad \text{(EQ 3-25)}$$

One consequence of this formula is that wire delay grows as the square of wire length, since n is proportional to wire length. Since the wire's delay also depends on its unit resistance and capacitance, it is imperative to use the material with the lowest RC product (which will almost always be metal) to minimize the constant factor attached to the n^2 growth rate.

Although the Elmore delay formula is widely used, we will need some results from the analysis of continuous transmission lines for our later discussion of crosstalk. The normalized voltage step response of the transmission line can be written as

$$V(t) = 1 + \sum_{k=1}^{\infty} K_k e^{-\sigma_k t/RC} \approx 1 + K_1 e^{-\sigma_1 t/RC}, \qquad \text{(EQ 3-26)}$$

where R and C are the total resistance and capacitance of the line. We will define R_t as the internal resistance of the driving gate and C_t as the load capacitance at the opposite end of the transmission line.

Sakurai [Sak93] estimated the required values for the first-order estimate of the step response as:

$$K_1 = \frac{-1.01(R_T + C_T + 1)}{R_T + C_T + \pi/4}, \qquad \text{(EQ 3-27)}$$

$$\sigma_1 = \frac{1.04}{R_T C_T + R_T + C_T + (2/\pi)^2}, \qquad \text{(EQ 3-28)}$$

where R_T and C_T are R_t/R and C_t/C, respectively.

So far, we have assumed that the wire has constant width. In fact, tapered wires provide lower delay. Consider the first resistance element in the transmission line—the current required to charge all the capacitance of the wire must flow through this resistance. In contrast, the resistance at the end of the wire handles only the capacitance at the end. Therefore, if we can decrease the resistance at the head of the wire, we can decrease the delay through the wire. Unfortunately, increasing the resistance by widening the wire also increases its capacitance, making this a non-trivial problem to solve.

Fishburn and Schevon [Fis95] proved that the optimum-shaped wire has an exponential taper. If the source resistance is R_0, the sink capacitance is C_0, and the unit resistance and capacitance are R_s and C_s, the width of the wire as a function of distance is

$$w(x) = \frac{2C_0}{C_sL}W\left(\frac{L}{2}\sqrt{\frac{R_sC_s}{R_0C_0}}\right)e^{2W\left(\frac{L}{2}\sqrt{\frac{R_sC_s}{R_0C_0}}\right)\frac{x}{L}}, \qquad \text{(EQ 3-29)}$$

where W is the function that satisfies the equality $W(x)e^{W(x)} = x$. The advantage of optimal tapering is noticeable. Fishburn and Schevon calculate that, for one example, the optimally tapered wire has a delay of 3.72 ns while the constant-width wire with minimum delay has a delay of 4.04 ns. In this example, the optimally tapered wire shrinks from 30.7 μm at the source to 7.8 μm at the sink.

Of course, exponentially-tapered wires are impossible to fabricate exactly, but it turns out that we can do nearly as well by dividing the wire into a few constant width sections. Figure 3-42 shows that a few segments of wire can be used to approximate the exponential taper reasonably well. This result also suggests that long wires which can be run on several layers should run on the lowest-resistance layer near the driver and can move to the higher-resistance layers as they move toward the signal sink.

Figure 3-42: *A step-tapered wire.*

3.7.2 Delay Through RC Trees

While analyzing a straight transmission line is straightforward, analyzing more complex networks is harder. We may not always need an exact answer, either—a good approximation is often enough considering the other uncertainties in IC design and manufacturing. In the case of RC trees, as shown in Figure 3-43, we can quickly compute accurate bounds on the delay through the wire [Rub83]. The wiring can be broken into an RC tree either by representing each branch by one RC lump or by breaking a branch into several lumps.

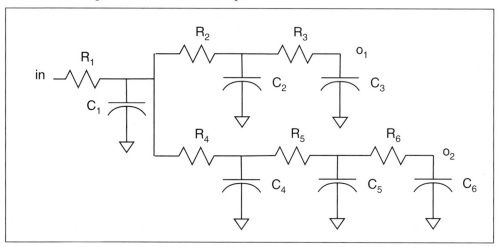

Figure 3-43: *An RC tree.*

When analyzing the RC tree, we assume the network has one input, which provides a voltage step, and several outputs. We can find the delay through the wire by analyzing the voltages at the output nodes and measuring the time between the 10% and 90% points. While an exact solution for the output voltages for an arbitrary RC network is complex, we can find accurate upper and lower bounds on the output voltage, and from those voltage bounds we can compute delay bounds. We won't perform a detailed derivation of the bounds formulas, but will only try to provide an intuitive explanation of their form.

The capacitance at a node k is called C_k. We are primarily concerned with resistances along paths, notably the resistances along shared paths. If o is an output node and k is an internal node, the resistance along the intersection of the paths from the input to o and to k is called R_{k0}. In Figure 3-43, $R_{1O_1} = R_1$ because R_1 is the only resistor shared by the paths to 1 and O_1. R_{00} is the total resistance from input to the output o and similarly, R_{kk} is the total resistance from input to the internal node k. The simplest time constant for the tree is

$$T_P = \sum_k R_{kk} C_k .$$ **(EQ 3-30)**

Each term in the summation is the time constant of the simple RC circuit built from the capacitance at k and all the resistance from the input to k. Two other time constants relative to the output o are important to the bounds:

$$T_{Do} = \sum_k R_{ko} C_k ; \qquad \text{(EQ 3-31)}$$

$$T_{Ro} = \left(\sum_k R_{ko}^2 C_k \right) / R_{oo} . \qquad \text{(EQ 3-32)}$$

The terms of T_{D0} compute the time constant of the capacitance at each node and the resistance shared by the paths to k and o available to charge C_k. The terms of T_{R0} weight the terms of T_{D0} against the total resistance along the path to the output, squaring R_{k0} to ensure the value has units of time. Although we won't prove it here, these inequalities relate the voltage at each output, $v_0(t)$, and the voltage at an interior node, $v_k(t)$, using the path resistances:

$$R_{oo}[1-v_k(t)] \geq R_{ko}[1-v_o(t)] \qquad \text{(EQ 3-33)}$$

$$R_{ko}[1-v_k(t)] \leq R_{kk}[1-v_o(t)] \qquad \text{(EQ 3-34)}$$

Some intermediate steps are required to find the $v_o(t)$'s; we will skip to the resulting bounds, shown in Table 3-2. The bounds are expressed both as the voltage at a given time and as the time required for the output to assume a specified voltage; the two formulas are, of course, equivalent.

Do these bounds match our intuition about the circuit's behavior? At $t=0$, the upper bound for the output voltage is $v_o(0) = 1 - T_{D0}$. T_{D0} is formed by the time constants of RC sections formed by all the resistance along the path to o that are also connected to the kth capacitor, such as the highlighted resistors at **a** in the figure. Some of the current through those resistors will go to outputs other than o, and so are not available to charge the capacitors closest to o; the upper bound assumes that *all* their current will be used to charge capacitors along the path from input to o. The lower bound is dominated by T_{R0}, which compares R_{k0} to the total resistance from the input to o; the ratio R_{k0}/R_{00} gives a minimum resistance available to charge the capacitor C_k.

	validity	bound
lower	$t \leq T_{Do}\text{-}T_{Ro}$ $T_{Do}\text{-}T_{Ro} \leq t \leq T_p\text{-}T_{Ro}$ $t \geq T_p\text{-}T_{Ro}$	$v_o(t) \geq 0$ $v_o(t) \geq 1\text{-}[T_{Do}/(t+T_{Ro})]$ $v_o(t) \geq 1 - \dfrac{T_{Do}}{T_p} e^{(T_p\text{-}T_{Ro})/T_P} e^{-(t/T_p)}$
upper	$t \leq T_{Do}\text{-}T_{Ro}$ $t \geq T_{Do}\text{-}T_{Ro}$	$v_o(t) \leq 1\text{-}((T_{Do}\text{-}t)/T_P)$ $v_o(t) \leq 1 - \dfrac{T_{Ro}}{T_p} e^{(T_{Do}\text{-}T_{Ro})/T_{Ro}} e^{-(t/T_p)}$

	validity	bound
lower	$v_o(t) \leq 1\text{-}(T_{Ro}/T_p)$ $v_o(t) \geq 1\text{-}(T_{Ro}/T_p)$	$t \geq T_{Do}\text{-}T_p[1\text{-}v_o(t)]$ $t \geq T_{Do}\text{-}T_{Ro} + T_{Ro}ln(\dfrac{T_{Ro}}{T_p[1\text{-}v_o(t)]})$
upper	$v_o(t) \leq 1\text{-}(T_{Do}/T_p)$ $v_o(t) \geq 1\text{-}(T_{Do}/T_p)$	$t \leq [T_{Do}/(1\text{-}v_o(t))]\text{-}T_{Ro}$ $t \leq T_P\text{-}T_{Ro} + T_Pln(\dfrac{T_{Do}}{T_p[1\text{-}v_o(t)]})$

Table 3-2 *Rubinstein-Penfield-Horowitz voltage and time bounds for RC trees.*

3.7.3 Buffer Insertion in RC Transmission Lines

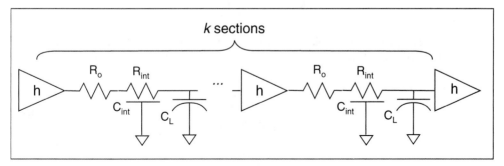

Figure 3-44: *An RC transmission line with repeaters.*

We do not obtain the minimum delay through an RC transmission line by putting a single large driver at the transmission line's source. Rather, we must put a series of buffers equally spaced through the line to restore the signal. Bakoglu [Bak90] derived the optimal number of repeaters and repeater size for an RC transmission line. As shown in Figure 3-44, we want to divide the line into k sections, each of length l. Each buffer will be of size h.

Let's first consider the case in which $h=1$ and the line is broken into k sections. R_{int} and C_{int} are the total resistance and capacitance of the transmission line. R_0 is the driver's equivalent resistance and C_0 its input capacitance. Then the 50% delay formula is

$$T_{50\%} = k\left[0.7R_0\left(\frac{C_{\text{int}}}{k} + C_0\right) + \frac{R_{\text{int}}}{k}\left(0.4\frac{C_{\text{int}}}{k} + 0.7C_0\right)\right] \qquad \textbf{(EQ 3-35)}$$

The various coefficients are due to the distributed nature of the transmission line. We find the minimum delay by setting $dT/dk = 0$. This gives the number of repeaters as

$$k = \sqrt{\frac{0.4R_{\text{int}}C_{\text{int}}}{0.7R_0C_0}}. \qquad \textbf{(EQ 3-36)}$$

When we free the size of the repeater to be an arbitrary value h, the delay equation becomes

$$T_{50\%} = k\left[0.7\frac{R_0}{h}\left(\frac{C_{int}}{k} + hC_0\right) + \frac{R_{int}}{k}\left(0.4\frac{C_{int}}{k} + 0.7hC_0\right)\right]. \qquad \textbf{(EQ 3-37)}$$

We solve for minimum delay by setting $\frac{dT}{dk} = 0$ and $\frac{dT}{dh} = 0$. This gives the optimal values for k and h as

$$k = \sqrt{\frac{0.4R_{int}C_{int}}{0.7R_0C_0}}, \qquad \textbf{(EQ 3-38)}$$

$$h = \sqrt{\frac{R_0C_{int}}{R_{int}C_0}}. \qquad \textbf{(EQ 3-39)}$$

The total delay at these values is

$$T_{50\%} = 2.5\sqrt{R_0C_0R_{int}C_{int}}. \qquad \textbf{(EQ 3-40)}$$

Example 3-3: *Buffer insertion in an RC line*

Let's calculate the buffers required when a minimum-size inverter drives a metal 1 wire that is 2000 λ x 3 λ. In this case, $R_0 = 3.9\text{k}\Omega$ and $C_0 = 0.68$ fF while $R_{int} = 53.3$ Ω and $C_{int} = 15$ fF + 90.1 fF = 105.1 fF. The optimal number of buffers is

$$k = \sqrt{\frac{0.4 \times 53.33 \times 105.1\times10^{-15}}{0.7 \times 3900 \times 0.68\times10^{-15}}} = 1.099.$$

The optimal buffer size is

$$h = \sqrt{\frac{3900 \times 105.1\times10^{-15}}{53.33 \times 0.68\times10^{-15}}} = 106.33.$$

The 50% delay is

$$T_{50\%} = 2.5\sqrt{3900 \times 0.68\times10^{-15} \times 53.33 \times 105.1\times10^{-15}} = 9.64\times10^{-12} .$$

If we increase the size of the driver by a factor of 4, reducing its resistance by 4X and increasing its capacitance by 4X, what happens? k and $T_{50\%}$ remain unchanged, but the buffer size drops by a factor of 4.

3.7.4 Crosstalk Between RC Wires

Crosstalk is important to analyze because it slows down signals—the crosstalk noise increases the signal's settling time. Crosstalk can become a major component of delay if wiring is not carefully designed.

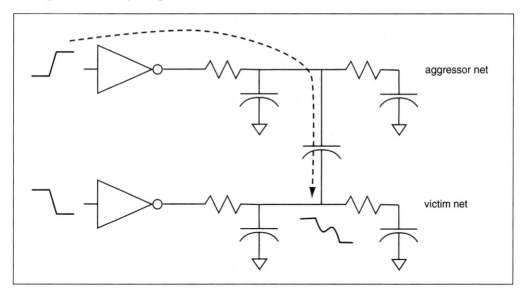

Figure 3-45: *Aggressor and victim nets.*

Figure 3-45 shows the basic situation in which crosstalk occurs. Two nets are coupled by parasitic capacitance. One net is the **aggressor net** that interferes with a **victim net** through that coupling capacitance. A transition in the aggressor net is transmitted to the victim net causing the victim to glitch. The glitch causes the victim net to take longer to settle to its final value. In static combinational logic, crosstalk increases the

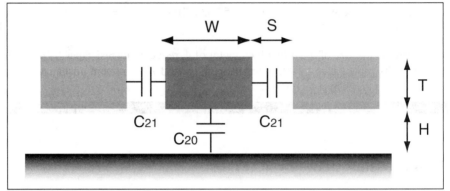

Figure 3-46: A simple crosstalk model (after Sakurai [Sak93], © 1993 IEEE).

delay across a net; in dynamic logic, crosstalk can cause the state of a node to flip, causing a permanent error.

In this section we will develop basic analytical models for crosstalk; in Section 4.5.4 we will learn how to minimize crosstalk through routing techniques. The simplest case to consider is a set of three wires [Sak93], as shown in Figure 3-46. The middle wire carries the signal of interest, while the other two capactively inject crosstalk noise. Each wire is of height T and width W, giving an aspect ratio of W/T. Each wire is height H above the substrate and the wires are spaced a distance S apart. We must consider three capacitances: C_{20} between the signal wire and the substrate, and two capacitances of equal value, C_{21}, to the two interfering wires. We denote the sum of these three capacitances as C_3. Sakurai estimates the RC delay through the signal wire in arbitrary time units as

$$t_r = \frac{(C_{20} + 4C_{21})}{W/H} .$$

(EQ 3-41)

Using this simple model, Figure 3-47 shows Sakurai's calculation of relative RC delay in arbitrary units for a 0.5 μm technology for the signal wire. This plot assumes that $T/H = 1$ and that the aspect ratio varies from near 0 through 4; the delay is shown for four different spacings between the wires, as given by the P/H ratio. This plot clearly shows two important results. First, there is an optimum wire width for any given wire spacing, as shown by the U shape of each curve. Second, the optimum width increases as the spacing between wires increases.

That analysis assumes that the signals on the surrounding wires are stable, which is the best case. In general, we must assume that the surrounding wires are in transition. Consider the model of Figure 3-48, in which we have two RC transmission lines with

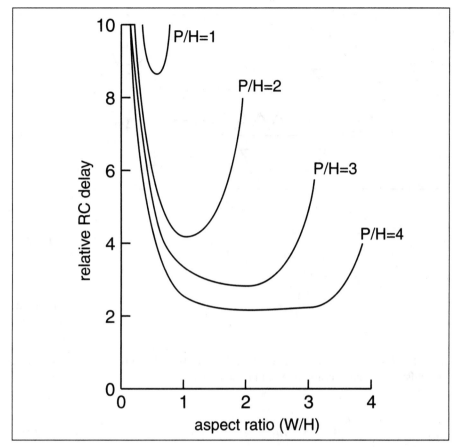

Figure 3-47:
Delay vs. wire
aspect ratio
and spacing
(after Sakurai
[Sak93] ©
1993 IEEE).

a coupling capacitance C_c between them. A step is applied to each wire at $t=0$, resulting in response waveforms at the opposite ends of the transmission lines [Sak93]. We assume that the unit resistances and capacitances of the two transmission lines are equal. Defining differential voltages between the two wires helps simplify the voltage response equations:

$$V_+ = \frac{V_1 + V_2}{\sqrt{2}}, V_- = \frac{V_1 - V_2}{\sqrt{2}}. \qquad \textbf{(EQ 3-42)}$$

The voltage responses of the transmission lines can then be written as

Figure 3-48:
Two coupled RC transmission lines.

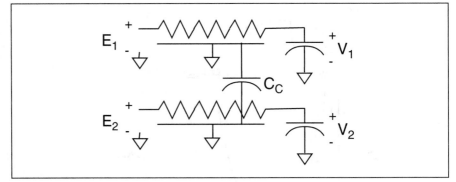

$$\frac{d^2 V_+}{dx^2} = rc\frac{dV_+}{dt},$$ (EQ 3-43)

$$\frac{d^2 V_-}{dx^2} = r(c + 2c_c)\frac{dV_-}{dt}.$$ (EQ 3-44)

If we let $R = rl$, $C = cl$, and $C_c = c_c l$, then the voltage responses V_1 and V_2 at the ends of the transmission lines can be written as

$$V_1(t) \approx E_1 + \frac{K_1}{2}[(E_1 + E_2)e^{-\sigma_1 t/RC} + (E_1 - E_2)e^{-\sigma_1 t/RC + 2RC_c}],$$ (EQ 3-45)

$$V_2(t) \approx E_2 + \frac{K_1}{2}[(E_1 + E_2)e^{-\sigma_1 t/RC} - (E_1 - E_2)e^{-\sigma_1 t/(RC + 2RC_c)}].$$ (EQ 3-46)

3.8 Delay Through Inductive Interconnect

Copper wiring provides much better performance, particularly for long wires. However, copper wires have significant inductance. Analyzing inductive wiring is more complicated than is analyzing RC transmission lines. RLC transmission lines have a more complex response that requires more subtle interpretation as well as more effort.

3.8.1 RLC Basics

First, let's review the basics of RLC circuits. A single RLC section is shown in Figure 3-49. The poles of the RLC section are at

$$\omega_0 \left[\xi \pm \sqrt{\xi^2 - 1} \right]$$ (EQ 3-47)

where the damping factor ξ is defined as

$$\xi = \frac{R}{2} \sqrt{\frac{C}{L}}.$$ (EQ 3-48)

If the damping factor is greater than 1, the circuit is **overdamped** and responds to an impulse or step by monotonically approaching the final voltage. If the damping factor is less than 1, the circuit is **underdamped** and oscillates as it converges to the steady-state voltage. Underdamped circuits create a new challenge for digital circuit analysis because it is harder to find their rise times. For an underdamped circuit, we simply have to find the first time the waveform crosses the desired voltage threshold, knowing that it will always remain above that level. To determine the rise time of an under-damped circuit, we must find the last time at which the waveform falls below the threshold.

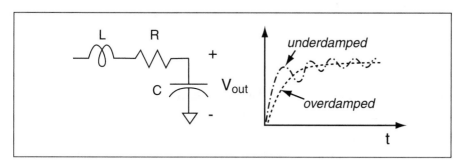

Figure 3-49: *An RLC circuit and its behavior.*

The simplest form of an RLC transmission is the lossless LC line with zero resistance. A signal propagates along an LC transmission line [Ram65] with velocity

$$v = \frac{1}{\sqrt{LC}}.$$ (EQ 3-49)

Therefore, the propagation delay through an LC transmission line of length l is $t_p = l\sqrt{LC}$. This value is a lower bound on the delay introduced by an RLC transmission line.

3.8.2 RLC Transmission Line Delay

In today's technology the resistance of the copper wiring cannot be ignored. Because we are designing digital systems, we are interested in an RLC transmission line that is being driven by a gate at one end and is connected to a receiving gate at the other end [Ism00]. We will model the driving gate as a resistance R_{tr} and the load gate as a capacitance C_L. We will use R, L and C for the unit resistance, inductance, and capacitance and R_t, L_t, and C_t for the total resistance, inductance, and capacitance of the line. The complete system is shown in Figure 3-50.

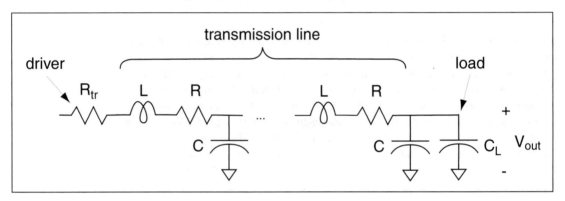

Figure 3-50: *An RLC transmission line with a driver and load.*

We can simplify our analysis by scaling time using the factor

$$\omega_n = \frac{1}{\sqrt{Ll(Cl + C_L)}} .$$
(EQ 3-50)

We normalize time by substituting $t = t'/\omega_n$. We also need two additional values:

$$R_T = \frac{R_{tr}}{R_t} ,$$
(EQ 3-51)

$$C_T = \frac{C_L}{C_t}.$$ (EQ 3-52)

where l is once again the length of the transmission line.

The complete derivation of the transmission line's response is rather complex, but we are most interested in the propagation delay through the wire to the load capacitance. Ismail and Friedman showed that propagation delay is primarily a function of ξ, which is defined as

$$\xi = \frac{R_t}{2}\sqrt{\frac{C_t}{L_t}}\left(\frac{Rl + Cl + RCl^2 + 0.5}{\sqrt{1 + Cl}}\right).$$ (EQ 3-53)

They used numerical techniques to approximate the 50% propagation delay of our RLC transmission line as

$$t_{pd} = \left(e^{-2.9\xi^{1.35}} + 1.48\xi\right)/\omega_n.$$ (EQ 3-54)

Figure 3-51 compares the response of RLC and RC wires for different values of ξ. These plots show that ignoring inductance results in very poor results for small values of ξ.

Figure 3-52 compares RC and RLC models for wires driven by inverters in a 0.25 μm technology. This figure shows that ignoring inductance results in serious errors in estimating delay for a variety of wire and driver configurations.

3.8.3 Buffer Insertion in RLC Transmission Lines

Ismail and Friedman also showed where to place buffers in an RLC transmission line [Ism00]. The circuit is shown in Figure 3-53. The transmission line is divided into k sections, each of length l/k. All the buffers are of the same size and are h times larger than a minimum-size buffer; we use R_0 and C_0 to represent the source resistance and load capacitance of a minimum-size buffer.

We can define

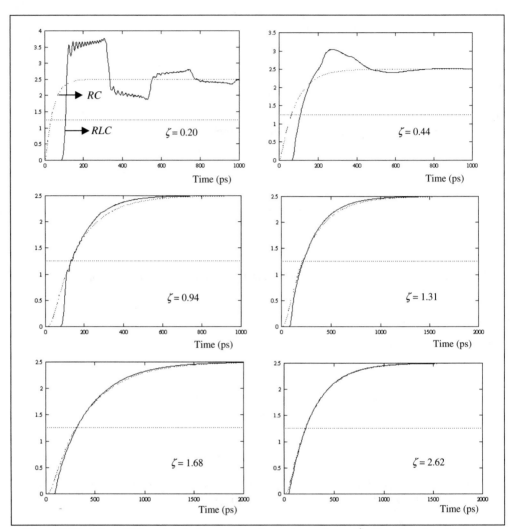

Figure 3-51: *RC vs. RLC models for interconnect for various values of* ξ *(from Ismail and Friedman [Ism00]).* © 2000 IEEE.

$$T_{L/R} = \sqrt{\frac{L_t/R_t}{R_0 C_0}}. \tag{EQ 3-55}$$

As in the RC case, we are interested in determining the optimum drive per stage h_{opt} and the optimum length of each stage's wire k_{opt}. This optimization problem cannot

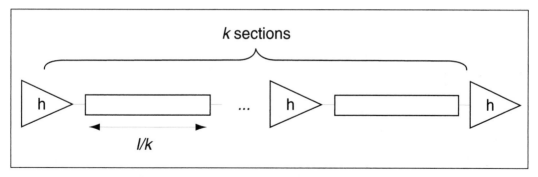

Figure 3-53: *Repeaters in an RLC transmission line.*

be solved analytically, but Ismail and Friedman fitted curves to the functions to provide these formula:

$$h_{opt} = \sqrt{\frac{R_0 C_t}{R_t C_0}} \frac{1}{[1 + 0.16(T_{L/R})^3]^{0.3}},$$ (EQ 3-56)

$$k_{opt} = \sqrt{\frac{R_t C_t}{2 R_0 C_0}} \frac{1}{[1 + 0.18(T_{L/R})^3]^{0.3}}.$$ (EQ 3-57)

3.9 References

Claude Shannon first described the relationship between Boolean logic and switching functions for his Master's thesis; his paper [Sha38] is still interesting reading. Hodges and Jackson [Hod83] give an excellent introduction to device characteristics and digital circuit design, showing how to analyze CMOS logic gates as well as design more complex digital circuits. Books by Rabaey [Rab96] and Uyemura [Uye92] are detailed presentations of digital logic circuits; Rabaey's book also covers bipolar circuits in detail. Geiger, Allen, and Strader [Gei90] give a good introduction to circuit simulation as well as a number of important topics in circuit and logic design. Shoji [Sho88] gives a very thorough analysis of delay through CMOS gates. Domino logic was introduced by Krambeck, Lee, and Law [Kra82]. De et al [De01] concentrate on leakage currents in CMOS logic.

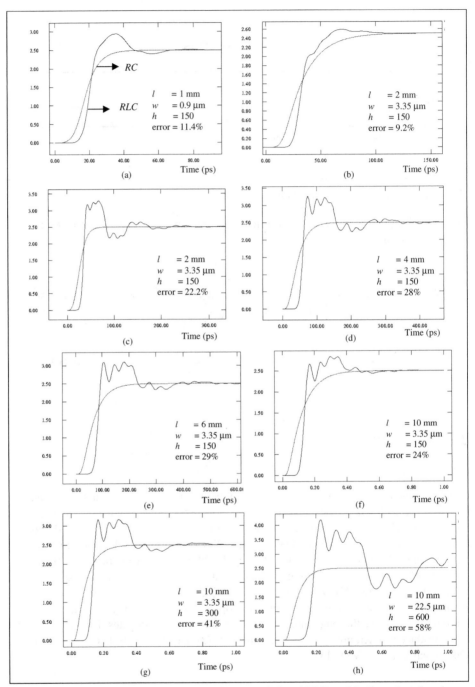

Figure 3-52: *CMOS gate driving a copper wire, using RC and RLC models (from Ismail and Friedman [Ism00]).* © 2000 IEEE.

3.10 Problems

Use the parameters for the 0.5 μ*m* process of Table 2-4 whenever process parameters are required, unless otherwise noted.

3-1. Design the static complementary pullup and pulldown networks for these logic expressions:

> a) $(a + b + c)'$.
>
> b) $[(a + b)c]'$.
>
> c) $(a + b)(c + d)$.

3-2. Write the defining logic equation and transistor topology for each complex gate below:

> a) AOI-22.
>
> b) OAI-22.
>
> c) AOI-212.
>
> d) OAI-321.
>
> e) AOI-2222.

3-3. Draw a layout for a two-input NOR gate using a static complementary circuit. The gate should have its inputs in metal on the cell's left hand side and its output in metal on the right hand side.

3-4. Size the transistors in a three-input, static complementary NOR gate such that the circuit's rise and fall times are approximately equal.

3-5. What is the difference in fall time of a two-input, static complementary NOR gate (assuming a minimum-size load capacitance) when one pulldown and when two pulldowns are activated?

3-6. Compute the low-to-high transition time and delay (at a power supply voltage of 3.3V) using the τ model through a two-input NAND gate which drives one input of a three-input NOR gate (both static complementary gates):

> a) Compute the load capacitance on the NAND gate, assuming the NOR gate's transistors all have $W=6\lambda$, $L=2\lambda$.
>
> b) Compute the equivalence resistance of appropriate transistors for the low-to-high transition, assuming the pulldown transistors have $W=6\lambda$, $L=2\lambda$ and the pullups have $W=6\lambda$, $L=2\lambda$.

c) Compute the transition time and delay.

3-7. Compute, using the τ model, the high-to-low transition time and delay (at a power supply voltage of 3.3V) of a two-input, static complementary NOR gate with minimum-sized transistors driving these loads:

a) An inverter with minimum-sized pullup and pulldown.

b) An inverter whose pullup and pulldown are both of size $W=10\lambda$, $L=10\lambda$.

c) A $2000\lambda \times 2\lambda$ poly wire connected to an inverter with minimum-sized pullup and pulldown (lump the wire and inverter parasitics into a single RC section).

d) Size the transistors in a two-input NOR gate such that the gate's rise time and fall time are approximately equal.

3-8. Redesign the layout of Figure 3-8 so that the inverter's output displays roughly symmetric rise and fall times.

3-9. Design a three-input, static complementary NAND gate, which implements this function:

a	b	c	NAND(a,b,c)
0	0	0	1
0	0	1	1
0	1	0	1
0	1	1	1
1	0	0	1
1	0	1	1
1	1	0	1
1	1	1	0

a) Draw a switch-level schematic.

b) Draw a stick diagram.

3-10. Here is a partial schematic for a two-input XOR gate:

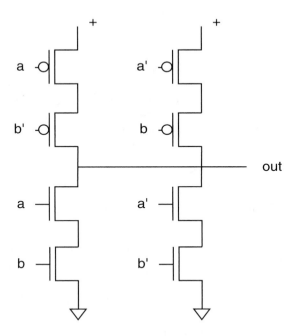

The gate's output is 1 when exactly one of its inputs is 1. The schematic is partial because transistors in the diagram require both the true (a) and complement (a') form of the inputs, but the inverters which generate a' and b' from a and b are not shown.

    a) Write the truth table for the two-input XOR.

    b) Complete the schematic for this gate—compute the complement of the inputs using inverters.

    c) Draw a stick diagram for the partial XOR schematic.

    d) Draw a stick diagram for the complete XOR schematic.

3-11. Draw a stick diagram for a three-input NOR gate.

3-12. Draw the stick diagram for a one-bit multiplexer cell built from static, complementary gates. The stick diagram should show the detail of the component NAND gates—don't draw the NAND gates as cells.

3-13. A two-input NAND gate drives a set of parallel inverters; all the logic gates use minimum-sized transistors. Compare the NAND gate driving n inverters to the delay through a two-stage network, in which two inverters each drive $n/2$ fanout inverters. What is the smallest value of n for which the buffered network has smaller delay than the unbuffered network?

3-14. At what length does a minimum-width poly wire present a capacitive load equal to a minimum-sized inverter? A minimum-width metal-1 wire?

3-15. Draw the circuit topology of a three-input NOR gate designed in pseudo-nMOS.

3-16. Design a two-input AND gate in domino logic:

   a) Draw a transistor schematic.

   b) Draw a stick diagram.

3-17. For a pseudo-nMOS inverter:

   a) What V_{OL} must be chosen such that the pulldown transistor of a subsequent pseudo-nMOS gate will be off when the inverter's output is logic 1?

   b) What is the ratio $W_p/L_p/W_n/L_n$ required to achieve this output voltage?

3-18. Draw a schematic for a 2-input NAND in MTCMOS logic.

3-19. Draw a schematic for these gates in DCVSL:

   a) 2 input NAND;

   b) 2 input NOR;

   c) 3 input NAND.

3-20. Draw schematic for these gates in DCSL:

   a) 2 input NAND;

   b) 2 input NOR;

   c) 3 input NAND;

3-21. Plot the Elmore delay for a metal 1 wire of size $2000\lambda \times 3\lambda$ using

   a) 2 sections;

   b) 4 sections;

   c) 8 sections.

3-22. Plot the Elmore delay for a metal 2 wire of size $3000\lambda \times 4\lambda$ using

   a) 2 sections;

   b) 4 sections;

c) 8 sections.

3-23. Compute the optimal number of buffers and buffer sizes for these RC (non-inductive) wires when driven by a minimum-size inverter:

    a) metal 1 $3000\lambda \times 3\lambda$.

    b) metal 1 $5000\lambda \times 3\lambda$.

    c) metal 2 $3000\lambda \times 4\lambda$.

    d) metal 2 $4000\lambda \times 4\lambda$.

4

Combinational Logic Networks

Highlights:

Layouts for logic networks.

Delay through networks.

Power consumption.

Switch logic networks.

Combinational logic testing.

4.1 Introduction

This chapter concentrates on the design of combinational logic functions. Building a single inverter doesn't justify a multi-billion VLSI fabrication line. We want to build complex systems of many combinational gates. To do so, we will study basic aspects of hierarchical design and analysis, especially delay and power analysis. The knowledge gained about fabrication is important for combinational logic design—technology-dependent parameters for minimum size, spacing, and parasitic values largely determine how big a gate circuit must be and how fast it can run. We will use our knowledge of logic gates, developed in the last chapter, to analyze the delay and testability properties of combinational logic networks, including both the interconnect and the gates.

4.2 Standard Cell-Based Layout

Many layout design methods are common to most subsystems. In this section we will cover general-purpose layout design methods for use in the rest of the chapter, largely by amplifying the lessons learned in Chapter 3.

CMOS layouts are pleasantly tedious, thanks to the segregation of pullups and pulldowns into separate tubs. The tub separation rules force a small layout into a row of p-type transistors stacked on top of a row of n-type transistors. On a larger scale, they force the design into rows of gates, each composed of their own p-type and n-type rows. That style makes layout design easier because it clearly marks the boundaries of the design space—in technologies with fewer layout restrictions, like nMOS, layout designers are often kept up at night wondering whether some unthought-of geometry could squeeze a few microns from the layout.

As has been mentioned before, a good way to attack the design of a layout is to divide the problem into placement, which positions components, and routing, which runs wires between the components. These two phases clearly interact: we can't route the wires until components are placed, but the quality of a placement is judged solely by the quality of the routing it allows. We separate layout design into these two phases to make each part more tractable. We generally perform placement using simple estimates of the quality of the final routing, then route the wires using that fixed placement; occasionally we modify the placement and patch up the routing to fix problems that weren't apparent until all the wires were routed. The primitives in placement are almost always logic gates, memory elements, and occasionally larger components like full adders. Transistors are too small to be useful as placement primitives—the transistors in a logic gate move as a clump since spreading them out would introduce huge parasitics within the gate. We generally place logic gates in single-row layouts and either gates or larger register-transfer components in multi-row layouts.

4.2.1 Single-Row Layout Design

We can design a one-row layout as a one-dimensional array of gates connected by wires. Changing the placement of logic gates (and as a result changing the wiring between the gates) has both area and delay effects. By sketching the wiring organization during placement, we can judge the feasibility of wiring, the size of the layout, and the wiring parasitics which will limit performance.

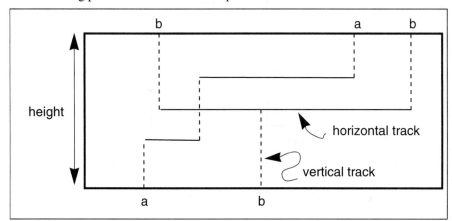

Figure 4-1:
Structure of a one-row layout.

The basic structure of a one-row layout is shown in Figure 4-1. The transistors are all between the **power rails** formed by the V_{DD} and V_{SS} lines. The major routing **channel** runs below the power rails (there is another channel above the row, of course, that can also be used by these transistors). The gate inputs and outputs are near the center of the row, so vertical wires connect the gates to the routing channel and outside world. Sometimes space is left in the transistor area for a feedthrough to allow a wire to be routed through the middle of the cell. Smaller areas within the transistor area—above the V_{SS} line, below the V_{DD} line, and between the n-type and p-type rows—are also available for routing wires.

We usually want to avoid routing wires between the p-type and n-type rows because stretching apart the logic gates adds harmful parasitics, as discussed in Section 3.3.7. However, useful routing areas can be created when transistor sizes in the row vary widely, leaving extra room around the smaller transistors, as shown in Figure 4-2. The intra-row wiring areas are useful for short wires between logic gates in the same row—not only is a routing track saved, but the wire has significantly less capacitance since it need not run down to the routing channel and back up. Intra-row routing is a method of last resort, but if it becomes necessary, the best way to take advantage of the available space is to first design the basic gate layout first, then look for interstitial

Figure 4-2: Intra-row wiring.

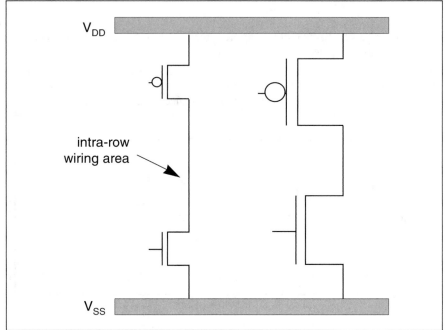

V_{DD}

intra-row
wiring area

V_{SS}

space around the small transistors where short wires can be routed, and finally to
route the remaining wires through the channel.

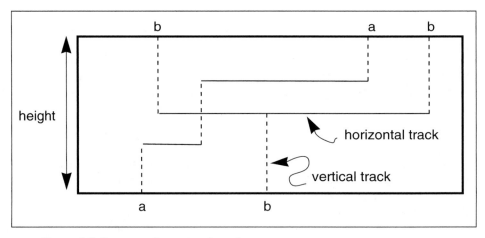

Figure 4-3: Structure of a routing channel.

The wiring channel's structure is shown in Figure 4-3. A channel has pins only along its top and bottom walls. We will discuss area routing, which allows pins on all four sides, in Chapter 7. The channel is divided into **horizontal tracks**, more typically called **tracks**, and **vertical tracks**. The horizontal and vertical tracks form a grid on which wire segments are placed. The distance between tracks is equal to the minimum spacing between a wire and a via. Using a standard grid greatly simplifies wiring design with little penalty—human or algorithmic routers need only place wires in the tracks to ensure there will be no design rule violations. Wire segments on horizontal and vertical tracks are on separate layers—some advanced routing programs occasionally violate this rule to improve the routing, but keeping vertical and horizontal wire segments separate greatly simplifies wiring design. Segregation ensures that vertical wires are in danger of shorting horizontal wires only at corners, where vias connect the horizontal and vertical layers. If we consider each horizontal segment to be terminated at both ends by vias, with longer connections formed by multiple segments, then the routing is completely determined by the endpoints of the horizontal segments.

The width of the routing channel is determined by the placement of pins along its top and bottom edges. The major variable in area devoted to signal routing is the height of the channel, which is determined by the **density**—the maximum number of horizontal tracks occupied on any vertical cut through the channel. Good routing algorithms work hard to minimize the number of tracks required to route all the signals in a channel, but they can do no better than the density: if three signals must go from one side of the channel to the other at a vertical cut, at least three tracks are required to accommodate those wires.

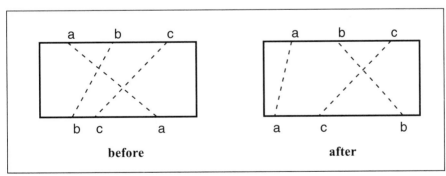

Figure 4-4:
Channel density changes with pin placement.

Changing the placement of pins can change both the density and the difficulty of the routing problem. Consider the example of Figure 4-4. The position of a pin along the top or bottom edge is determined by the position of the incoming vertical wire that connects the channel to the appropriate logic gate input or output; the transistor rows above and below the wiring channel can both connect to the channel, though at opposite edges. In this case, swapping the a and b pins reduces the channel density from three to two.

Density is a measure that can be used to evaluate the wirability of a channel before we have actually completed the wiring. It is very important to be able to estimate the results of routing so that we can provide for adequate space in the design. It is sometimes valuable to leave extra space in the channel to make it easier to route the wires, as well as to be able to change the wiring to accommodate logic design changes. Not all blocks of logic are equally performance-critical, and it may be worth spending some area to make a logic block easier to layout and to modify.

The next example walks through the design process for a one-row layout.

Example 4-1: Layout of a full adder

A full adder illustrates the techniques used to design single-row layouts. The full adder computes two functions: $s_i = a_i \oplus b_i \oplus c_i$ and $c_{i+1} = a_i \cdot b_i + a_i \cdot c_i + b_i \cdot c_i$. We will compute s_i using two two-input XORs:

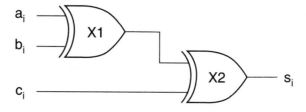

We will use a two-level NAND network to compute c_{i+1}:

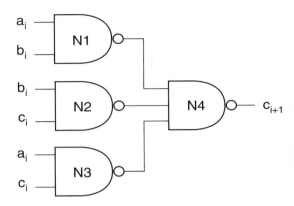

(An AOI gate is a better choice for the carry computation, but we have more cells to use to illustrate placement options by choosing this implementation.) We have a total of six gates to place and route. In this case, we won't use intra-row wiring—all of the wires will go into the wiring channel below the gates. Our layout job is to place the gates such that the wiring channel below the row of gates has as few tracks as possible.

We can use a three-step process to generate and evaluate a candidate layout for the adder:

1. Place the gates in a row by any method.

2. Draw the wires between the gates and the primary inputs and outputs.

3. Measure the density of the channel.

Once we have evaluated a placement, we can decide how to improve it. We can generate the first placement by almost any means; after that, we use the results of the last routing to suggest changes to the placement.

The full adder has four gates to place, two for each function. We will start by keeping together the gates for each function:

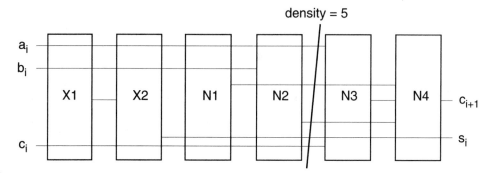

Though the final layout will have its wires in the wiring channel below the cell, we have drawn the wires over the gates here for clarity—drawing the paths of the wires down to the channel and then back up to the gate inputs is too confusing. The channel density for this placement is five.

We can try to reduce the channel density by interchanging gates. Two opportunities suggest themselves:

- **Swap the gates within each function**. A simple test shows that this doesn't reduce the density.

- **Swap the XOR pair with the NAND network**. This doesn't help either, because we must still drag a_i, b_i, and c_i, to the XORs and send c_{i+1} to the right edge of the cell.

This placement seems to give the minimum-height routing channel, which means that the channel's area will be as small as possible. Gate placement can affect transistor size in larger layouts—we may be able to reduce the sizes of the transistors in some critical gates by placing those gates closer together. But in a layout this small, if we use metal wiring as much as possible, the sizes of the gate cells are fixed. So minimizing wiring channel density minimizes total layout area.

Systems with more than six gates provide more opportunities for placement optimization. If we are more concerned about parasitics on some critical wires (such as the carry), we can choose a placement to make those wires as short as possible. If those wires are sufficiently critical, we may even want to increase density beyond the minimum required to make those critical wires shorter.

We also need to know how to route the wires in the channel. Channel routing is NP-complete [Szy85], but simple algorithms exist for special cases, and effective heuristics exist that can solve many problems. We will defer the description of sophisticated routing algorithms to Chapter 10; here, we will identify what makes each problem difficult and identify some simple algorithms and heuristics that can be applied by hand.

The **left-edge algorithm** is a simple channel routing algorithm which uses only one horizontal wire segment per net. The algorithm sweeps the channel from left to right; imagine holding a ruler vertically over the channel and stopping at each pin, whether it is on the top or bottom of the channel. If the pin is the first pin on a net, that net is assigned its lone horizontal wire segment immediately. The track assignment is greedy—the bottommost empty track is assigned to the net. When the last pin on a net is encountered, the net's track is marked as empty and it can be reused by another net farther to the right. The vertical wire segments that connect the pins to the horizontal segment, along with the necessary vias, can be added separately, after assignment of horizontal segments is complete. The next example shows how to use this algorithm to route a channel.

Example 4-2: Left-edge channel routing

The channel has three nets:

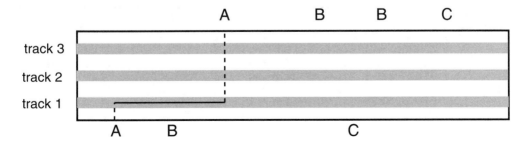

The left-most net is A; we route it in the first empty track, which is track 1. We run a wire segment from A's left-most pin to its right-most:

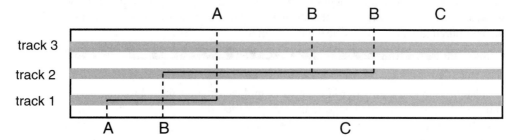

Moving to the right, the next pin is B. Track 1 is occupied, so we route B in track 2:

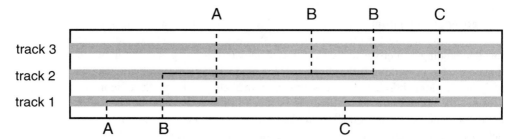

The third and final net is C. At this position, A no longer occupies track 1, so we can reuse it to route C:

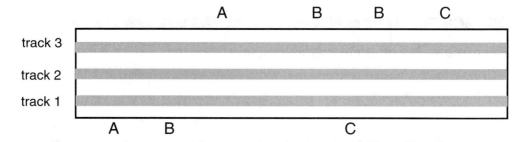

Once the horizontal wire segments have all been placed, we can add the vertical wire segments to connect the tracks to the pins and, of course, the vias needed to connect the horizontal and vertical segments. Since the channel needs only two tracks, its height can be reduced appropriately.

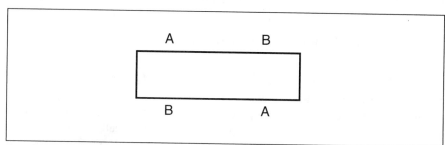

Figure 4-5: A channel that cannot be routed by the left-edge algorithm.

The left-edge algorithm is exact for the problems we have encountered so far—it always gives a channel with the smallest possible height. But it fails in an important class of problems illustrated in Figure 4-5. Both ends of nets A and B are on the same vertical tracks. As a result, we can't route both nets using only one horizontal track each. If only one of the pins were moved—for instance, the right pin of B—we could route A in the first track and B in the second track. But pins along the top and bottom of the track are fixed and can't be moved by the router—the router controls only the placement of horizontal segments in tracks. Vertically aligned pins form a **vertical**

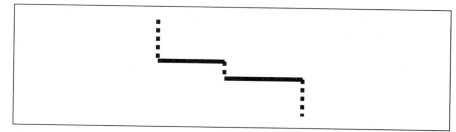

Figure 4-6: A dogleg wire.

constraint on the routing problem: on the left-hand side of this channel, the placement of A's pin above B's constrains A's horizontal segment to be above B's at that point; on the right-hand side, B's horizontal segment must be above A's at that point in the channel. We obviously can't satisfy both constraints simultaneously if we restrict each net to one horizontal segment.

The natural solution is to allow a net to move from track to track as it travels along the channel [Deu76]. Figure 4-6 shows a **dogleg**—those who can see Greek gods in the constellations should also be able to identify this wire as a dog's outstretched hind leg. We can use one single-track net and one dogleg to route the channel of Figure 4-5. Dogleg channel routing algorithms are much more sophisticated than the left-edge algorithm; we will survey dogleg routing algorithms in Section 10.4. If you want to route a channel with a few cyclic constraints by hand, a good strategy is to route the nets that require doglegs first, then route the remaining nets using the left-edge algorithm, avoiding the regions occupied by the previously routed nets.

4.2.2 Standard Cell Layout Design

Large layouts are composed from several rows. We introduced standard cell layout in Chapter 2; we are now in a position to investigate standard cell layout design in more detail. A standard cell layout is composed of cells taken from a library. Cells include combinational logic gates and memory elements, and perhaps cells as complex as full adders and multiplexers. A good standard cell library includes many variations on logic gates: NANDs, NORs, AOIs, OAIs, etc., all with varying number of inputs. The more complete the library, the less that is wasted when mapping your logic function onto the available components.

Figure 4-7 shows how the layout of a typical standard cell is organized. All cells in the library must have the same **pitch** (the distance between two points, in this case height) because they will be connected by abutment and their V_{DD} and V_{SS} lines must match up. Wires which must be connected to other cells are pulled to the top and bottom edges of the cell and placed to match the grid of the routing channel. The wire must be presented at the cell's edge on the layer used to make vertical connections in the channel. Most of the cell's area cannot be used for wiring, but some cells can be designed with a feedthrough area. Without feedthroughs, any wire going from one channel to another would have to be run to the end of the channel and around the end of the cell row; feedthroughs provide shortcuts through which delay-critical wires can be routed.

Transistors in standard cells are typically much larger than those in custom layouts. The designer of a library cell doesn't know how it will be used. In the worst case, a cell may have to drive a wire from one corner of a large chip to the other. To ensure that even worst-case delays are acceptable, the cells are designed with large transistors. Some libraries give two varieties of cells: high-power cells can be used to drive long wires, while low-power cells can be used to drive nodes with lower capacitive loads. Of course, the final selection cannot be made until after placement; we usually make an initial selection of low- or high-power based on the critical path of the gate network, then adjusting the selection after layout. Furthermore, both low-power and

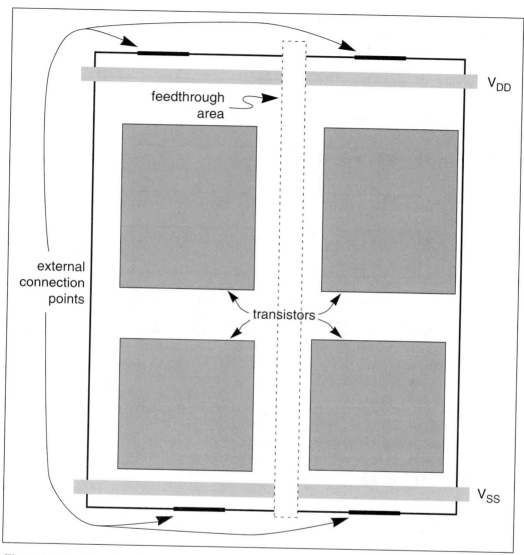

Figure 4-7: *Configuration of a typical standard cell.*

high-power cells must be the same height so that they can be mixed; the smaller transistor sizes of low-power cells may result in narrower cells.

The interaction between area and delay in a multi-row layout can be complex. Generally we are interested in minimizing area while satisfying a maximum delay through the combinational logic. One good way to judge the wirability of a placement is to

Figure 4-8: A rat's nest plot of wires.

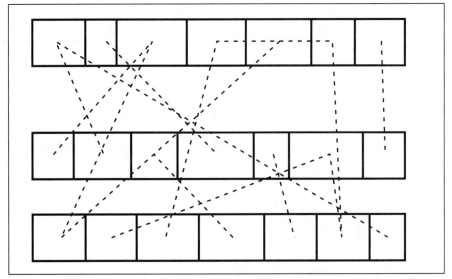

write a program to generate a **rat's nest** plot. (Use a program to generate the plot—it is too tedious to construct by hand for examples of interesting size.) An example is shown in Figure 4-8. The plot shows the position of each component, usually as a point or a small box, and straight lines between components connected by a wire. The straight line is a grossly simplified cartoon of the wire's actual path in the final routing, but for medium-sized layouts it is sufficient to identify congested areas. If many lines run through a small section, either the routing channel in that area will be very tall, or wires will have to be routed around that region, filling up other channels. Individual wires also point to delay problems—a long line from one end of the layout to the other indicates a long wire. If that wire is on the critical delay path, the capacitance of the wire will seriously affect performance.

4.3 Simulation

It's very easy to make a mistake during layout and not draw the circuit you intended. It is also surprisingly easy to write down a logic expression which doesn't perform the function you intended. **Simulation** is a key design validation tool for both checking a circuit's function and other parameters, such as performance. Simulation tools are available for all levels of design abstraction. Four types of simulator are most commonly used for combinational logic design:

- **circuit simulation** (which we introduced in Section 2.3.8) performs a detailed analysis of voltages and currents in the circuit;

- **timing simulation** uses slightly less accurate models of transistors to allow simulation of larger circuits than is possible with circuit simulation;

- **switch simulation** treats the transistors as ideal or semi-ideal switches;

- **gate simulation** uses the logic gate as the fundamental unit of simulation.

Simulation can and should occur at two stages: you can enter gate or transistor circuits (either textually or using a graphical schematic editor) before layout is complete; or you can extract a circuit from the layout and simulate it. Both are critical. You certainly want to find logic errors before you commit to layout. Simulation of an extracted layout both verifies that you made no mistakes translating the circuit into rectangles and checks that layout parasitics don't destroy the circuit's behavior.

Circuit simulation is important when delay information is critical. The simple delay models introduced in Section 3.3.4 may not be accurate enough to give good delay values for the circuits which determine the system clock rate. Circuit simulation is also invaluable for testing whether the circuit will work not only at nominal process values, but at the process corners as well. As a result, cells designed for libraries, which will be used in many different chips, undergo extensive circuit simulation before they are added to the library.

Circuit simulators are very accurate but slow. We may be able to simulate a few hundred transistors on a computer, but since even simple VLSI chips have thousands of transistors, we can't use circuit simulation exclusively. A timing simulator trades accuracy for speed; while the waveforms it generates are not always accurate, a timing simulator can handle circuits with many hundreds of transistors. The standard explanation of how to build a timing simulator is to start with a circuit simulator, throw out its component algorithms one-by-one until the simulator breaks, then put that last component back in. Timing simulators use simpler transistor models and solution methods which generally give good results for digital circuits. A timing simulator is very useful for scrutiny of performance-critical circuits, such as the carry chain of a large adder.

A switch simulator models transistors as switches. Most represent node values only as logic values: 0, 1, and X for unknown. The next example illustrates switch-level simulation of a simple logic circuit.

Example 4-3: *Switch simulation*

We want to simulate this circuit:

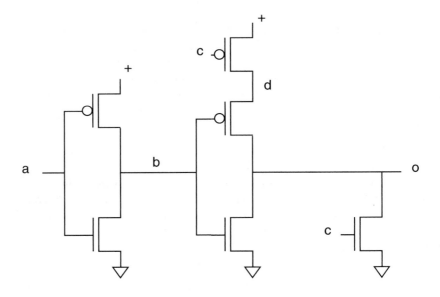

(The two *c* inputs are connected; we've omitted the wire to simplify the drawing.)

We will need initial values for the nodes. A safe assumption is that all non-primary input nodes start out at X, the unknown value. After we've simulated for enough cycles to exercise all nodes, X's should have been replaced by proper logic values. If they aren't, we know the circuit has a problem.

Let's start with *a*=1, *c*=0. A simulation step solves for the new internal node values given the present primary input values. However, that one step requires several iterations, because values may propagate through the circuit. At each iteration, we examine the nodes connected by a closed (that is, on) switch: if the source and drain nodes have the same value, no change is required. If one of them is a 1 and the other is a 0, the result depends in part on the relative capacitances of the two nodes. If each has about the same capacitance, both nodes change to X, because the voltage resulting from the fighting capacitors will probably not be a valid logic value. If, however, one capacitance is much larger than the other, that capacitance's voltage will dominate because it can share charge with the smaller capacitor without significantly changing its own voltage. Nodal capacitances are usually assigned to only a few sizes: the power supplies must be assigned the greatest strength, since they can override any node in the circuit; typical nodes are assigned a smaller strength; and sometimes long wires are assigned an intermediate strength.

The first iteration propagates two values: b is set to 0 because the inverter's n-type transistor is closed; and d is set to 1 because the c-controlled pullup is closed. In the next iteration, the values at b and d propagate to node o: since the b-controlled pullup is now on, the 1 from the d node propagates to o. If we try another iteration, no node values change, so we are done with this step.

After one step, here are the node values:

a	1
b	0
c	0
d	1
o	1

If we change any primary inputs, we can execute another simulation step and propagate the new values through the circuit. If we set a=0, the first iteration sets b to 1. In the next iteration, the b-controlled pullup is off and the pulldown is on; since the power supply is larger than any internal node, o's new value is 0. The new node values at the end of the iteration are:

a	0
b	1
c	0
d	1
o	0

Note that d's value hasn't changed because it is still driven to 1 through the upper pullup and protected from the o node by the lower pullup.

Switch simulation is the workhorse simulation algorithm for functional verification: a switch simulator can simulate tens or hundreds of thousands of transistors, while a circuit simulator can simulate only tens or perhaps a few hundred transistors. Switch simulation can also detect some kinds of circuit problems, since it uses a simple model of node capacitance. Switch-level simulation algorithms are discussed in more detail in Section 10.3. Switch-level simulators can be extended to provide timing information. IRSIM [Sal89] is a popular example of a switch-level timing simulator. It uses RC delay models to predict delays in the circuit.

Logic simulation is, like switch simulation, a digital simulation because it assigns discrete values rather than continuous voltages. However, rather than deal with the complete switch network, it uses NANDs, AOIs, and other logic gates as the primitives. The higher-level model speeds up simulation, since the logic gate's output can be computed directly. But logic simulators cannot simulate behaviors which depend on the properties of switches and capacitors.

4.4 Combinational Network Delay

We know how to analyze the speed of a single logic gate, but that isn't sufficient to know the delay through a complex network of logic gates. The delay through one or two gates may in fact limit a system's clock rate—transistors that are too small to drive the gate's load, particularly if the gate fans out to a number of other gates may cause one gate to run much more slowly than all the other gates in the system. However, the clock rate may be limited by delay on a path through a number of gates. The delay through a combinational network depends in part on the number of gates the signal must go through; if some paths are significantly longer than others, the long paths will determine the maximum clock rate. The two problems must be solved in different ways: speeding up a single gate requires modifying the transistor sizes or perhaps the layout to reduce parasitics; cutting down excessively long paths requires redesigning the logic at the gate level. We must consider both to obtain maximum system performance.

In this section, we'll assume that the wires between the gates are ideal. In the next section we will extend our techniques to take into account the characteristics of real interconnect.

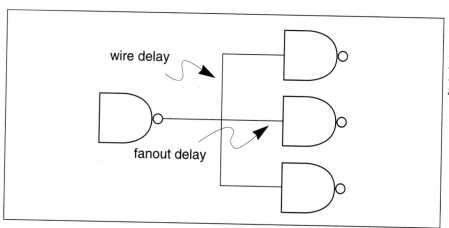

Figure 4-9:
Sources of delay
through a single
gate.

4.4.1 Fanout

Let's first consider the problems that can cause a single gate to run too slowly. A gate runs slowly when its pullup and pulldown transistors have W/Ls too small to drive the capacitance attached to the gate's output. As shown in Figure 4-9, that capacitance may come from the transistor gates or from the wires to those gates. The gate can be sped up by increasing the sizes of its transistors or reducing the capacitance attached to it.

Logic gates which have large **fanout** (many gates attached to the output) are prime candidates for slow operation. Even if all the fanout gates use minimum-size transistors, presenting the smallest possible load, they may add up to a large load capacitance. Some of the fanout gates may use transistors that are larger than they need, in which case those transistors can be reduced in size to speed up the previous gate. In many cases this fortuitous situation does not occur, leaving two possible solutions:

- The transistors of the driving gate can be enlarged, in severe cases using the buffer chains of Section 3.3.8.

- The logic can be redesigned to reduce the gate's fanout.

An example of logic redesign is shown in Figure 4-10. The driver gate now drives two inverters, each of which drives two other gates. Since inverters were used, the fanout gates must be reversed in sense to absorb the inversion; alternatively, non-inverting buffers can be used. The inverters/buffers add delay themselves but cut down the load capacitance on the driver gate. In the case shown in the figure, adding

Figure 4-10:
Fanout reduction
by buffer inser-
tion.

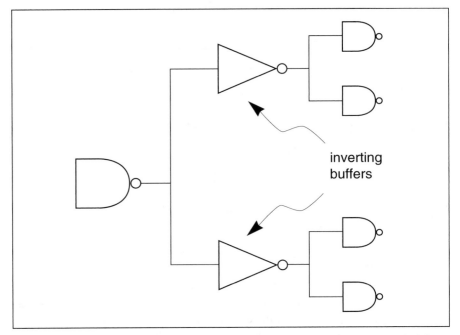

inverting
buffers

the inverters probably slowed down the circuit because they added too much delay; a gate which drives more fanout gates can benefit from buffer insertion.

Excess load capacitance can also come from the wires between the gate output and its fanout gates. We saw in Section 3.7.3 how to optimally add buffers in RC transmission lines.

4.4.2 Path Delay

In other cases, performance may be limited not by a single gate, but by a **path** through a number of gates. To understand how this can happen and what we can do about it, we need a concise model of the combinational logic that considers only delays. As shown in Figure 4-11, we can model the logic network and its delays as a directed graph. Each logic gate and each primary input or output is assigned its own node in the graph. When one gate drives another, an edge is added from the driving gate's node to the driven gate's node; the number assigned to the edge is the delay required for a signal value to propagate from the driver to the input of the driven gate. (The delay for $0 \rightarrow 1$ and $1 \rightarrow 0$ transitions will in general be different; since the wires in the network may be changing arbitrarily, we will choose the worst delay to represent the delay along a path.)

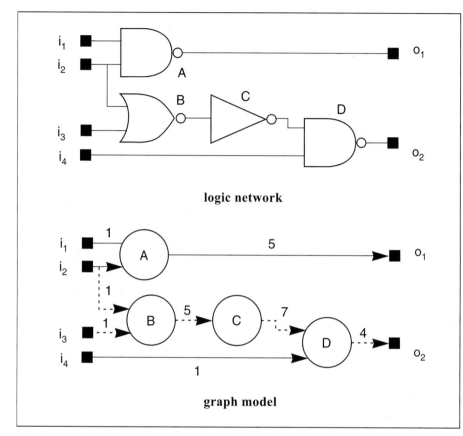

logic network

graph model

Figure 4-11: A graph model for delay through combinational logic.

In building the graph of Figure 4-11, need to know the gate along each edge in the graph. We use a **delay calculator** to estimate the delay from one gate's input through the gate and its interconnect to the next gate's input. The delay calculator may use a variety of models ranging from simple to complex. We will consider the problem of calculating the delay between one pair of gates in more detail in Section 4.5.1.

The simplest delay problem to analyze is to change the value at only one input and determine how long it takes for the effect to be propagated to a single output. (Of course, there must be a path from the selected input to the output.) That delay can be found by summing the delays along all the edges on the path from the input to the output. In Figure 4-11, the path from $i4$ to $o2$ has two edges with a total delay of 5 ns.

We could use a logic simulator which models delays to compute the delays through various paths in the logic. However, system performance is determined by the *maximum* delay through the logic—the longest delay from any input to any output for any possible set of input values. To determine the maximum delay by simulation, we would have to simulate all 2^n possible input values to the combinational logic. It is possible, however, to find the logic network's maximum delay without exhaustive simulation. **Timing analysis** [McW80,Ost83] builds a graph which models delays through the network and identifies the longest delay path. Timing analysis is also known as *static timing analysis* because it determines delays statically, independent of the values input to the logic gates.

The longest delay path is known as the **critical path** since that path limits system performance. We know that the graph has no cycles, or paths from a node back to itself—a cycle in the graph would correspond to feedback in the logic network. As a result, finding the critical path isn't too difficult. In Figure 4-11, there are two paths of equal length: $i2 \rightarrow B \rightarrow C \rightarrow D \rightarrow o2$ and $i3 \rightarrow B \rightarrow C \rightarrow D \rightarrow o2$ both have total delays of 17 ns. Any sequential system built from this logic must have a total delay of 17 ns, plus the setup time of the latches attached to the outputs, plus the time required for the driving latches to switch the logic's inputs (a term which was ignored in labeling the graph's delays).

Figure 4-12: A cutset through a critical timing path.

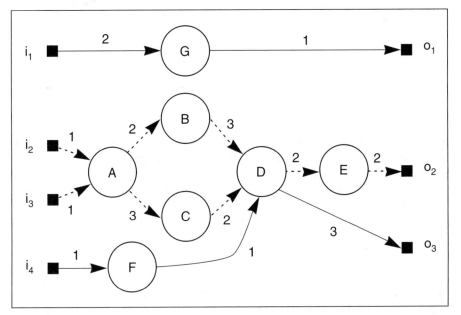

The critical path not only tells us the system cycle time, it points out what part of the combinational logic must be changed to improve system performance. Speeding up a gate off the critical path, such as A in the example, won't speed up the combinational logic. The only way to reduce the longest delay is to speed up a gate on the critical path. That can be done by increasing transistor sizes or reducing wiring capacitance. It can also be done by redesigning the logic along the critical path to use a faster gate configuration.

Speeding up the system may require modifying several sections of logic since the critical path can have multiple branches. The circuit in Figure 4-12 has a critical path with a split and a join in it. Speeding up the path from B to D will not speed up the system—when that branch is removed from the critical path, the parallel branch remains to maintain its length. The system can be improved only by speeding up both branches [Sin88]. A **cutset** is a set of edges in a graph that, when removed, break the graph into two unconnected pieces. Any cutset that separates the primary inputs and primary outputs identifies a set of speedups sufficient to reduce the critical delay path. The set b-d and c-d is one such cutset; the single edge d-e is another. We probably want to speed up the circuit by making as few changes to the network as possible. It may not be possible, however, to speed up every connection on the critical path. After selecting a set of optimization locations identified by a cutset, you must analyze them to be sure they can be sped up, and possibly alter the cutset to find better optimization points.

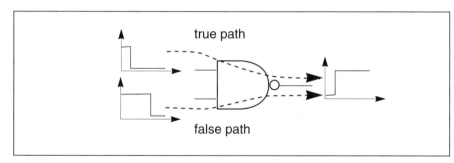

Figure 4-14:
Boolean gates cre-
ate false delay
paths.

However, not all paths in the timing analysis graph represent changes propagating through the circuit which limit combinational delay. Because logic gates compute Boolean functions, some paths through the logic network are cut short. Consider the example of Figure 4-14—the upper input of the NAND gate goes low first, followed by the lower input. Either input going low causes the NAND's output to go low, but after one has changed, the high-to-low transition of the other input doesn't affect the gate's output. If we know that the upper input changes first, we can declare the path through the lower input a **false path** for the combination of primary input values

Figure 4-13:
*Using Boolean
identities to
reduce delay.*

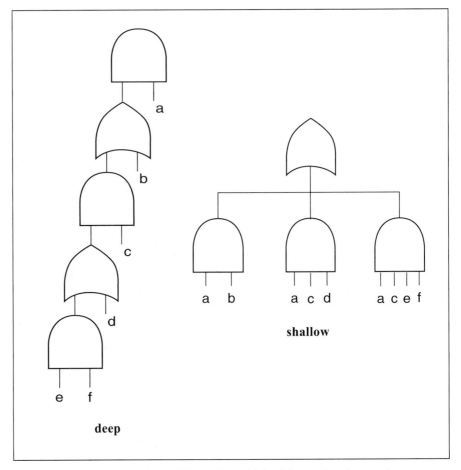

which cause these internal transitions. Even if the false path is longer than any true path, it won't determine the network's combinational delay because the transitions along that path don't cause the primary outputs to change. Note, however, that to identify false paths we must throw away our previous, simplifying assumption that the delay between two gates is equal to the worst of the rise and fall times.

Redesigning logic to reduce the critical path length requires rewriting the function to reduce the number of logic levels. Consider the logic of Figure 4-13. The critical path is clearly from e and f to the primary output. Writing the Boolean expression for this network both illustrates its depth and suggests a solution. The function is $a(b + c(d + ef))$; by eliminating parentheses we can reduce the depth of the equivalent logic network. The logic corresponding to $ab + acd + acef$ has only two levels. Care must be taken, however—flattening logic leads to gates with higher fanin. Since adding inputs to a gate slows it down (due to the delay through series transistors), all of the delay gained by flattening may be eaten up in the gates.

4.4.3 Transistor Sizing

One of the most powerful tools available to the integrated circuit designer is transistor sizing. By varying the sizes of transistors at strategic points, a circuit can be made to run much faster than when all its transistors have the same size. Transistor sizing can be chosen arbitrarily in full-custom layout, though it will take extra time to construct the layout. But transistor sizing can also be used to a limited extent in standard cells if logic gates come in several versions with differently-sized transistors.

The next example illustrates the effects of transistor sizing on one of the most important circuits, the adder.

Example 4-4: *Transistor sizing in an adder carry chain*

We will concentrate on the carry chain of the adder, since the longest delay follows that path. We will use a ripple-carry adder made of full adders (the full adder is described in more detail in Section 6.4). For this adder, we will use an AOI gate to implement the carry computation because it is faster and more compact than a

NAND-NAND network. The AOI must compute the function $c_{i+1} = a_i b_i + (a_i + b_i) c_i$. Here is the schematic for the AOI gate:

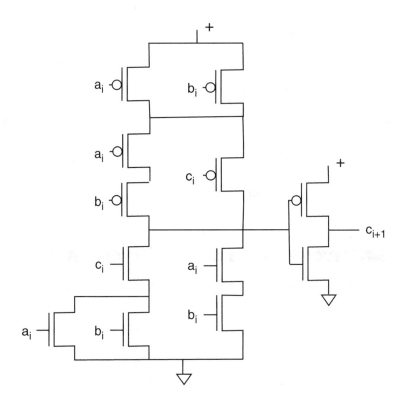

We arranged the order of transistors to put the early-arriving signals, ai and b_i, closer to the power supplies, as was discussed in Section 3.3.4. We will build a four-bit carry chain using four of these AOI gates, with the four outputs being c_1, c_2, c_3, and c_4.

The worst case for delay is that the a or b are 1 and the carry-in to the zero-th stage c_0 is 1. We will make the simplifying assumption that $a_i=1$ and $b_i=0$ for all bits, since other combinations only add a small delay which is independent of the stage. The carry of the ith stage, on the other hand, must wait for the i-1th stage to complete.

The simplest layout is to use small transistors of the same size. Using the 0.5 μm technology, we have made all n-type transistors with $W/L = 0.75/0.5$ (in μm, not λ) and all p-type transistors with $W/L = 1.5/0.5$. Here are the waveforms for the four carry outputs in this case:

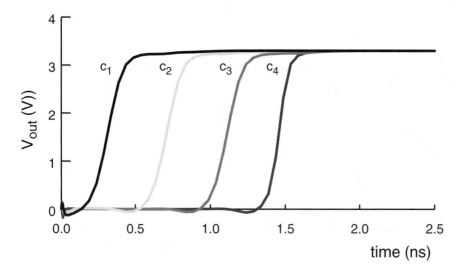

You can verify for yourself that uniformly increasing the sizes of all the transistors in the carry chain does not decrease delay—all the gates have larger loads to drive, negating the effect of higher drive.

The worst-case delay scenario for the pulldown network is the same in every stage: c_i is the latest-arriving signal and is rising. We can therefore widen these highlighted transistors in the AND-OR pulldown network:

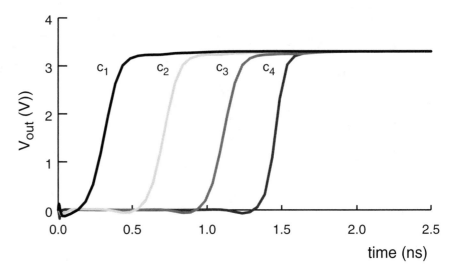

These transistors include the pulldown controlled by c_i and the a_i and b_i transistors on the same path, which also must be widened to ensure that they do not become a bottleneck. We must also increase the size of the output inverter.

We will first try making the a, b, and c pulldowns with $W/L = 1.5/0.5$, the first-stage inverter pullup with $W/L = 4.5/0.5$ and the pulldown with $W/L = 1.5/0.5$.

The inverters in the subsequent stages have pullup of size $W/L = 3/0.5$ and pull-down of size $W/L = 1.5/0.5$. Here is the result:

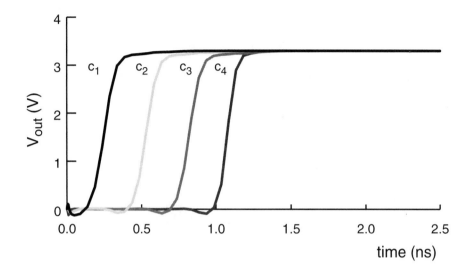

The adder is now considerably faster. The slope of c_1 is now steeper and the all the c's are spaced closer together.

We can try increasing the sizes of the transistors in the second and third stages, also increasing the first-stage transistors somewhat. In this case, the first-stage a, b, and c pulldowns in the first stage $W/L = 3/0.5$ with the first-stage inverter the same size as before. The second- and third-stage a, b, and c pulldowns have been increased in size to $W/L = 3/0.5$. The other transistors are left the same as in the last case. Here are the results:

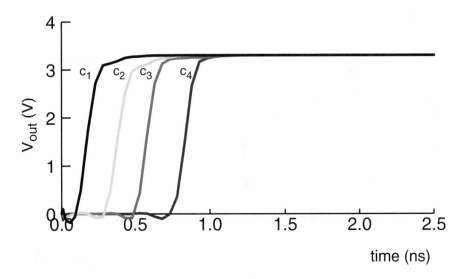

The carry chain has been sped up again, though not quite as much as before.

The theory of **logical effort** [Sut99] provides a clear and useful foundation for transistor sizing. Logical effort uses relatively simple models to analyze the behavior of chains of gates in order to optimally size all the transistors in the gates. Logical effort works best on tree networks and less well on circuits with reconvergent feedback, but the theory is both widely useful and intuitively appealing. Logical effort not only lets us easily calculate delay, it shows us how to size transistors to optimize delay along a path.

Logical effort computes d, the delay of a gate, in units of τ, the delay of a minimum-size inverter. We start with a model for a single gate. A gate's delay consists of two components:

$$d = f + p.$$ **(EQ 4-58)**

The **effort delay** f is related to the gate's load, while the **parasitic delay** p is fixed by the gate's structure. We can express the effort delay in terms of its components:

$$f = gh.$$
(EQ 4-59)

The **electrical effort** h is determined by the gate's load while the logical effort g is determined by the gate's structure. Electrical effort is given by the relationship between the gate's capacitive load and the capacitance of its own drivers (which is related to the drivers' current capability):

$$h = \frac{C_{out}}{C_{in}}.$$
(EQ 4-60)

	1 input	2 inputs	3 inputs	4 inputs	n inputs
inverter	1				
NAND		4/3	5/3	6/3	(n+2)/3
NOR		5/3	7/3	9/3	(2n+1)/3
mux		2	2	2	2
XOR		4	12	32	

Table 4-3 Logical effort for several types of static CMOS gates.

The **logical effort** g for several different gates is given in Table 4-3. The logical effort can be computed by a few simple rules.

We can rewrite Equation 3-58 using our definition of f to give

$$d = gh + p.$$
(EQ 4-61)

We are now ready to consider the logical effort along a path of logic gates. The path logical effort of a chain of gates is

$$G = \prod_{i=1}^{n} g_i.$$
(EQ 4-62)

The electrical effort along a path is the ratio of the last stage's load to the first stage's input capacitance:

$$H = \frac{C_{out}}{C_{in}}.$$

(EQ 4-63)

Branching effort takes fanout into account. We define the branching effort b at a gate as

$$b = \frac{C_{onpath} + C_{offpath}}{C_{onpath}}.$$

(EQ 4-64)

The branching effort along an entire path is

$$B = \prod_{i=1}^{n} b_i.$$

(EQ 4-65)

The path effort is defined as

$$F = GBH.$$

(EQ 4-66)

The path delay is the sum of the delays of the gates along the path:

$$D = \sum_{i=1}^{N} d_i = \sum_{i=1}^{N} g_i h_i + \sum_{i=1}^{N} p_i = D_F + P.$$

(EQ 4-67)

We can use these results to choose the transistor sizes that minimize the delay along that path. We know from Section 3.3.8 that optimal buffer chains are exponentially tapered. When recast in the logical effort framework, this means that each stage exerts the same effort. Therefore, the optimal stage effort is

$$\hat{f} = F^{1/N}.$$

(EQ 4-68)

We can determine the ratios of each of the gates along the path by starting from the last gate and working back to the first gate. Each gate i has a ratio of

$$C_{in,i} = \frac{g_i C_{out,i}}{\hat{f}} .$$
(EQ 4-69)

The delay along the path is

$$\hat{D} = NF^{1/N} + P .$$
(EQ 4-70)

Example 4-5 illustrates the use of logical effort in transistor sizing.

Example 4-5: Sizing transistors with logical effort

Let us apply logical effort to an a chain of three two-input NAND gates. the first NAND gate is driven by a minimum-size inverter and the output of the last NAND gate is connected to an inverter that is 4X the minimum size.

The logical effort for the chain of three NAND gates is

$$G = \prod_{i=1}^{3} \frac{4}{3} .$$

The branching effort along the path is $B = 1$ since there is no fanout. The electrical effort is the ratio of input to output capacitances, which was given as 4. Then

$$F = GBH = \left(\frac{4}{3}\right)^3 \times 1 \times 4 = 9.5 .$$

The optimum effort per stage is

$$\hat{f} = \sqrt[3]{9.5} = 2.1 .$$

Since all the stages have the same type of gate, we can compute the output-to-input capacitance ratio for the stages as

$$\frac{C_{\text{in,i}}}{C_{\text{out,i}}} = \frac{g_i}{\hat{h}} = \frac{4/3}{2.1} = 0.6 \, .$$

4.4.4 Automated Logic Optimization

Logic design—turning a logic function into a network of gates—is tedious and time-consuming. While we may use specialized logic designs for ALUs, **logic optimization** programs are often used to design random logic. Logic optimization programs have two goals: area minimization and delay satisfaction. Logic optimizers typically minimize area subject to meeting the designer's specified maximum delay. Logic optimizers like *misII* [Bra87], as well as several commercial tools from Synopsys, Cadence, Mentor, and other CAD vendors, can generate multi-level logic using a variety of methods: simplification, which takes advantage of don't-cares; common factor extraction; and structure collapsing, which eliminates common factors by reducing logic depth.

Finding good common factors is one of the most important steps in multi-level logic optimization. There are two particularly useful types of common factors: a cube is a product of literals; a kernel is a sum-of-products expression. A factor for a function f must be made of literals found in f. One way to factorize logic is to generate potential common factors and test each factor k to see whether it divides f—that is, whether there is some function g such that $g = f/k$. Once we have found a set of candidate factors for f, we can evaluate how they will affect the network's costs. A factor that can be used in more than one place (a common factor) can help save gate area, though at the cost of some additional wiring area. But factors increase the delay and power consumption of the logic network. The effects of introducing a factor can be evaluated in several ways with varying levels of accuracy.

Logic optimization algorithms are described in detail in Section 10.7. The important point to remember at this point is that logic optimization along with place-and-route algorithms give us an automated path from Boolean logic equations to a complete layout.

4.5 Logic and Interconnect Design

In this section we will consider how to design logic networks using realistic interconnect models. Interconnect comes in all shapes and sizes. Not only do nets vary in the number of gates they connect, but they can be laid out in a number of different topologies as well. Figure 4-15 shows the two basic forms of interconnection trees. Think of the gate inputs and outputs as nodes in a graph and the wires connecting them as edges in the graph. A **spanning tree** uses wire segments to directly connect the gate inputs and outputs. A **Steiner tree** adds nodes to the graph so that wires can join at a **Steiner point** rather than meeting at a gate input or output.

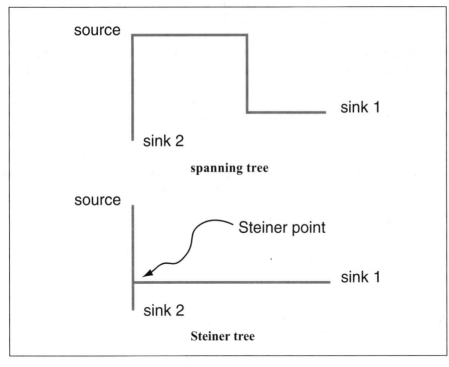

Figure 4-15: *Varieties of wiring trees.*

In order to make the problem tractable, we will generally assume that the logic structure is fixed. This still leaves us many degrees of freedom:

- we can change the topology of the wires connecting the gates;

- we can change the sizes of the wires;

- we can add buffers;

- we can size transistors.

We would like to solve all these problems simultaneously but in practice we solve either one at a time or a few in combination. Even this careful approach leaves us with quite a few opportunities for optimizing the implementation of our combinational network.

4.5.1 Delay Modeling

We saw in Section 4.4.2 that timing analysis consists of two phases: using a delay calculator to determine the delay to each gate's output; and using a path analyzer to determine the worst-case critical timing path. The delay calculator's model should take into account the wiring delay as well as the driving and driven gates. When analyzing large networks, we want to use a model that is accurate but that also can be evaluated quickly. Quick evaluation is important in timing analysis but even more important when you are optimizing the design of a wiring network. Fast analysis lets you try more wiring combinations to determine the best topology.

The Elmore model is well-known because it is computationally tractable. However, it works only for single RC sections. In some problems, such as when we are designing wiring tree topologies, we can break the wiring tree into a set of RC sections and use the Elmore model to evaluate each one independently. In other cases, we want to evaluate the entire wiring tree, which generally requires numerical techniques.

Figure 4-16: The effective capacitance model.

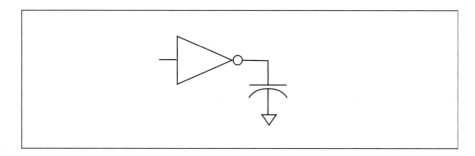

One model often used is the effective capacitance model shown in Figure 4-16. This model considers the interconnect as a single capacitance. While this is a simplified model, it allows us to separate the calculation of gate and interconnect delay. We then model the total delay as the sum of the gate and interconnect delays. The gate delay is determined using the total load capacitance and numerically fitting a set of parameters that characterize the delay. Qian et al. developed methods for determining an effective capacitance value [Qia94]. **Asymptotic waveform evaluation (AWE)** [Pil90] is a well-known numerical technique that can be used to evaluate the interconnect delay. AWE uses numerical techniques to find the dominant poles in the response of the network; those poles can be used to characterize the network's response.

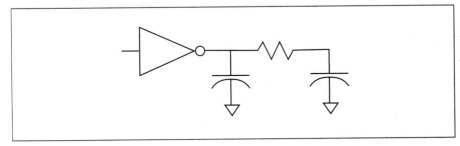

Figure 4-17: A π model for RC interconnect.

The π model, shown in Figure 4-17, is often used to model RC interconnect. The π model consists of two capacitors connected by a resistor. The values of these components are determined numerically by analyzing the characteristics of the RC network. The waveform at the output of the π model (the node at the second capacitor) does not reflect the wire's output waveform—this model is intended only to capture the effect of the wire's load on the gate. This model is chosen to be simple yet capture the way that resistance in an RC line shields downstream capacitance. Capacitance near the driver has relatively little resistance between it and the driver, while wire capacitance farther from the driver is partially shielded from the driver by the wire's resistance. The π model divides the wire's total capacitance into shielded and unshielded components.

4.5.2 Wire Sizing

We saw in Section 3.7.1 that the delay through an RC line can be reduced by tapering it. The formulas in that section assumed a single RC section. Since many wires connect more than two gates, we need methods to determine how to size wires in more complex wiring trees.

Cong and Leung [Con93] developed CAD algorithms for sizing wires in **wiring trees**. As shown in Figure 4-15, wires that connect a source to several sinks generally form trees. A pure **spanning tree** has distinct paths from endpoint to endpoint. If two paths intersect at an intermediate point, that point is called a **Steiner point** and the

Figure 4-18: *A tree with sized segments.*

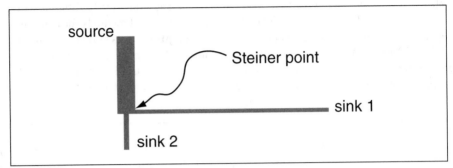

tree is referred to as a **Steiner tree**. (Although the upper branch of the spanning tree has several segments, it joins the other branch only at the source, not at some intermediate point.) In a tree, the sizing problem is to assign wire widths to each segment in the wire, with each segment having constant width; since most paths require several turns to reach their destinations, most trees have ample opportunities for tapering. Their algorithm also puts wider wires near the source and narrower wires near the sinks to minimize delay, as illustrated in Figure 4-18.

4.5.3 Buffer Insertion

We saw in Section 3.7.3 how to insert buffers in a single RC transmission line. However, in practice we must be able to handle RC trees. Not only do the RC trees have more complex topologies, but different subtrees may have differing sizes and arrival time requirements.

van Ginneken [van90] developed an algorithm for placing buffers in RC trees. The algorithm is given the placement of the sources and sinks and the routing of the wiring tree. It places buffers within the tree to minimize the departure time required at the source that meets the delay requirements at the sinks:

$$T_{\text{source}} = min_i(T_i - D_i) \qquad \text{(EQ 4-71)}$$

where T_i is the arrival time at node i and D_i is the required delay between the source and sink i. This ensures that even the longest delay in the tree satisfies its arrival time requirement.

This algorithm uses the Elmore model to compute the delay through the RC network. As shown in Figure 4-19, when we want to compute the delay from the source to sink

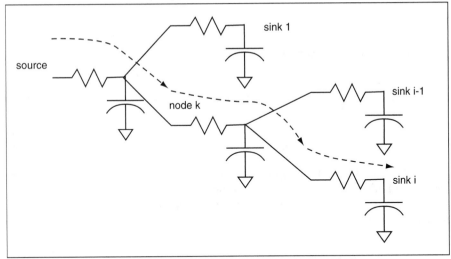

source
sink 1
node k
sink i-1
sink i

Figure 4-19:
Recursively computing delay in the van Ginneken algorithm.

i, we apply the R and C values along that path to the Elmore formula. If we want to compute the delay from some interior node *k* to sink *i*, we can use the same approach, counting only the resistance and capacitance on the path from *k* to *i*.

This formulation allows us to recursively compute the Elmore delay through the tree starting from the sinks and working back to the source. Let *r* and *c* be the unit resistance and capacitance of the wire and L_k be the total capacitive load of the subtree rooted at node *k*. As we walk the tree, we need to compute the required time T_k of the signal at node *k* assuming the tree is driven by a zero-impedance buffer. When we add a wire of length *l* at node *k*, then the new delay at node *k* is

$$T_k' = T_k - rlL_k - \frac{1}{2}rcl^2, \qquad \text{(EQ 4-72)}$$

$$L_k' = L_k + cl. \qquad \text{(EQ 4-73)}$$

When node *k* is buffered the required time becomes

$$T_k' = T_k - D_{\text{buf}} - R_{\text{buf}}L_k, \qquad \text{(EQ 4-74)}$$

$$L_k' = C_{\text{buf}}, \qquad \text{(EQ 4-75)}$$

where D_{buf}, R_{buf}, and C_{buf} are the delay, resistance, and capacitance of the buffer, respectively.

When we join two subtrees m and n at node k, the new values become

$$T_k = min(T_m, T_n),$$ (EQ 4-76)

$$L_k = L_m + L_n.$$ (EQ 4-77)

We can then use these formulas to recursively evaluate buffering options in the tree. The algorithm's first phase moves bottom-up to calculate all the buffering options at each node in the tree. The second phase chooses the best buffering strategy at each node in the tree.

4.5.4 Crosstalk Minimization

Coupling capacitances between wires can introduce crosstalk between signals. Crosstalk at best increases the delay required for combinational networks to settle down; at worst, it causes errors in dynamic circuits and memory elements. We can, however, design logic networks to minimize the crosstalk generated between signals.

Figure 4-20: *Interleaved ground signals for crosstalk minimization.*

We can use basic circuit techniques as a first line of defense against crosstalk. One way to minimize crosstalk is to introduce a larger capacitance to ground (or to V_{DD}, which is also a stable voltage). Since ground is at a stable voltage, it will not introduce noise into a signal. The larger the capacitance to ground relative to the coupling capacitance, the smaller the effect of the coupling capacitance, since the amount of charge on each capacitance is proportional to the value of the capacitance. In that case, the ground capacitance is said to *swamp out* the coupling capacitance. One way to add capacitance to ground is to interleave V_{SS} or V_{DD} wires between the signal wires as shown in Figure 4-20. This method is particularly well-suited to signals which must run together for long distances. Adding ground wires works best for groups of signals which travel together for long distances.

If we cannot provide shielding, minimizing coupling capacitance will help to reduce the effects of crosstalk. A simple example shows how we can redesign wire routes to reduce crosstalk.

━━━

Example 4-6: Crosstalk minimization

We need to route these signals in a channel:

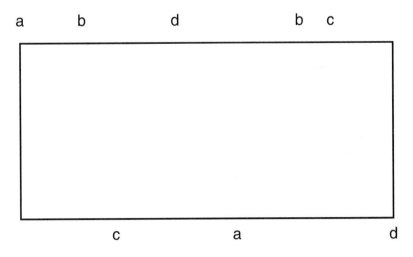

Let us assume for the moment that we can measure the total crosstalk in the wiring by examining only the horizontal wires. The vertical wires can introduce coupling, but they are generally shorter and we can also arrange to put them on a layer with lower coupling capacitance.

Here is one routing for these wires which minimizes channel height (assuming one horizontal segment per wire) but which has significant capacitive coupling:

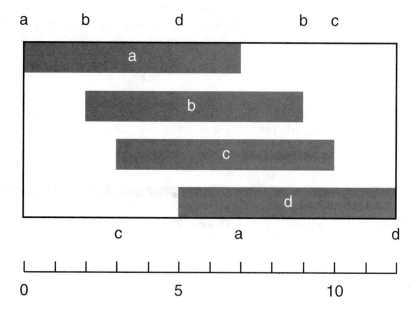

If we assume that non-adjacent wires have no coupling, then the total coupling capacitance for this route is:

a-b	6
b-c	6
c-d	5

for a total coupling capacitance of 17 units.

By rearranging the track assignments, we can significantly reduce the total coupling without changing the channel height or total wire length:

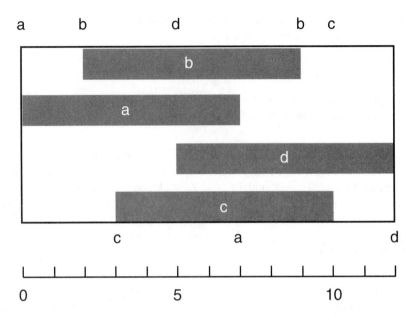

This routing has a much smaller coupling:

a-b	5
a-d	2
c-d	5

for a total coupling capacitance of 12 units.

However, in practice, minimizing crosstalk requires estimating the delays induced by crosstalk, not just minimizing coupling capacitance. Routing problems are sufficiently complex that it may not be obvious how to balance coupling capacitance against other criteria in the absence of information about how much critical path delay that coupling capacitance actually induces. Detailed analytical models are too complex and slow to be used in the inner loop of a routing algorithm. Sapatnekar [Sap00] developed an efficient crosstalk model that can be used during routing.

The effect of the coupling capacitance depends on the relative transitions of the aggressor and victim nets:

- When the aggressor changes and the victim does not, the coupling capacitance takes its nominal value C_c.

- When the aggressor and victim switch in opposite directions, the coupling capacitance is modeled as $2C_c$.

- When the aggressor and victim switch in the same direction, the coupling capacitance is modeled as 0.

The major problem in modeling the effect of coupling is that those effects depend on the relative switching times of the two nets. If the driver inputs of two nets switch in the intervals $[T_{min,1}, T_{max,1}]$ and $[T_{min,2}, T_{max,2}]$ and the propagation delays for those two signals are $[d_{1,min}, d_{1,max}]$ and $[d_{2,min}, d_{2,max}]$, then the lines can switch during the intervals $[T_{min,1} + d_{1,min}, T_{max,1} + d_{1,max}]$ and $[T_{min,2} + d_{2,min}, T_{max,2} + d_{2,max}]$. We can write the above observations on the coupling capacitance more precisely in terms of these intervals:

- $max(T_{min,1} + d_{1,min}, T_{min,2} + d_{2,min}) < t < min(T_{max,1} + d_{1,max}, T_{max,2} + d_{2,max})$: coupling capacitance is 0 or $2C_c$, depending on whether the aggressor and victim nets switch in the same or opposite directions.

- $min(T_{min,1} + d_{1,min}, T_{min,2} + d_{2,min}) < t < max(T_{min,1} + d_{1,min}, T_{min,2} + d_{2,min})$: coupling capacitance is C_c.

- $min(T_{max,1} + d_{1,max}, T_{max,2} + d_{2,max}) < t < max(T_{max,1} + d_{1,max}, T_{max,2} + d_{2,max})$: coupling capacitance is C_c.

Furthermore, the values for the ds depend on the values chosen for the coupling capacitance, which of course depends on the ds. As a result, an iterative algorithm must be used to solve for the transition times and coupling capacitances. The effective coupling capacitance's value changes over the course of the signal propagation and the order in which the transition times and coupling capacitances are updated affect the speed at which the solution converges.

Sapatnekar's algorithm iteratively finds the delays through the signals; since only a few iterations are generally required, the algorithm can be used in the inner loop of a router. This allows the router to exchange nets to reduce the actual crosstalk between the wires, not just their coupling capacitance.

There are several other ways to redesign the layout to reduce the amount of coupling capacitance between wires. One method is to increase the spacing between critical signals [Cha93]. Since the coupling capacitance decreases with distance, this technique can reduce the coupling capacitance to an acceptable level. However, this may require significant space when applied to signals which are coupled over a long distance. Alternatively, signals may be swapped in their tracks[Gao94] or a more global view may be taken to assign signals to tracks to minimize total crosstalk risk [Kir94]. Xue et al. [Xue96] developed an algorithm which tries to minimize the total crosstalk risk across the chip. It starts with crosstalk risk values for signal pairs, based on an assessment of the criticality of a signal, etc. It then selects nets for rip-up and reroute in order to minimize the total crosstalk risk.

4.6 Power Optimization

4.6.1 Power Analysis

We saw in Section 3.3.5 how to optimize the power consumption of an isolated logic gate. One important way to reduce a gate's power consumption is to make it change its output as few times as possible. While the gate would not be useful if it never changed its output value, it is possible to design the logic network to reduce the number of *unnecessary* changes to a gate's output as it works to compute the desired value.

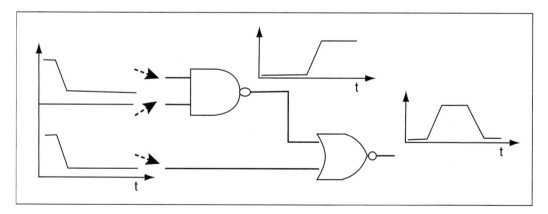

Figure 4-21: *Glitching in a simple logic network.*

Figure 4-21 shows an example of power-consuming glitching in a logic network. Glitches are more likely to occur in multi-level logic networks because the signals arrive at gates at different times. In this example, the NOR gate at the output starts at 0 and ends at 0, but differences in arrival times between the gate input connected to the primary input and the output of the NAND gate cause the NOR gate's output to glitch to 1.

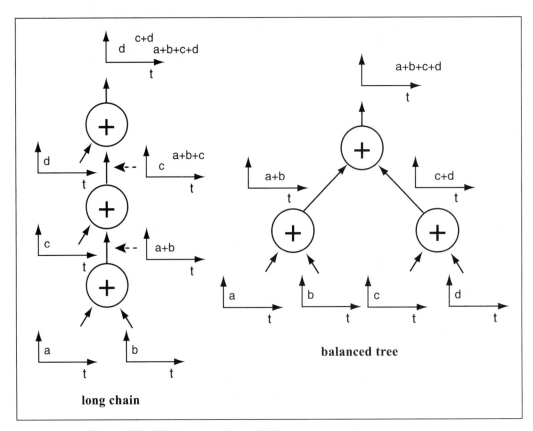

Figure 4-22: *Glitching in a chain of adders.*

Some sources of glitches are more systematic and easier to eliminate. Consider the logic networks of Figure 4-22, both of which compute the sum $a+b+c+d$. The network on the left-hand side of the figure is configured as a long chain. The effects of a change in any signal—either a primary input or an intermediate value—propagate through the successive stages. As a result, the output of each adder assumes multiple values as values reach its inputs. For example, the last adder first takes on the value of the d input (assuming, for simplicity, that all the signals start at 0), then computes $c+d$

as the initial value of the middle adder arrives, and finally settles at $a+b+c+d$. The right-hand network, on the other hand, is more balanced. Intermediate results from various subnetworks reach the next level of adder at roughly the same time. As a result, the adders glitch much less while settling to their final values.

We cannot in general eliminate glitches in all cases. We may, however, be able to eliminate the most common kinds of glitches. To do so, we need to be able to estimate the **signal probabilities** in the network. The signal probability P_s is the probability that signal s is 1. The probability of a transition $P_{tr,s}$ can be derived from the signal probability, assuming that the signal's values on clock cycles are independent:

$$P_{tr,s} = 2P_s(1-P_s).$$ **(EQ 4-78)**

The first matter to consider is the probability distribution of values on primary inputs. The simplest model is that a signal is equally likely to be 0 or 1. We may, however, have some specialized knowledge about signal probabilities. Some control signals may, for example, assume one value most of the time and only occasionally take on the opposite value to signal an operation. Some sets of signals may also have correlated values, which will in turn affect the signal probabilities of logic gate outputs connected to those sets of signals.

Signal probabilities are generally computed by **power estimation tools** which take in a logic network, primary input signal probabilities, and perhaps some wiring capacitance values and estimate the power consumption of the network. There are two major ways to compute signal probabilities and power consumption: **delay-independent** and **delay-dependent** [Ped96]. Analysis based on delay-independent signal probabilities is less accurate than delay-dependent analysis but delay-independent values can be computed much more quickly. The signal probabilities of primitive Boolean functions can be computed from the signal probabilities of their inputs. Here are the formulas for NOT, OR, and AND:

$$P_{NOT} = 1-P_{in};$$ **(EQ 4-1)**

$$P_{OR} = 1- \prod_{i \in in} (1-P_i);$$ **(EQ 4-2)**

$$P_{AND} = \prod_{i \in in} P_i.$$ **(EQ 4-3)**

When simple gates are combined in networks without reconvergent fanout, the signal probabilities of the network outputs can easily be computed exactly. More sophisticated algorithms are required for networks that include reconvergent fanout.

Delay-independent power estimation, although useful, is subject to errors because it cannot predict delay-dependent glitching. The designer can manually assess power consumption using a circuit simulator or, in the case of larger circuits, a switch-level simulator such as IRSIM [Sal89]. This technique, however, suffers the same limitation as does simulation for delay in that the user must manually evaluate the combinations of inputs which produce the worst-case behavior. Power estimation tools may rely either directly on simulation results or on extended techniques that use simulation-style algorithms to compute signal probabilities. The time/accuracy trade-offs for power estimation track those for delay estimation: circuit-level methods are the most accurate and costly; switch-level simulation is somewhat less accurate but more efficient; logic-based simulation is less powerful but can handle larger networks.

Figure 4-23:
Logic factorization for low power.

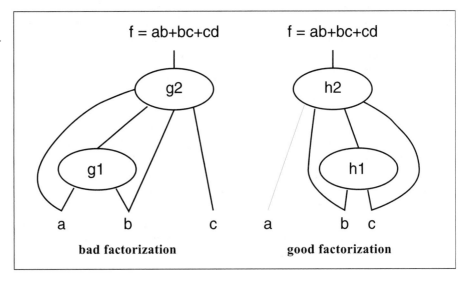

Given the power estimates from a tool, the designer can choose to redesign the logic network to reduce power consumption as required. Logic synthesis algorithms designed to minimize power can take advantage of signal probabilities to redesign the network [Roy93]. Figure 4-23 shows two factorizations of the function $f = ab + bc + cd$ [Ped96]. If a glitches much more frequently than b and c, then the right-hand network exhibits lower total glitching: in the left-hand network, both $g1$ and $g2$ glitch when a changes; in the right-hand network, glitches in a cause only $h2$ to glitch.

Glitch analysis can also be used to optimize placement and routing. Nodes that suffer from high glitching should be laid out to minimize their routing capacitance. The capacitance estimates from placement and routing can be fed back to power estimation to improve the results of that analysis.

Of course, the best way to make sure that signals in a logic block do not glitch is to not change the inputs to the logic. Of course, logic that is never used should not be included in the design, but when a block of logic is not used on a particular clock cycle, it may be simple to ensure that the inputs to that block are not changed unnecessarily. In some cases, eliminating unnecessary register loads can eliminate unnecessary changes to the inputs. In other cases, logic gates at the start of the logic block can be used to stop the propagation of logic signals based on a disable signal.

4.7 Switch Logic Networks

We have used MOS transistors to build logic gates, which we use to construct combinational logic functions. But MOS transistors are good switches—a switch being a device which makes or breaks an electrical connection—and switches can themselves be used to directly implement Boolean function [Sha38]. Switch logic isn't universally useful: large switch circuits are slow and switches introduce hard-to-trace electrical problems; and the lack of drive current presents particular problems when faced with the relatively high parasitics of deep-submicron processes. But building logic directly from switches can help save area and parasitics in some specialized cases.

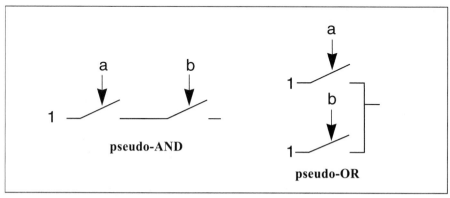

Figure 4-24:
Boolean functions built from switches.

Figure 4-24 shows how to build AND and OR functions from switches. The control inputs control the switches—a switch is closed when its control input is 1. The switch drains are connected to constants (V_{DD} or V_{SS}). A pseudo-AND is computed by series switches: the output is a logic 1 if and only if both inputs are 1. Similarly, a pseudo-OR is computed by parallel switches: the output is logic 1 if either input is 1. We call these functions *pseudo* because when none of the switches is turned on by the input variables, the output is not connected to any constant source and its value is not defined. As we will see shortly, this property causes havoc in real circuits with parasitic capacitance. Switch logic is not complete—we can compute AND and OR but we cannot invert an input signal. If, however, we supply both the true and complement forms of the input variables, we can compute any function of the variables by combining true and complement forms with AND and OR switch networks.

Figure 4-25: A switch network with non-constant source inputs.

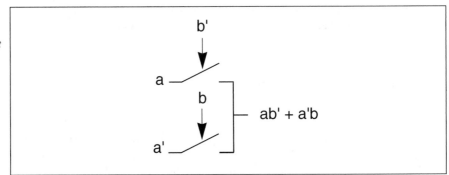

We can reduce the size of a switch network by applying some of the input variables to the switches' gate inputs. The network of Figure 4-25, for example, computes the function $ab' + a'b$ using two switches by using one variable to select another. This network's output is also defined for all input combinations. Switch networks which apply the inputs to both the switch gate and drain are especially useful because some functions can be computed with a very small number of switches.

Example 4-7: Switch implementation of a multiplexer

We want to design a multiplexer (commonly called a mux) with four data inputs and four select inputs—the two select bits s_1, s_0 and their complements are all fed into the switch network. The network's structure is simple:

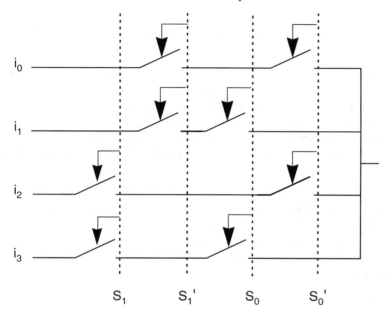

In practice, the number of select lines limits the useful size of the multiplexer.

Computing delay through a switch network is similar to computing pullup or pulldown delay through a logic gate—the switch transistor is in series with the pullup or pulldown network. However, the resistive approximation to the transistor becomes less accurate as more transistors are placed in series. For accurate delay analysis, you should perform a more accurate circuit or timing simulation. Switch networks with long paths from input to output may be slow. Just as no more than four transistors should be in series in a logic gate, switch logic should contain no paths with more than four switches for minimum delay (though long chains of pass transistors may be useful in some situations, as in the Manchester carry chain of Section 6.4).

Figure 4-26:
Charge sharing.

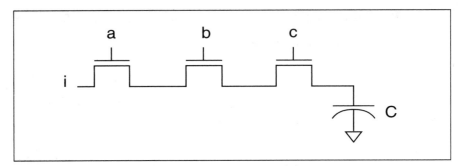

The most insidious electrical problem in switch networks is **charge sharing**. Switches built from MOS transistors have parasitic capacitances at their sources and drains thanks to the source/drain diffusion; capacitance can be added by wires between switches. While this capacitance is too small to be of much use (such as building a memory element), it is enough to cause trouble. We saw in Example 4-3 that the relative capacitances of nodes must be considered to properly compute node values.

Consider the circuit of Figure 4-26. Initially, $a = b = c = i = 1$ and the output o is driven to 1. Now set $a = b = c = i = 0$—the output remains one, at least until substrate resistance drains the parasitic capacitance, because the parasitic capacitance at the output stores the value. The network's output should be undefined, but instead it gives us an erroneous 1.

Figure 4-27:
*Charge division
across a switch.*

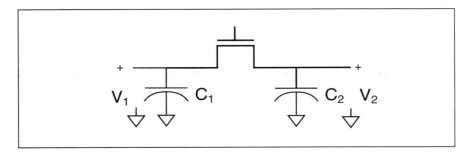

When we look at the network's behavior over several cycles, we see that much worse things can happen. As shown in Figure 4-27, when a switch connects two capacitors not driven by the power supply, current flows to place the same voltage across the two capacitors. The final amounts of charge depend on the ratio of the capacitances. Charge division can produce arbitrary voltages on intermediate nodes. These bad logic values can be propagated to the output of the switch network and wreak havoc on the logic connected there. Consider the value of each input and of the parasitic capacitance between each pair of switches/terminals over time:

time	i	C_{ia}	a	C_{ab}	b	C_{bc}	c	C_{co}
0	1	1	1	1	1	1	1	1
1	0	0	1	0	0	1	0	1
2	0	0	0	1/2	1	1/2	0	1
3	0	0	0	1/2	0	3/4	1	3/4
4	0	0	1	0	0	3/4	0	3/4
5	0	0	0	3/8	1	3/8	0	3/4

The switches can shuttle charge back and forth through the network, creating arbitrary voltages, before presenting the corrupted value to the network's output. Charge sharing can be easily avoided—design the switch network so that its output is always driven by a power supply. There must be a path from V_{DD} or V_{SS} through some set of switches to the output for every possible combination of inputs. Since charge can be divided only between undriven capacitors, always driving the output capacitance ensures that it receives a valid logic value.

The severity of charge sharing suggests that strong measures be used to ensure the correct behavior of switch logic networks. One way to improve the reliability of transmission gates is to insert buffers before and after them.

4.8 Combinational Logic Testing

Once we have designed our logic, we must develop tests to allow manufacturing to separate faulty chips from good ones. A **fault** is a manifestation of a manufacturing defect; faults may be caused by mechanisms ranging from crystalline dislocations to lithography errors to bad etching of vias.

It is worth considering our goals for testing. Can we ensure that the chips coming off the manufacturing line are totally defect-free? No—it is impossible to predict all the ways a chip can fail, let alone test for them all. A somewhat more realistic goal is to choose one or several fault models, such as the stuck-at-0/1 model, and test for all possible modeled faults. Even this goal is hard to achieve because it considers multiple faults. An even more modest goal is to test for all *single faults*—assume that only one gate is faulty at any time. Single-fault coverage for stuck-at-0/1 faults is the most common test; many multiple faults are discovered by single-fault testing, since many of the fault combinations are independent.

The simulation vectors used for design verification typically cover about 80% of the single-stuck-at-0/1 faults in a system. While it may be tempting to leave it at that, 80% fault coverage lets an unacceptable number of bad parts slip into customers' hands. Williams and Brown analyzed the field reject rate as a function of the yield of the manufacturing process (called Y) and the coverage of manufacturing defects (called T) [Wil81]. They found, using simple assumptions about the distribution of manufacturing errors, that the percentage of defective parts allowed to slip into the customers' hands was

$$D = 1 - Y^{(1-T)}$$ (EQ 4-4)

What does this equation mean in practice? Let's be generous for a moment and assume that testing for single stuck-at-0/1 covers all manufacturing defects. If we use our simulation vectors for testing, and our process has a yield of 50%, then the defect rate is 13%—that is, 13% of the chips that pass our tests are found by our customers to be bad. If we increase our fault coverage to 95%, the defect rate drops to 3.4%— better, but still unacceptably large. (How would you react if 3.4% of all the quarts of milk you bought in the grocery store were spoiled?) If we increase the fault coverage to 99.9%, the defect rate drops to 0.07%, which is closer to the range we associate with high quality.

But, in fact, single stuck-at-0/1 testing is not sufficient to catch all faults. Even if we test for all the single stuck-at faults, we will still let defective chips slip through. So how much test coverage is sufficient? Testing folklore holds that covering 99-100% of the single stuck-at-0/1 faults results in low customer return rates, and that letting fault coverage slip significantly below 100% quickly leads to unacceptably poor quality.

Manufacturing tests also help catch **infant mortalities**. One commonly-used model for chip reliability is an exponential probability for failure {Mur93}:

$$R(t) = e^{-\lambda_0 t}.$$ (EQ 4-5)

This model assumes that the failure rate starts high and rapidly decreases. The testing process exercises the chips enough to cause very marginal chips to fail, which is likely to be detected in the tests which remain after the failure occurs.

We first need to first understand what types of faults we need to test; we can then consider how to test for these faults when the fault is not easily located but is embedded in a combinational logic network.

4.8.1 Gate Testing

Testing a logic gate requires a **fault model**. The simplest fault model considers the entire logic gate as one unit; more sophisticated models consider the effects of faults in individual transistors in the gate.

The most common fault model is the **stuck-at-0/1** model. Under this model, the output of a faulty logic gate is 0 (or 1), independent of the value of its inputs. The fault does not depend on the logic function the gate computes, so any type of gate can exhibit a stuck-at-0 (S-A-0) or stuck-at-1 (S-A-1) fault. Detecting a S-A-0 fault simply requires applying a set of inputs that sets a fault-free gate's output to 1, then examining the output to see if it has the true or faulty value.

a	b	fault-free	S-A-0	S-A-1		a	b	fault-free	S-A-0	S-A-1
0	0	1	0	1		0	0	1	0	1
0	1	1	0	1		0	1	0	0	1
1	0	1	0	1		1	0	0	0	1
1	1	0	0	1		1	1	0	0	1

NAND   **NOR**

Table 4-1 *True and faulty behavior for stuck-at-0/1 faults.*

Table 4-1 compares the proper behavior of two-input NAND and NOR gates with their stuck-at-0 and stuck-at-1 behavior. While the output value of a gate stuck at 0 isn't hard to figure out, it is instructive to compare the difficulty of testing for S-A-0 and S-A-1 faults for each type of gate. A NAND gate has three input combinations which set a fault-free gate's output to 1; that gives three ways to test for a stuck-at-0 fault. There is only one way to test for stuck-at-1—set both inputs to 0. Similarly, there are three tests for stuck-at-1 for a NOR gate, but only one stuck-at-0 test.

Figure 4-28: A simple logic network that requires two tests.

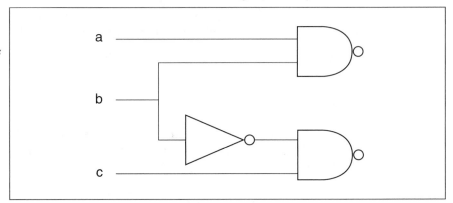

The number of input combinations that can test for a fault becomes important when gates are connected together. Consider testing the logic network of Figure 4-28 for stuck-at-0 and stuck-at-1 faults in the two NAND gates, assuming, for the moment, that the inverter is not faulty. We can test both NAND gates for stuck-at-0 faults simultaneously, using, for example, abc = 011. (A set of values simultaneously applied to the inputs of a logic network is called a **vector**.) However, there is no way to test both NAND gates simultaneously for stuck-at-1 faults: the test requires that both NAND gate inputs are 1, and the inverter assures that only one of the NAND gates can receive a 1 from the b input at a time. Testing both gates requires two vectors: abc = 00- (where - means the input's value is a don't-care, so that doesn't matter) and abc = -10.

The stuck-at-0/1 model for faults doesn't correspond well to real physical problems in CMOS processing. While a gate's output may be stuck at 0 by a short between the gate's output and V_{SS}, for example, that manufacturing error is unlikely to occur. The stuck-at-0/1 model is still used for CMOS because many faults from a variety of causes are exposed by testing vectors designed to catch stuck-at faults. The stuck-at model, however, does not predict all faults in a circuit; it is comforting to have a fault model which corresponds more closely to real processing errors.

One such model is the **stuck-open** model [Gal80], which models faults in individual transistors rather than entire logic gates. A stuck-open fault at a transistor means that the transistor never conducts—it is an open circuit. As Figure 4-29 shows, a stuck-open transistor in a logic gate prevents the gate from pulling its output in one direc-

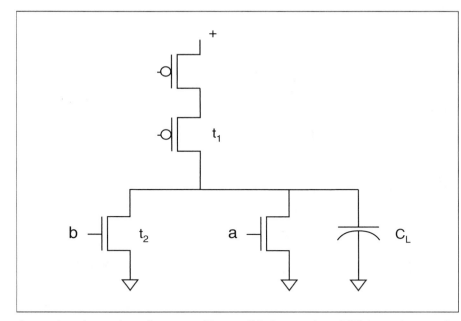

Figure 4-29: A circuit model for stuck-open faults.

tion or the other, at least for some of its possible input values. If t1 is stuck open, the gate cannot pull its output to V_{DD} for any input combination which should force the gate's output to 1. In contrast, if t2 is stuck open, the gate can pull its output to V_{SS} when $a = 1$ but not when $b = 1$.

This example also shows why reliably catching a stuck-open fault requires a *two cycle* test. If the gate's output is not driven to V_{DD} or V_{SS} due to a stuck-open fault, the gate's output value depends on the charge stored on the parasitic capacitance at its output. If we try setting $b = 1$ to test for a stuck-open fault at t_2, for example, if the last set of inputs applied to the gate was $ab = 01$, the gate charged its output to logic 0; when b is set to 0 to test t_2, the output will remain at 0, and we can't tell if the gate's output is due to a fault or not. Testing the stuck-open fault at t_2 requires setting the logic gate's output to one value with one vector, then testing with another vector whether the gate's output changes. In this case, we must first apply $ab = 00$ to set the gate's output to 1; then, when we apply $ab = 01$, the gate's output will be pulled down to 0 if t_2 is not faulty but will remain at 1 if t_2 is stuck open.

Both stuck-at and stuck-open faults check for function. We can also treat delay problems as faults: a **delay fault** [Lin87] occurs when the delay along a path falls outside specified limits. (Depending on the circuit, too-short paths may cause failures as well as too-long paths.) Delay faults can be modeled in either of two ways: a **gate delay fault** assumes that all the delay errors are lumped at one gate along the path; a **path**

delay fault is the result of accumulation of delay errors along the entire path. Detecting either type of fault usually requires a large number of tests due to the many paths through the logic. However, since delay faults reduce yield, good testing of delay faults is important. If delay faults are not adequately caught in the factory, the bad chips end up in customers' hands, who discover the problems when they plug the chips into their systems.

4.8.2 Combinational Network Testing

Just as network delay is harder to compute than the delay through an individual gate, testing a logic gate in a network is harder than testing it in isolation. Testing a gate inside a combinational network requires exercising the gate in place, without direct access to its inputs and outputs. The problem can be split into two parts:

- **Controlling** the gate's inputs by applying values to the network's primary inputs.

- **Observing** the gate's output by inferring its value from the values at the network's primary outputs.

Figure 4-30:
Testing for combi-
national faults.

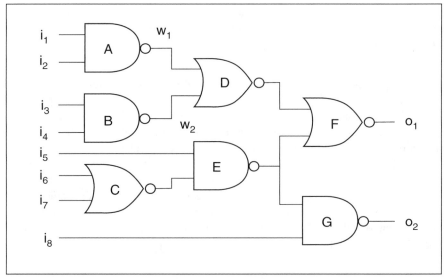

Consider testing gate D in Figure 4-30 for a stuck-at-0 fault. The first job is to control D's inputs to set both to 0, also called **justifying** 0 values on the inputs. We can justify the required values by working backward from the pins to the primary inputs. To set wire w1 to 0, we need to make gate A's output 0, which we can do by setting both its inputs to 1. Since those wires are connected to primary inputs, we have succeeded in justifying w1's value. The other required 0 can be similarly controlled through B.

The second job is to set up conditions that let us observe the fault at the primary outputs—one or more of the primary outputs should have different values if D is stuck-at 0. Observing the fault requires both working forward and backward through the network. D's faulty behavior can be observed only through F—we need to find some combination of input values to F that gives one output value when D is 1 or 0. Setting F's other input to 0 has the desired result: if D's output is good, the input combination 10 results in a 0 at F's output; if D is faulty, the 00 inputs give a 1 at the output. Since F is connected to a primary output, we don't have to propagate any farther, but we do have to find primary input values that make F's other input 0. Justification tells us that $i_5 = 1$, $i_6 = 0$, $i_7 = 0$ provides the required value; i_8 's value doesn't matter for this test. Many tests may have more than one possible sequence. Testing D for stuck-at-1 is relatively easy, since three input combinations form a test. Some tests may also be combined into a single vector, such as tests for F and G.

Finding a test for a combinational fault is NP-complete [Gar79]—finding the test will, in the worst case, require checking every possible input combination. However, much random logic is relatively easy to test, and many harder-to-test structures have well-known tests. In practice, programs do a relatively good job of generating combinational test patterns.

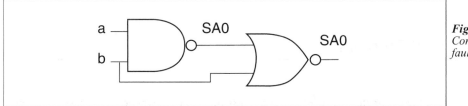

Figure 4-31: *Combinational fault masking.*

Not all faults in a combinational network can be tested. In Figure 4-31, testing the NOR gate for stuck-at-0 requires setting both its inputs to 0, but the NAND gate ensures that one of the NOR's inputs will always be 1. Observing the NAND gate's stuck-at-0 fault requires setting the other input of the NOR gate to 0, but that doesn't allow the NAND gate's fault to be exercised. In both cases, the logic is untestable because it is redundant. Simplifying the logic gives:

$$f = [\overline{ab} + \bar{b}]$$
$$= [\bar{a} + \bar{b} + b]$$
$$= 0$$

The entire network could be replaced by a connection to V_{SS}. Any irredundant logic network can be completely tested. While it may seem dumb to introduce redundancies in a network—they make the logic larger and slower as well as less testable—it often isn't easy to recognize redundancies.

4.9 References

Dogleg channel routing was introduced by Deutsch [Deu76]. The seminal paper on low-power CMOS design is by Chandrakasan, Sheng, and Brodersen[Cha92]. Devadas *et al.* describe algorithms for eliminating false paths during combinational logic delay analysis [Dev91]. Lawrence Pillage and Lawrence Pileggi are two well-known researchers in interconnect analysis who are in fact the same person; Prof. Pileggi reverted to his family's traditional name. Algorithms for technology-independent performance optimization of combinational logic have been developed by Singh, *et al.* [Sin88]. Jha and Kundu [Jha90] discuss CMOS testing methods in detail.

4.10 Problems

Use the parameters for the 0.5 μm process of Chapter 2 whenever process parameters are required, unless otherwise noted.

4-1. If we route horizontal tracks in metal1 and vertical tracks in metal2, what are the horizontal and vertical grid spacings for channels in our SCMOS process?

4-2. For the channel shown below (dotted lines show vertically aligned pins):

    a) Can this channel be routed using the left-edge algorithm? Explain.

    b) Route the channel.

4-3. For each example, rewrite the functions in multi-level form, introducing where possible additional functions for common factors.

    a) $f = ab + ac + bc$; $g = ad + c' + bd$.

    b) $f = abd + c'e$; $g = de' + abc$.

    c) $f = ab + de'$; $g = ab + abc$; $h = a + bc + d'e$.

4-4. Using the assumptions and techniques of Example 4-6, find minimum-crosstalk routings for these channels. Use the distances between terminals shown on the ruler.

a)

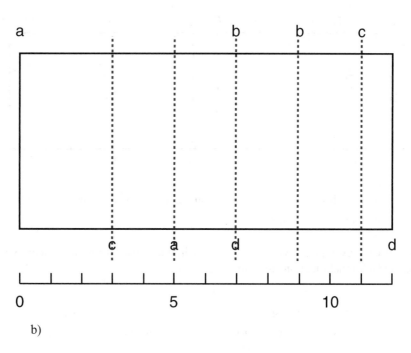

b)

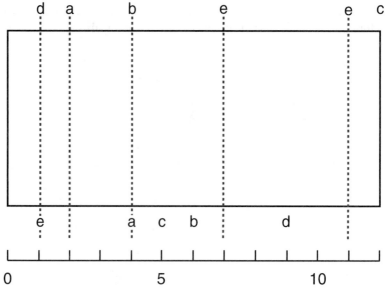

c)

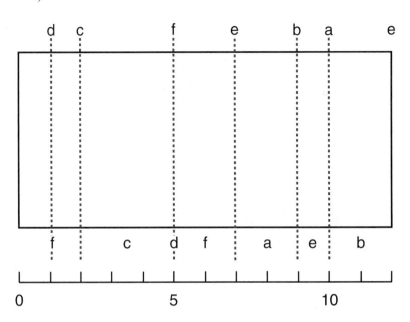

4-5. Find the combination of input transitions which introduces maximum glitching at the primary output o, assuming that all gates have a delay of one time unit.

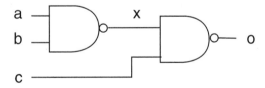

4-6. Compute the zero-delay signal probabilities for all signals in these networks assuming the signal probabilities for the primary inputs as shown.

a)

$P_a = 0.5$
$P_b = 0.1$

$P_a = 0.5$

b)

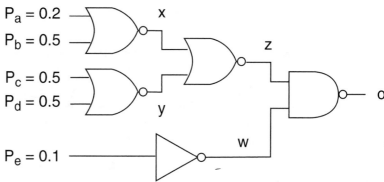

$P_a = 0.5$

$P_b = 0.1$

x

$P_c = 0.4$

$P_d = 0.05$

y

o

c)

$P_a = 0.2$

$P_b = 0.5$

x

$P_c = 0.5$

$P_d = 0.5$

y

z

$P_e = 0.1$

w

o

4-7. Give at least one test for stuck-at-0 and stuck-at-1 faults for each of these static gates:

    a) $(a + b + c)'$.

    b) $[(a + b)c]'$.

    c) $(a + b)(c + d)$.

4-8. Give tests for stuck-open faults for each transistor in a two-input static NOR gate.

5

Sequential Machines

Highlights:

Latches and flip-flops.

Clocking structures and timing disciplines.

Sequential system design.

Sequential system testing.

5.1 Introduction

A **sequential machine** is a machine whose output values depend not only on the present input values, but also the history of previous inputs. The sequential machine's memory lets us build much more sophisticated functions; it also complicates design, validation, and testing. In this chapter we will learn the design methods common to all sequential systems: memory element design, construction rules which guarantee that our sequential system operates properly, specification techniques, state assignment, performance optimization, and testing.

5.2 Latches and Flip-Flops

5.2.1 Categories of Memory Elements

Building a sequential machine requires **memory elements** which read a value, save it for some time, and then can write that stored value somewhere else, even if the element's input value has subsequently changed. A Boolean logic gate can compute values, but its output value will change shortly after its input changes. Each alternative circuit used as a memory element has its own advantages and disadvantages.

A generic memory element has an internal memory and some circuitry to control access to the internal memory. In CMOS circuits the memory is formed by some kind of capacitance or by positive feedback of energy from the power supply. Access to the internal memory is controlled by the *clock* input—the memory element reads its data input value when instructed by the clock and stores that value in its memory. The output reflects the stored value, probably after some delay. Memory elements differ in many key respects:

- exactly what form of clock signal causes the input data value to be read;

- how the behavior of *data* around the read signal from *clock* affects the stored value;

- when the stored value is presented to the output;

- whether there is ever a combinational path from the input to the output.

Introducing a terminology for memory elements requires caution—many terms are used in slightly or grossly different ways by different people. We choose to follow Dietmeyer's convention [Die78] by dividing memory elements into two major types:

- **Latches** are transparent while the internal memory is being set from the data input—the (possibly changing) input value is transmitted to the output.

- **Flip-flops** are not transparent—reading the input value and changing the flip-flop's output are two separate events.

Within these types, many subclasses exist. But the latch vs. flip-flop dichotomy is most important because, as we will see in Section 5.3, the decision to use latches or flip-flops dictates substantial differences in the structure of the sequential machine.

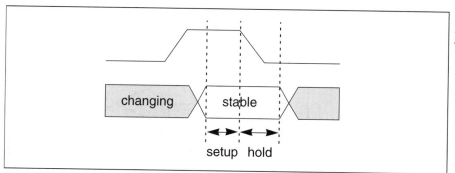

Figure 5-1: Setup
and hold times.

Memory elements can also be categorized along another dimension, namely the types
of data inputs they present.

- The most common data input type in VLSI design is the **D-type** memory
  element. Think of "D" as standing for data—the Q output of the memory
  element is determined by the D input value at the clocking event.

- The **T-type** memory element toggles its state when the T input is set at the
  clocking event.

- The **SR-type** memory element is either set by the S input or reset by the R
  input (the S and R inputs are not allowed to be 1 simultaneously).

- The **JK-type** is similar but its J and K inputs can both be 1. The other memory element types can be built using the JK-type as a component.

The two most commonly quoted parameters of a memory element are its **setup time**
and **hold time**, which define the relationship between the clock and input data signals. The data value to be latched must remain stable around the time the clock signal
changes value to ensure that the memory element retains the proper value. In Figure
5-1, the memory element stores the input value around the clock's falling edge. The
setup time is the minimum time the data input must be stable before the clock signal
changes, while the hold time is the minimum time the data must remain stable after
the clock changes. The setup and hold times, along with the delay times through the
combinational logic, determine how fast the system can run. The **duty cycle** of a
clock signal is the fraction of the clock period for which the clock is active.

5.2.2 Latches

***Figure 5-2:** A dynamic latch circuit.*

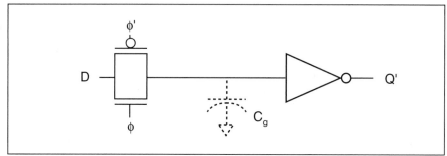

The simplest memory element in MOS technology is the **dynamic latch** shown in Figure 5-2. It is called dynamic because the memory value is not refreshed by the power supply and a latch because its output follows its input under some conditions. The latch is a D-type, so its input is D and its output is Q'. The inverter connected to the output should be familiar. The **storage capacitance** has been shown in dotted lines since it is a parasitic component; this capacitance has been named C_g since most of the capacitance comes from the gates of the transistors in the inverter.

The latch capacitance is guarded by a fully-complementary transmission gate. Dynamic latches generally use p-n pair switches because they transmit logic 0 and 1 equally well and provide better storage on the capacitor. (The inverter requires both p- and n-type transistors, so the area savings from an n-type transmission gate would be small.) The transmission gate is controlled by two clock signals, ϕ and ϕ'—a complementary switch requires both true and false forms of the control signal.

The latch's operation is straightforward. When the transmission gate is closed, whatever logic gate is connected to the D input is allowed to charge or discharge C_g. As the voltage on C_g changes, Q' follows in complement—as C_g goes to low voltages, Q' follows to high voltages, and vise versa. When the transmission gate opens, C_g is disconnected from any logic gate that could change its value. Therefore, the value of latch's output Q' depends on the voltage of the storage capacitor: if the capacitor has been discharged, the latch's output will be a logic 1; if the storage capacitor has been charged, the latch's output will be a 0. Note that the value of Q' is the logical complement of the value presented to the latch at D; we must take this inversion into account when using the latch. To change the value stored in the latch, we can close the transmission gate by setting $\phi = 1$ and $\phi' = 0$ and change the voltage on C_g.

When operating the latch, we must be sure that the final voltage stored on C_g is high enough or low enough to produce a valid logic 1 or 0 voltage at the latch's output. The storage capacitance adds delay, just as does any other parasitic capacitance; we must be sure that the logic gate connected to the latch's input has time to drive C_g to its final value before ϕ is set to 0 and the latch is closed. This latch does not keep its value forever. Parasitic resistances on the chip conspire to leak away the charge stored in the capacitor. A latch's value can usually be maintained for about a millisecond (10^{-3} s). Since gate delays range in the fractions of a nanoseconds (10^{-9} s), however, memory degradation doesn't present a significant problem, so long as the clock ticks regularly. The memory's value is restored when a new value is written to the latch, and we generally want to write a new value to the latch as quickly as possible to make maximum use of the chip's logic.

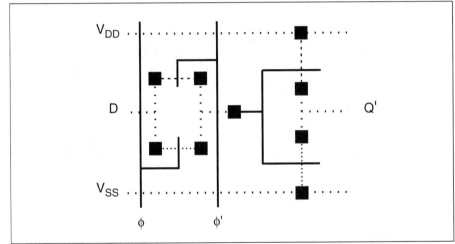

Figure 5-3: A stick diagram of a dynamic latch.

Figure 5-3 shows one possible stick diagram for a dynamic latch. The familiar inverter is on the right-hand side of the cell and the transmission gate is on the left. Figure 5-4 shows a layout of the latch. This latch's setup time is determined by the time required to charge the storage capacitance. The hold time is dominated by the time required to turn off the transistors in the transmission gate.

We should consider one simple but useful extension of the basic dynamic latch, the **multiplexed latch** shown in Figure 5-5. This latch has two data inputs, D1 and D2; the control signals A and B (and their complements) control which value is loaded into the latch; A and B should be the AND of the clock and some control signal. To ensure that a valid datum is written into the latch, A and B must never simultaneously be 1. This latch, which can be extended to more inputs, is useful because either of two

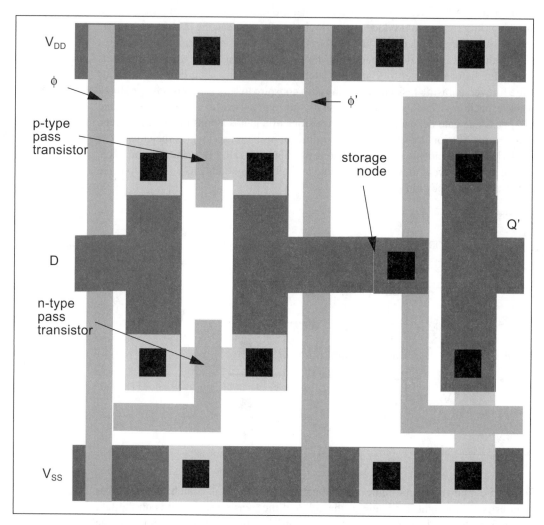

Figure 5-4: *A layout of a dynamic latch.*

different pieces of data may be loaded into a latch, depending on the value of an independently computed condition.

The dynamic latch has a small layout, but the value stored on the capacitor leaks away over time. The **recirculating latch** eliminates this problem by supplying current to constantly refresh the stored value. A recirculating latch design is shown in Figure 5-6. This latch is called **quasi-static** because the latched data will vanish if the clocks are stopped, but as long as the clocks are running the data will be recirculated and refreshed. The latch is also said to be **static on one phase** because the stored data

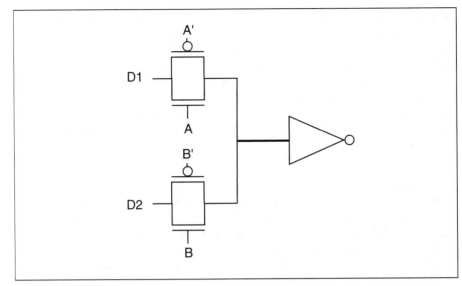

Figure 5-5: *A multiplexed dynamic latch.*

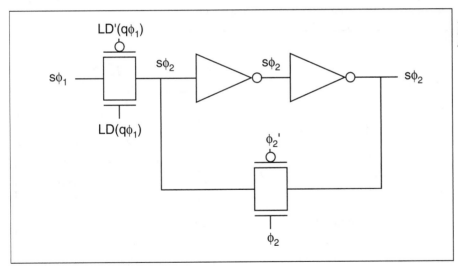

Figure 5-6: *A recirculating quasi-static latch.*

will be saved so long as the clock controlling the feedback connection remains high. During ϕ_1, if the latch is to be loaded with a new value, **LD** is set to 1, turning on the transmission gate and changing the value stored on the first inverter's gate capacitance. During ϕ_2, the two inverters are connected in a cycle; since there is an even number of inverters in the cycle, their values reinforce each other. So long as the ϕ_2 clock ticks, the latch will be repeatedly refreshed.

This latch can suffer from charge sharing when placed in a larger layout. The latch's value is stored on node A. When ϕ_2 is high, the storage node is connected to the latch's output; if the output node has a large capacitance, the charge stored there will redistribute itself to the storage node, destroying its value. Another way to look at this problem is that the output inverter won't be able to drive the large capacitance to its final value in the clock period, and the storage node's value will be destroyed as a side effect since it is connected to the output capacitance by the transmission gate. If you need to drive a large capacitance with this latch (for example, when the latch drives a long wire), you can add a buffer to the output to present an insignificant capacitance to the latch's feedback loop.

The **clocked inverter**, shown in Figure 5-7, lets us build more sophisticated latch circuits. As implied by its schematic symbol, the clocked inverter is controlled by its clock input ϕ. When $\phi = 1$, both n1 and p1 are turned on and the circuit acts as a normal inverter. When $\phi = 0$, both transistors are turned off and the output is disconnected from the rest of the inverter circuit. The control transistors p1 and n1 are closest to the output to ensure that the output is disconnected as quickly as possible when ϕ goes low. The clocked inverter is a clever way of combining transmission gate and logic gate functions into a single, compact circuit.

Figure 5-7: *A clocked inverter.*

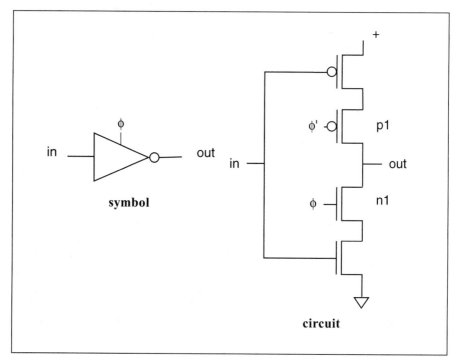

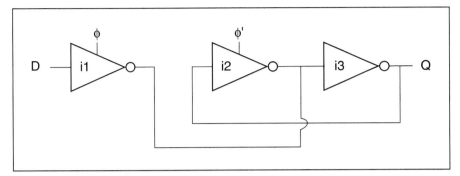

Figure 5-8: *A D-latch built from clocked inverters.*

A latch circuit built from clocked inverters is shown in Figure 5-8. This latch takes a clock and its complement and has a non-inverting output. When $\phi = 0$, $\phi' = 1$ and the inverters i2 and i3 form a positive feedback loop which retains the latch's value. When $\phi = 1$, i2 is turned off, breaking the feedback loop, and i1 is turned on to inject a new value into the loop. The hold time depends on the time required to turn off the clocked inverter. An alternative design uses a weak inverter rather than a clocked inverter for i2; if i1 is much stronger than i2, i1 can flip the state of the inverter pair when it is enabled.

Figure 5-9 shows a latch with feedback that was used in a the DEC Alpha 21064 [Dob92]. Both stages of the latch use the clocked inverter structure in their pulldown networks. The two p-type transistors in the latch's second stage form the feedback loop for the latch. When the latch value is set to 1 by a 0 value from the first stage while the clock is high, the output also turns off the upper p-type transistor, reinforcing the output value. If the first stage output goes to 1 and then subsequently to 0 without a clock event, the clocked inverter structure prevents the p-type feedback pair from flipping state.

Cells that require both true and complement forms of the clock usually generate *CLK'* internally. Such circuits often require close synchronization between the true and complement clocks, which is very difficult to achieve when the two are distributed from a distant driver. In many standard cells, the cell's clock input is connected to two inverters; this is largely a bookkeeping measure to ensure that the load presented to the clock signal by the cell is independent of the cell's internals. One inverter delivers *CLK'* while the other feeds another inverter to regenerate *CLK*. These two chains obviously don't have the same delay. One way to equalize their delays is to insert a transmission gate before the single-inverter to slow down that

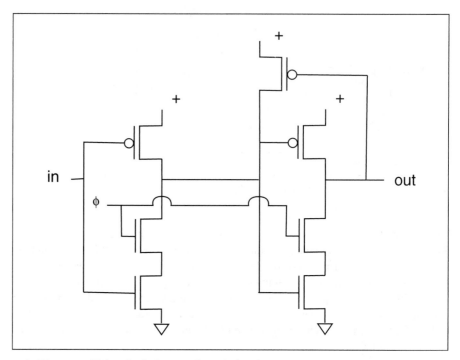

Figure 5-9: *A regenerative latch.*

path. However, if the clock duty cycle and circuit are such that overlapping phases are a serious concern, circuit simulation with accurate parasitics may be warranted.

5.2.3 Flip-Flops

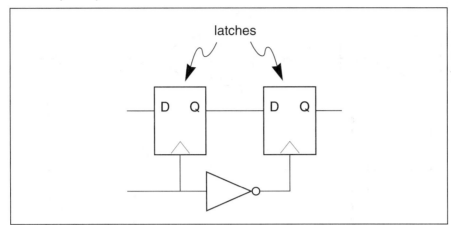

latches

Figure 5-10: *A master-slave flip-flop built from latches.*

There are two major types of flip-flops: **master-slave** and **edge-triggered**. The structure of a master-slave flip-flop is shown in Figure 5-10. It is built from two back-to-back latches called, naturally enough, the master and the slave. The master latch reads the data input when the clock is high. Meanwhile, the internal inverter assures that the slave latch's clock input is low, insulating the slave latch from changes in the master's output and leaving the flip-flop's output value stable. After the clock has gone low, the slave's clock input is high, making it transparent, but a stable value is presented to the slave by the master. When the clock moves back from 0 to 1, the slave will save its value before the master's output has a chance to change. An edge-triggered flip-flop uses additional circuitry to change the flip-flop's state only at a clock edge; a master-slave flip-flop, in contrast, is sensitive to the input as long as the clock remains active.

Figure 5-11 shows a D-type master-slave flip-flop built from the D-type quasi-static latch. This circuit follows the basic structure shown in Figure 5-10, using the quasi-static latch structure for each of the component latches.

Figure 5-12 shows the circuit diagram for an SR-type clocked flip-flop [Rab96]. (The traditional SR-type flip-flop, built from cross-coupled NOR gates, does not have a clock input.) This circuit uses a pair of cross-coupled inverters to implement the storage nodes. The additional transistors flip the state according to the SR protocol. This flip-flop is fully static and consumes no quiescent power. It can be used as a building block for more complex flip-flops.

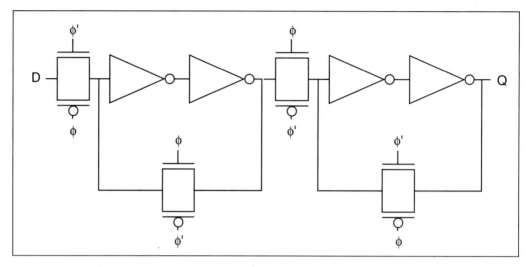

Figure 5-11: A quasi-static D-type flip-flop.

Figure 5-12: An
SR-type flip-flop.

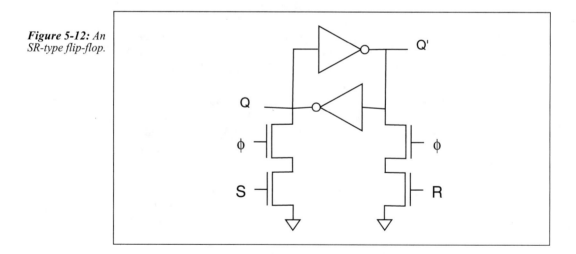

5.3 Sequential Systems and Clocking Disciplines

It is now time to study large sequential systems built from combinational networks
and memory elements. We need to understand how to build a sequential system that

performs a desired function, paying special attention to the clocks that run the memory elements to ensure that improper values are never stored; we also need to understand how to build a testable sequential machine.

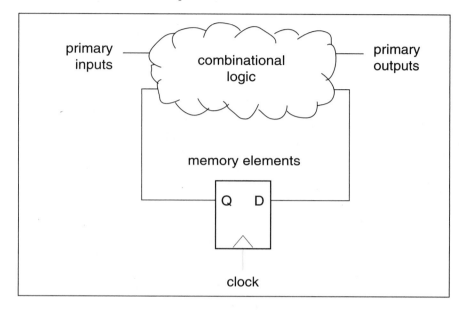

Figure 5-13:
Structure of a generic sequential system.

The structure of a generic sequential system—also known as a finite-state machine—is shown in Figure 5-13. Memory elements hold the machine's state; the machine's inputs and outputs are also called **primary inputs** and **primary outputs**. If the primary outputs are a function of both the primary inputs and state, the machine is known as a **Mealy machine**; if the primary outputs depend only on the state, the machine is called a **Moore machine**. A properly interconnected set of sequential systems is also a sequential system. It is often convenient to break a large system into a network of communicating machines: if decomposed properly, the system can be much easier to understand; it may also have a smaller layout and run faster.

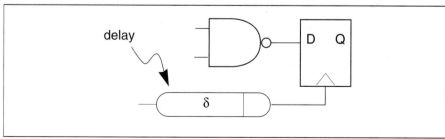

Figure 5-15: *A circuit which can suffer from clock skew.*

Figure 5-14: A
circuit which
introduces signal
skew.

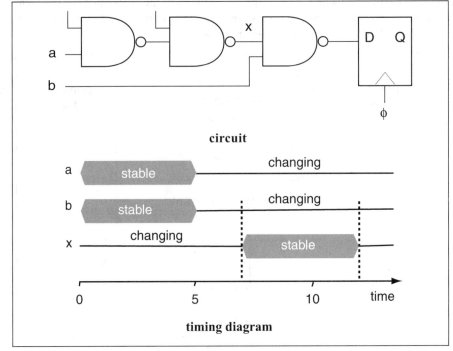

A sequential machine will work properly only if its inputs obey certain restrictions on their time behavior and the memory elements are properly clocked, notably **signal skew** and **clock skew**. In Figure 5-14, the latches that provide inputs a and b produce valid signals over the range [0,5 ns]. At those a and b inputs, the two signals are aligned in time to be simultaneously valid. By the time a's signal has propagated to point x, however, the combinational logic has delayed that signal so that it does not become valid until after b has ceased to be valid. As a result, the gate that combines b and x produces garbage during the time window marked by the dotted lines. Figure 5-15 illustrates clock skew: the signal provided to the latch is valid from 0 to 5 ns, but the clock does not close the latch until 6 ns; as a result, the latch stores a garbage value. We will describe the allowable limits on clock skew in Section 5.3.3.

We need reliable rules that tell us when a circuit acts as a sequential machine—we can't afford to simulate the circuit thoroughly enough to catch the many subtle problems which can occur. A **clocking discipline** is a set of rules that tell us:

• how to construct a sequential system from gates and memory elements;

• how to constrain the behavior of the system inputs over time.

Adherence to the clocking discipline ensures that the system will work at *some* clock frequency. Making the system work at the required clock frequency requires additional analysis and optimization.

The constraints on system inputs are defined as **signal types**, which define both how signals behave over time and what signals can be combined in logic gates or memory elements. By following these rules, we can ensure that the system will operate properly at some rate; we can then worry about optimizing the system to run as fast as possible while still functioning correctly. Different memory element types require different rules, so we will end up with a family of clocking disciplines. All disciplines have two common rules, however. The first is simple:

> **Clocking rule 1**: *Combinational logic gates cannot be connected in a cycle.*

Gates connected in a cycle form a primitive memory element and cease to be combinational—the gates' outputs depend not only on the inputs to the cycle but the values running around the cycle. In fact, this rule is stronger than is absolutely necessary. It is possible to build a network of logic gates which has cycles but is still combinational—the values of its outputs depend only on the present input values, not past input values. However, careful analysis is required to ensure that a cyclic network is combinational, whereas cycles can be detected easily. For most practical circuits, the acyclic rule is not overly restrictive.

The second common rule is somewhat technical:

> **Clocking rule 2**: *All components must have bounded delay.*

This rule is easily satisfied by standard components, but does rule out synchronizers for asynchronous signals.

5.3.1 One-Phase Systems for Flip-Flops

The clocking discipline for systems built from flip-flops is simplest, so let's consider that first. A flip-flop system looks very much like that of the generic sequential system, with a single rank of memory elements.

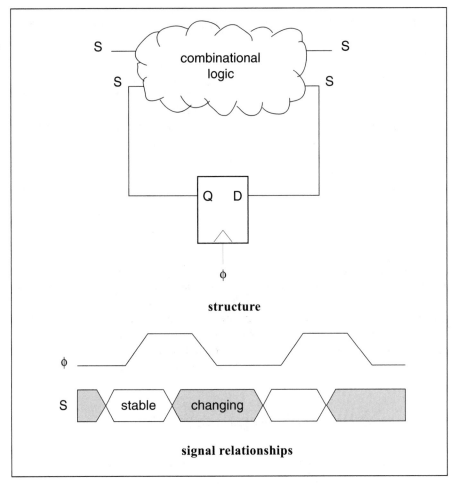

Figure 5-16: *Signal types in a flip-flop system.*

structure

signal relationships

We can define conditions that the clock and data signals must satisfy which are conservative but safe. A flip-flop system has one type of clock signal, ϕ, and one type of data signal, S, as shown in Figure 5-16. The figure assumes that the flip-flops read their inputs on the positive $(0 \to 1)$ clock edge. The data inputs must have reached stable values at the flip-flop inputs on the rising clock edge, which gives this requirement on the primary inputs:

Flip-flop clocking rule 1: *All primary inputs can change only in an interval just after the clock edge. All primary inputs must become stable before the next clock edge.*

The length of the clock period is adjusted to allow all signals to propagate from the primary inputs to the flip-flops. If all the primary inputs satisfy these conditions, the flip-flops will latch the proper next state values. The signals generated by the flip-flops satisfy the clocking discipline requirements.

5.3.2 Two-Phase Systems for Latches

A single rank of flip-flops cutting the system's combinational logic is sufficient to ensure that the proper values will be latched—a flip-flop can simultaneously send one value to its output and read a different value at its input. Sequential systems built from latches, however, are normally built from two ranks of latches. To understand why, consider the relationships of the delays through the system to the clock signal which controls the latches, as illustrated in Figure 5-17. The delay from the present state input to the next state output is very short. As long as the latch's clock is high, the latch will be transparent. If the clock signal is held high long enough, the signal can make more than one loop around the system: the next state value can go through the latch, change the value on the present state input, and then cause the next state output to change.

In such a system, the clock must be high long enough to securely latch the new value, but not so long that erroneous values can be stored. That restriction can be expressed as a **two-sided** constraint on the relative lengths of the combinational logic delays and the clock period:

- the latch must be open less than the shortest combinational delay;
- the period between latching operations must be longer than the longest combinational delay.

It is possible to meet two-sided constraint, but it is very difficult to make such a circuit work properly.

A safer architecture—the **strict two-phase** clocking discipline system—is shown in Figure 5-18. Each loop through the system is broken by two ranks of latches:

Two-phase clocking rule 1: *Every cycle through the logic must be broken by n ϕ_1 latches and n ϕ_2 latches.*

Figure 5-17: Sin-
gle latches may let
data shoot
through.

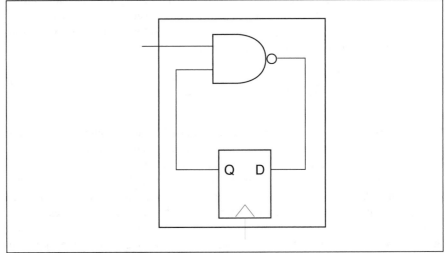

Figure 5-17: Sin-
gle latches may let
data shoot
through.

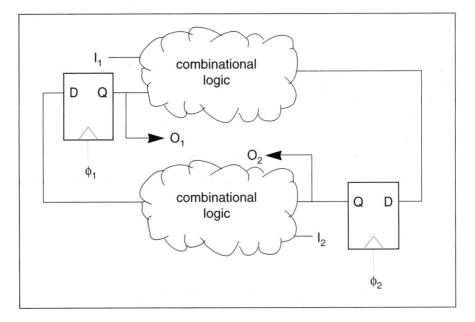

Figure 5-18: The
strict two-phase
system.

The latches are controlled by the **non-overlapping** clock phases clocks shown in
Figure 5-19. A ϕ_1-high, ϕ_2-high sequence forms a complete clock cycle. The non-
overlapping clocks ensure that no signal can propagate all the way from a latch's output back to its output. When ϕ_1 is high, the ϕ_2-controlled latches are disabled; when
ϕ_2 is high, the ϕ_1 latches are off. As a result, the delays through combinational logic

and clocks in the strict two-phase system need satisfy only a **one-sided** timing constraint: each phase must be longer than the longest combinational delay through that phase's logic. A one-sided constraint is simple to satisfy—if the clocks are run slow enough, the phases will be longer than the maximum combinational delay and the system will work properly. (A chip built from dynamic latches that is run so slowly that the stored charge leaks away won't work, of course. But a chip with combinational logic delays over a millisecond wouldn't be very useful anyway.)

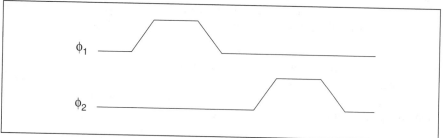

Figure 5-19: A two-phase, non-overlapping clock.

It is easy to see that we can stretch the clock phases and inter-phase gaps to ensure that the strict two-phase system works. The inputs to the combinational logic block at the ϕ_1 latch outputs are guaranteed to have settled by the time ϕ_1 goes low; the outputs of that block must have settled by the time ϕ_2 goes low for the proper values to be stored in the ϕ_2 latches. Because the block is combinational there is an upper bound on the delay from settled inputs to settled outputs. If the time between the falling edges of ϕ_1 and ϕ_2 is made longer than that maximum delay, the correct state will always be read in time to be latched. A similar argument can be made for the ϕ_1 latches and logic attached to their outputs. Therefore, if the clock cycle is properly designed, the system will function properly.

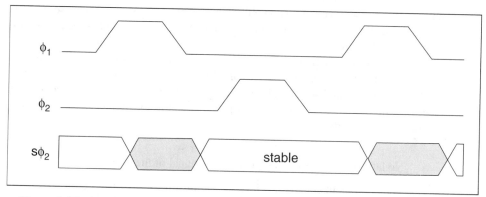

Figure 5-20: A stable ϕ_2 signal.

The strict two-phase system has two clock types, ϕ_1 and ϕ_2. Each clock type has its own data type [Noi82], the **stable** signal, which is equivalent to a **valid** signal on the opposite clock phase. Figure 5-20 shows the two clock phases and the output of a ϕ_1-clocked latch. The latch output changes only during a portion of the ϕ_1 phase. It therefore meets the setup and hold requirements of the succeeding ϕ_2 latch once the clock phase durations are properly chosen. Because the signal's value is settled during the entire ϕ_2 portion of the clock cycle, it is called **stable** ϕ_2, abbreviated as $s\phi_2$. The output of a ϕ_2-clocked latch is stable ϕ_1. A $s\phi_2$ signal is also called **valid** ϕ_1, abbreviated as $v\phi_1$, since it becomes valid around the time the ϕ_1 latch closes. Similarly, a signal that is stable during the entire ϕ_1 portion of the clock is known as **stable** ϕ_1 or $s\phi_1$.

Figure 5-21 summarizes how clocking types combine. Combinational logic preserves signal type: if all the inputs to a gate are $s\phi_1$ then its output is $s\phi_1$. Clocking types cannot be mixed in combinational logic: a gate cannot have both $s\phi_1$ and $s\phi_2$ inputs. The input to a ϕ_1-controlled latch is $s\phi_1$ and its output is $s\phi_2$.

Figure 5-21:
How strict two-phase clocking types combine.

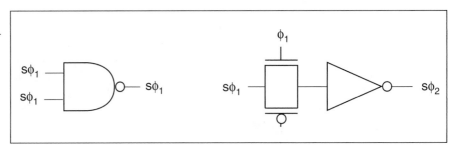

Figure 5-22 shows how signal types are used in the strict two-phase system. The system can have inputs on either phase, but all inputs must be stable at the defined times. The system can also have outputs on either phase. When two strict two-phase systems are connected, the connected inputs and outputs must have identical clocking types. Assigning clocking types to signals in a system ensures that signals are properly combined, but it will not guarantee that all loops are broken by both ϕ_1 and ϕ_2 latches. This check can be performed by **two-coloring** the block diagram. To two-color a schematic, color ϕ_1 and all signals derived from it red, and all ϕ_2-related signals green. For the system to satisfy the two-phase clocking discipline, the two-colored diagram must satisfy these rules:

•   No latch may have an input and output signal of the same color.

•   The latch input signal and clock signal must be of the same color.

•   All signals to a combinational logic element must be of the same color.

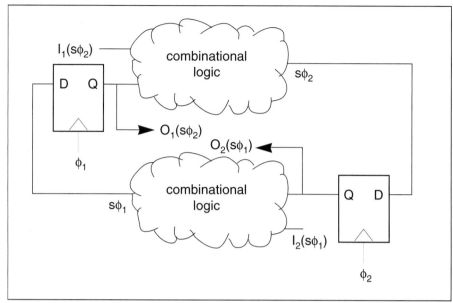

Figure 5-22:
Clocking types in the strict two-phase system.

The two-coloring check is a simple way to ensure that the rules of the clocking discipline are satisfied.

The next example builds a shift register from latches operated by a two-phase clock.

Example 5-1: Shift register design

The simplest machine we can build with the dynamic latch is a **shift register**. An n-bit shift register has a one-bit input and a one-bit output; the value at the input on a clock cycle appears at the output n clock cycles later. We can save design time by building a single cell and replicating it to create the complete shift register. We will design a component which stores a value for one clock cycle, then connect together n copies of the component so the value is shifted from one to the next for n clock cycles.

The basic bit storage component is built from a pair of latches. The schematic for a two-bit shift register looks like this:

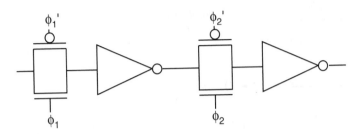

The stick diagram for a single shift register cell is identical to the dynamic latch cell, though we want to be sure that the input and output are both in poly so they can be directly connected. To build a shift register, we simply **tile** or **abut** the cells to make a linear array:

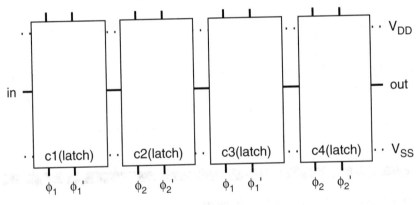

This arrangement gives us a large number of clock phase lines through the cell. Eventually, the ϕ_1s will be connected together, *etc*. Exactly how we do that depends on the design of the other components around this shift register.

The shift register's operation over one cycle looks like this:

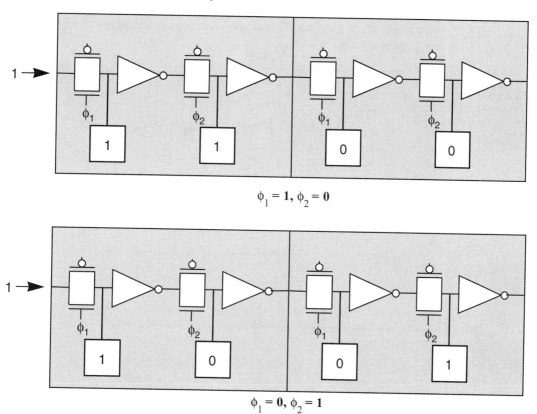

$$\phi_1 = 1, \phi_2 = 0$$

$$\phi_1 = 0, \phi_2 = 1$$

In the first phase, when $\phi_1 = 1$, the first latch in each bit is loaded with the value input to that bit. In the second phase, when $\phi_2 = 1$, that value is transferred to the second bit in the latch. So after one cycle of operation, the bit which appeared at the output of the first bit has been transferred to the output of the second and last bit of the shift register.

The multiplexed latch of Figure 5-5 needs a new type of control signal, the **qualified clock**. In the strict two-phase system there are two qualified clock types, qualified ϕ_1 (qϕ_1) and qualified ϕ_2 (qϕ_2). Qualified clocks may be substituted for clocks at latches. Since a static latch controlled by a qualified clock is no longer refreshed on every clock cycle, the designer is responsible for ensuring that the latch is reloaded often enough to refresh the storage node and to ensure that at most one transmission gate is on at a time. Qualified clocks are generated from the logical AND of a stable signal and a clock signal. For instance, a $q\phi_1$ signal is generated from a $s\phi_1$ signal and the ϕ_1 clock phase. When the clock is run slowly enough, the resulting signal will be a stable 0 or 1 through the entire ϕ_1 period.

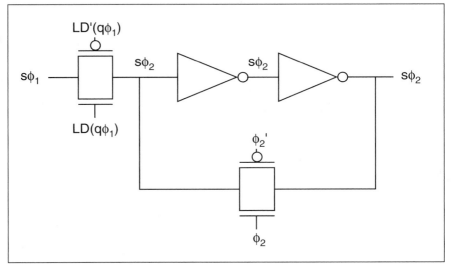

Figure 5-23:
Clocking types in the recirculating latch.

The quasi-static latch of Figure 5-6 does not satisfy the strict clocking discipline. First, it requires qualified clocks to operate; second, its feedback loop makes the type of its output problematic. If the clocking types shown in Figure 5-23 are used, the latch will operate properly in a strict two-phase system. The latch itself works even though it does not internally obey the clocking discipline—synchronous machine design hides many circuit difficulties in the memory elements.

Qualified clocks should be used carefully because they introduce clock skew, which can invalidate the assumptions made about the relationships between combinational logic delays and clock values. Consider the circuit of Figure 5-24: the ϕ_1 latch is run by a qualified clock while the ϕ_2 latch is not. When the ϕ_1 signal falls at the system input, the clock input to the latch falls δ_{clk} time later. In the worst case, if δ_{clk} is large enough, the ϕ_1 and ϕ_2 phases may both be 1 simultaneously. If that occurs, signals can propagate completely through latches and improper values may be stored. Qualified clocks also play havoc with dynamic latches, since the latch may be disabled for long enough to allow the stored charge to leak away. This constitutes another good reason to be careful when using qualified clocks.

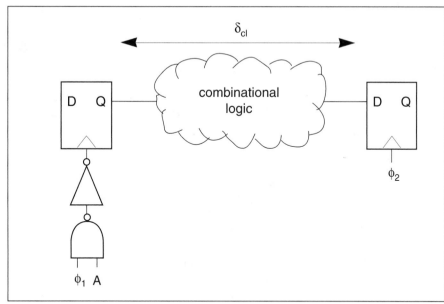

Figure 5-24:
Skew in quali-fied clock signals.

5.3.3 Advanced Clocking Analysis

In practice, it is difficult to design a large, high-performance chip which satisfies the clocking discipline. The non-overlapping phase requirement is hard to meet for at least two reasons: non-overlapping phases add dead time to the clock period which designers may not be willing to give away; and it is difficult to ensure that the clock phases are non-overlapping at all points on the chip. Many large chips are built with two-phase overlapping clock schemes. A single clock can be used to generate two phases (one for the upward edge and another for the downward edge) but these phases will overlap. When more advanced clocking schemes are used, we need to be more careful about the timing requirements on the system to ensure that it will work reliably.

If we assume that all the memory elements in the system act as flip-flops, then the combinational delay from the output of one memory element to the input of the next memory element is the critical measurement to determine the clock period: the clock period must be no smaller than the combinational delay plus the delay through the memory element. The techniques of Section 4.4 can be used to determine the worst-case combinational delay. (Of course, if the clock period is split into phases, the length of each phase is determined by looking at the logic and memory element delay of the section of the machine active on that phase.)

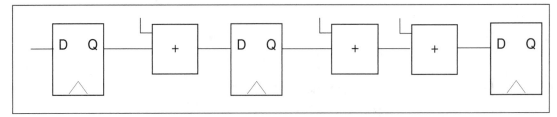

Figure 5-25: *A sequential system with unbalanced delays.*

Because the clock period is determined by the *longest* combinational delay, having some very long and some very short combinational paths causes part of the system logic to sit idle during each clock period. Consider the system of Figure 5-25: one path from flip-flop to flip-flop has one adder while the other has two. The system's clock period will be limited by the two-adder path. In some cases, the result of a long path is not used every cycle—for example, the flip-flop may be conditionally loaded. The system's clock period is still limited by the delay through this occasionally-used path, since we can't predict the cycles on which that path will be used.

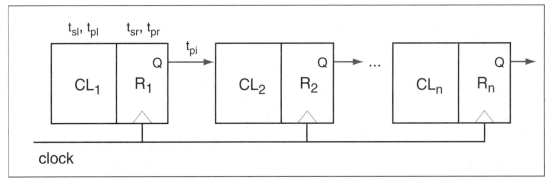

Figure 5-26: *Analysis of clocking requirements in a pipeline.*

The clock must meet requirements on skew at all points in the circuit—since the clock skew varies by position, clock skew must be checked at all registers. Hatamian and Cash [Hat87] derived the conditions for proper operation of flip-flop-based sequential machines in the presence of clock skew. As shown in Figure 5-26, they analyzed a linear chain of sequential stages in which each stage has combinational logic and a register. The clock period is T and the skew between two adjacent cells is δ. The combinational logic has a settling delay t_{sl} and a propagation delay t_{pl}; the register also has settling and propagation delays. The delay introduced by the interconnect between stages is t_{pi}. For data to be correctly transferred between adjacent pipe stages, the various delays must satisfy two relationships. The first constraint requires that the clock skew be less than the sum of the propagation delays from the input of one register to the input of the next:

$$\delta < t_{pr} + t_{pi} + t_{pl}.$$ (EQ 6-6)

The second constraint requires that the period be greater than the settling times through the logic plus the inter-stage propagation time, adjusted by the clock skew:

$$T > t_{sr} + t_{pi} + t_{sl} - \delta.$$ (EQ 6-7)

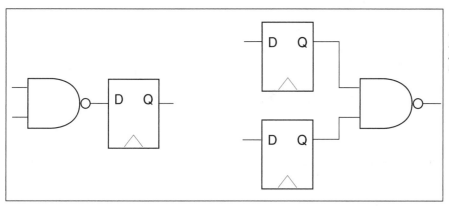

Figure 5-27:
Retiming pre-
serves combina-
tional function.

An alternative way to view the constraints on the clock for flip-flop-based systems was provided by Fishburn [Fis90]. This formulation places no restrictions on the system's interconnections but is more difficult to solve. Fishburn formulates two linear programs. The first minimizes the clock period P subject to clocking correctness constraints. In these formulas, the ith flip flop receives the clock after a delay of length x_i; the arrival time of the clock may vary within the range $[\alpha x_i, \beta x_i]$. The minimum and maximum logic delays between a the flip flops i and j are $MIN(i, j)$ and $MAX(i, j)$, respectively. The minimum delay that can be introduced by logic is $MINDELAY$. The linear program to be solved can be written as:

$$\text{minimize } P \text{ subject to}$$

$$\alpha x_i - \beta x_j \geq HOLD\text{-}MIN(i, j)$$

$$\alpha x_i - \beta x_j + P \geq SETUP + MAX(i, j) \qquad \textbf{(EQ 4-8)}$$

$$x_i \geq MINDELAY$$

The second problem maximizes the minimum margin for error in clocking constraints—this formulation distributes the slack in clock arrival times more evenly. This problem can be written as a linear program by introducing a variable M which is added to each of the main inequalities; when M is maximized, it will be the minimum slack over all the inequalities. In this formulation, P is given and therefore constant. The problem can be written as:

$$\text{maximize } M \text{ subject to}$$

$$\alpha x_i - \beta x_j - M \geq HOLD\text{-}MIN(i, j)$$

$$\alpha x_i - \beta x_j - M \geq SETUP + MAX(i, j) - P \qquad \textbf{(EQ 4-9)}$$

$$x_i \geq MINDELAY$$

Either of these problems can be solved by a standard linear programming package.

In many cases, we can move memory elements to balance combinational delays. A simple example of **retiming** [Lei83] is shown in Figure 5-27. Moving the memory element from the output of the NAND to its inputs doesn't change the combinational function computed, only the time at which the result is available. We can often move memory elements within the system to balance delays without changing the times of the signals at the primary inputs and outputs. In the example of Figure 5-25, we could move the middle flip-flop to split the middle addition in two. Retiming may also expose new logic optimization opportunities as logic is moved to the other side of a memory element [Mal90].

Sequential systems built from latches and flip-flops require different topologies—two-phase vs. single-phase systems—but how to do the speeds of the two systems compare? While taking advantage of these differences should be done only after experience and with care, the two systems do have distinct advantages. A system built from flip-flops uses all its logic on every cycle, while a latch-based system lets part of the logic sit idle on each phase. On the other hand, a latch-based machine can average delays across the two phases.

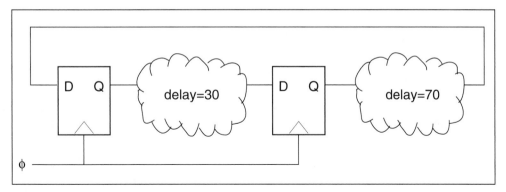

Figure 5-28: *A two-rank flip-flop system.*

Finding the maximum clock rate for latch-based systems is more difficult than the flip-flop clock-rate problem due to the transparency of latches. To understand the differences, consider first the flip-flop-based system shown in Figure 5-28. The computation done by the system requires recirculating values through two ranks of registers, and the delays through the two combinational logic blocks are not equal. The minimum clock period of the flip-flop based system is 70 ns. Completing the two-step operation requires 140 *ns*, since the machine requires a 70 ns clock period even for the operation of the 30 ns combinational block.

In the latch-based system, however, we can equalize the length of each phase to 50 ns, as shown in Figure 5-29, by taking advantage of the transparency of latches. Ignore for a moment the setup and hold times of the latches to simplify the explanation. Signals that become valid at the end of ϕ_2 propagate through the short-delay combinational logic. If the clock phases are arranged so that $\phi_1 = 1$ when they arrive, those signals can shoot through the ϕ_1 latch and start the computation in the long-delay combinational block. When the ϕ_1 latch closes, the signals are kept stable by the ϕ_1 latch, leaving the ϕ_2 latch free to open and receive the signals at the end of the next 50 *ns* interval. This technique is rarely used in practice, however: not only is it hard to debug, but it is hard to initialize the system state, since some of that state is stored on wires in the combinational logic.

Sakallah, Mudge, and Olukotun developed a set of constraints which must be obeyed by a latch-controlled synchronous system [Sak92]. Their formulation allows an arbitrary number of phases and takes into account propagation of signals through the latches. While the constraints must be solved by an algorithm for problems of reasonable size, studying the form of the constraints helps us understand the constraints which must be obeyed by a latch-controlled system.

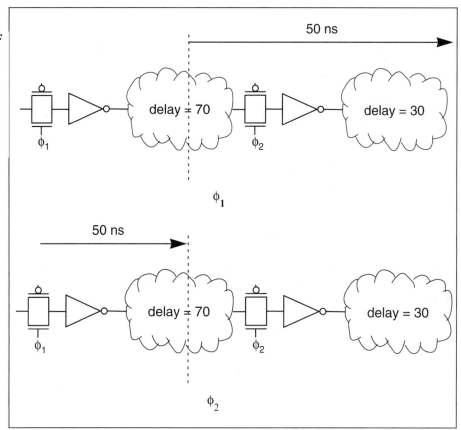

Figure 5-29:
*Spreading a
computation
across two
phases in a
latch-based
system.*

The system clock period is T_c. The clock is divided into k phases ϕ_1, ..., ϕ_k, each of which is specified by two values: the start time s_i, relative to the beginning of the system clock period, of the ith phase; and T_i, the duration of the active interval of the ith phase. Connectivity is defined by two $k \times k$ matrices. $C_{ij} = 1$ if $i \geq j$ and 0 otherwise; it defines whether a system clock cycle boundary must be crossed when going from phase i to phase j. $K_{ij} = 1$ if any latch in the system takes as its input a signal from phase ϕ_i and emits as its output a signal of phase ϕ_j and is 0 otherwise. We can easily write basic constraints on the composition of the clock phases:

- periodicity requires that $T_i \leq T_c$, $i = 1, ..., k$ and $s_i \leq T_c$, $i = 1, ..., k$;

- phase ordering requires that $s_i \leq s_{i+1}$, $i = 1, ..., k\text{-}1$;

- the requirement that phases not overlap produces the constraints
  $s_i \geq s_j + T_j \text{-} C_{ji} T_c$, $\forall(i,j) \ni K_{ij} = 1$;

- clock non-negativity requires that $T_c \geq 0, T_i \geq 0, i = 1, ..., k,$ and $s_i \geq 0, i = 1, ..., k$.

We now need constraints imposed by the behavior of the latches. The latches are numbered from 1 to l for purposes of subscripting variables which refer to the latches. The constraints require these new constraints and parameters:

- p_i is the clock phase used to control latch i; we need this mapping from latches to phases since we will in general have several latches assigned to a single phase.

- A_i is the **arrival time**, relative to the beginning of phase p_i, of a valid signal at the input of latch i.

- D_i is the **departure time** of a signal at latch i, which is the time, relative to the beginning of phase p_i, when the signal at the latch's data input starts to propagate through the latch.

- Q_i is the earliest time, relative to the beginning of phase p_i, when latch i's data output starts to propagate through the combinational logic at i's output.

- Δ_{DCi} is the setup time for latch i.

- Δ_{DQi} is the propagation delay of latch i from the data input to the data output of the latch while the latch's clock is active.

- Δ_{ij} is the propagation delay from an input latch i through combinational logic to an output latch j. If there is no direct, latch-free combinational path from i to j, then $\Delta_{ij} = -\infty$. The Δ array gives all the combinational logic delays in the system.

The latches impose setup and propagation constraints:

- Setup requires that $D_i + \Delta_{DCi} \leq T_{pi}, i = 1, ..., l$. These constraints ensure that a valid datum is setup at the latch long enough to let the latch store it.

- Propagation constraints ensure that the phases are long enough to allow signals to propagate through the necessary combinational logic. We can use a time-zone-shift equation to move a latch variable from one clock phase to another: $S_{ij} \equiv s_i - (s_j + C_{ij}T_c)$. A signal moving from latch j to latch i propagates in time $Q_j + \Delta_{ij}$, relative to the beginning of phase p_j. We can use the time-zone-shift formula to compute the arrival time of the signal at latch i measured in the time zone p_i, which is $Q_j + \Delta_{ji} + S_{pipj}$. The signal at the input of latch i is not valid until the latest signal has arrived at that latch: the time $A_i = max_i(Q_j + \Delta_{ij} + S_{pipj})$. To make sure that propagation delays

are non-negative, we can write the constraints as
$D_i = max(0, A_i), i = 1, ..., l$.

- Solving the constraints also requires that we constrain all the D_i's to be non-negative.

Optimizing the system cycle time requires minimizing T_c subject to these constraints.

5.3.4 Clock Generation

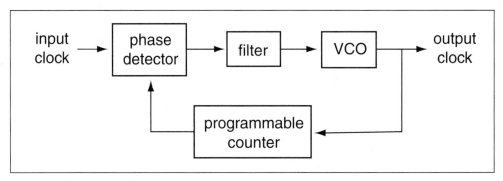

Figure 5-30: *Block diagram of a phase-locked loop for clock generation.*

Generating a high-speed clock is a non-trivial problem in itself. Many chips require clock signals of frequencies much too high to be driven onto the chip from the pads. As a result, the high-frequency clock must be generated on-chip from a lower-frequency input. Furthermore, the on-chip clock must be aligned in phase with the external reference—multiple chips are usually driven from the same external reference clock, and they will not communicate properly if their internal clocks are not phase-locked.

The **phase-locked loop (PLL)** shown in Figure 5-30 is commonly used to generate the on-chip clock signal. The higher-frequency output clock is generated by a voltage-controlled oscillator (VCO). The VCO's frequency is controlled by the feedback loop of the PLL. The signal generated by the PLL is divided down to the frequency of the input reference clock; the phase detector compares the difference in phases between the input and output clocks; a filter is imposed between the phase detector and VCO to ensure that the PLL is stable. The PLL is designed to quickly lock onto the input clock signal and to follow the input clock with low jitter. The phase-locked loop compares the input clock to the internal clock to keep the internal clock in the proper phase relationship. The circuit design of PLLs is beyond the scope of this book, but several articles [You92, Bow95, Man96] and a book [Raz98] describe PLL circuits used in high-speed chips.

5.4 Sequential System Design

5.4.1 Structural Specification of Sequential Machines

Now that we know how to construct reliable sequential machines, we can experiment with building real sequential machines, starting with a specification of function and finishing with a layout. We have already designed the simplest sequential machine— the shift register with no combinational logic. The shift register is relatively boring, not only because it has no combinational logic, but for another reason as well: there is no feedback, or closed path, between the latches. A binary counter is a simple system which exhibits both properties. We will define the counter as a *structure*: an *n*-bit counter will be defined in terms of interconnected one-bit counters.

Example 5-2: A counter

A one-bit counter consists of two components: a specialized form of adder, stripped of unnecessary logic so that it can only add 1; and a memory element to hold the value. We want to build an *n*-bit binary counter from one-bit counters.

What logical function must the one-bit counter execute? The truth table for the one-bit counter in terms of the present count stored in the latch and the carry-in is shown below. The table reveals that the next value of the count is the exclusive-or (XOR) of the current count and C_{in}, while the carry-out is the AND of those two values.

count	C_{in}	next count	C_{out}
0	0	0	0
0	1	1	0
1	0	1	0
1	1	0	1

Here is a logic schematic for the one-bit counter:

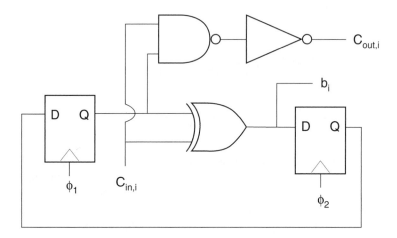

The AND function is built from a NAND gate and an inverter. The latches in this counter have the same basic connections as the latches in the shift register, except that logic is added between the ϕ_1 and ϕ_2 latches to compute the count. The next count is loaded into one latch while ϕ_2 is high, then transferred to the other latch during ϕ_1, allowing the next count cycle to be computed.

The n-bit counter's structure looks like this:

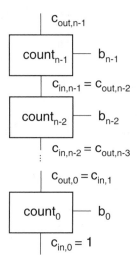

Each bit has one input and two outputs: the input $C_{in,i}$ is the carry into the ith bit; the output b_i is the current value of the count for that bit; and $C_{out,i}$ is the carry out of the bit. The carry-in value for the 0th bit is 1; on each clock cycle this carry value causes the counter to increment itself. (The counter, to be useful, should also have a reset input which forces all bits in the counter to 0; we have omitted it here for simplicity.)

Here is a hierarchical stick diagram for the one-bit counter:

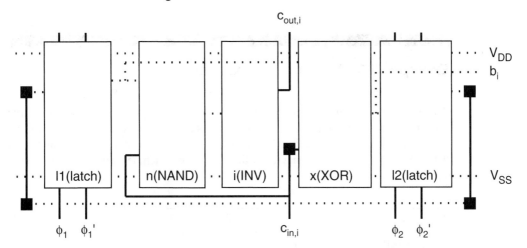

It has been designed to tile vertically to form an n-bit counter. All the one-bit counter's components are arranged in one long row. The ϕ_1 and ϕ_2 latches are on opposite ends of the cell, so a long metal wire must be used to connect them. The connections between the logic gates are relatively simple, though care must be taken to route the wires over cells so they do not create inadvertent shorts.

5.4.2 State Transition Graphs and Tables

To build complex sequential systems, we need powerful specification techniques. We described the counter of Example 5-2 as a structure. A more abstract and powerful specification is **functional**—describing the next-state and output functions directly, independent of the structure used to compute them. We can then use programs to generate the Mealy or Moore structure of Figure 5-13 to generate an initial structure, which can be optimized by CAD tools. Some behaviors are cumbersome to specify as state transition tables or graphs and are best described as structures—a register file is a good example of a sequential machine best described structurally—but functional descriptions of FSMs occur in nearly every chip design.

An FSM can be specified in one of two equivalent ways: as a **state transition table** or a **state transition graph**. Either is a compact description of a sequential machine's behavior. The next example shows how to design a simple machine from a state transition table.

Example 5-3: *A 01-string recognizer*

Consider as an example a very simple FSM with one input and one output. If the machine's inputs are thought of as a string of 0's and 1's, the machine's job is to recognize the string "01"—the FSM's output is set to 1 for one cycle as soon as it sees "01." This table shows the behavior of the recognizer machine over time for a sample input:

time	0	1	2	3	4	5
input	0	0	1	1	0	1
present state	bit1	bit2	bit2	bit1	bit1	bit2
next state	bit2	bit2	bit1	bit1	bit2	bit1
output	0	0	1	0	0	1

We can describe the machine's behavior as either a state transition graph or a state transition table. The machine has one input, the data string, and one output, which signals recognition. It also has two states: bit1 is looking for "0", the first bit in the string; bit2 is looking for the trailing "1". Both representations specify, for each possible combination of input and present state, the output generated by the FSM and the next state it will assume.

Here is the state transition table:

input	present state	next state	output
0	bit1	bit2	0
1	bit1	bit1	0
0	bit2	bit2	0
1	bit2	bit1	1

And here is the equivalent state transition graph:

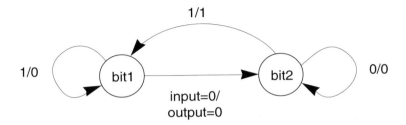

Assume that the machine starts in state bit1 at time $t=0$. The machine moves from bit1 to bit2 when it has received a 0 and is waiting for a 1 to appear on the next cycle. If the machine receives a 0 in state bit2, the "01" string can still be found if the next bit is a 1, so the machine stays in bit2. The machine recognizes its first "01" string at $t=2$; it then goes back to state bit1 to wait for a 0. The machine recognizes another "01" string at $t=5$.

Translating the state transition graph/table into a chip layout requires several steps, most of which are familiar from the counter design. The first step is to **encode** the machine's states into binary values, a step also known as **state assignment**. We didn't discuss the encoding of the counter machine because we already knew a good encoding, namely, two's-complement binary numbers. All the counter's signals were specified as binary values, which mapped directly into the 0s and 1s produced by logic gates.

The present and next state values of a machine specified as a state transition graph, however, are **symbolic**—they may range over more than two values, and so do not map directly into Boolean 0s and 1s. This string-recognizer machine has only two states, but even in this simple case we don't know which state to code as 0 and which as 1. The encoding problem is difficult and important because the choice of which

Boolean value is associated with each symbolic state can change the amount of logic required to implement the machine.

Encoding assigns a binary number, which is equivalent to a string of Boolean values, to each symbolic state. By substituting the state codes into the state transition table, we obtain a **truth table** which specifies the combinational logic required to compute the machine's output and next state. If we choose the encoding bit1 = 0, bit2 = 1 for the 01-string recognizer, we obtain this truth table:

input	present state	next state	output
0	0	1	0
1	0	0	0
0	1	1	0
1	1	0	1

From the encoded state transition table we can design the logic for to compute the next state and the output, either manually or by using logic optimization. Here is one logic network for the 01-string recognizer:

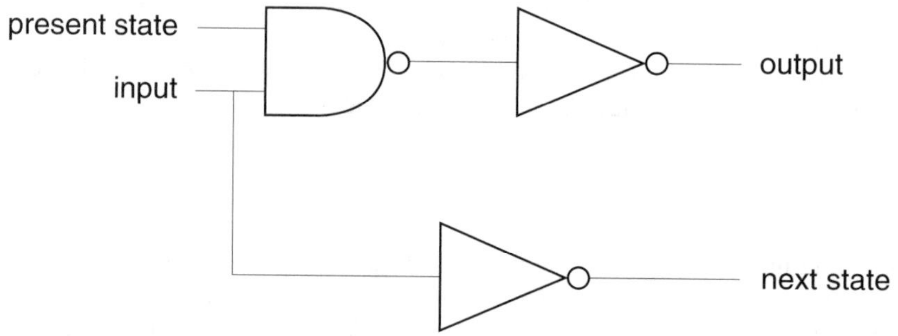

Inspection shows that the gates in fact implement the functions described in the truth table. Creating a logic network for this machine is easy but the task is more difficult for machines with larger state transition tables. Luckily, programs can design small, fast logic networks for us from encoded state transition tables. For example, we can use a set of synthesis tools which will take a truth table, optimize the logic, then create a standard-cell layout. The resulting layout looks somewhat different than our hand-designed examples because the standard cells's transistors are are designed to drive larger loads and so are much wider than the ones we have been drawing by hand, but the layout is still two rows of CMOS gates with wiring in between:

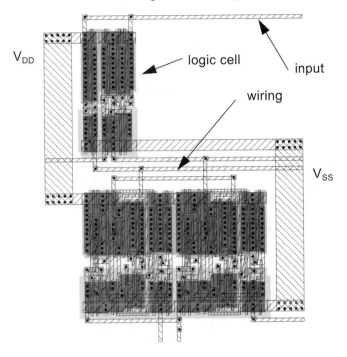

If necessary, we can use a layout editor to examine the layout and determine exactly how the logic functions were designed and where the transistors for each gate were placed. However, one of the nicest things about synthesis tools (well-debugged tools, at least) is that you don't have to worry about how they did their job. All we need to know is the position of each input and output around the edge of the layout cell.

A slightly more complex example of finite-state machine designs is a controller for a traffic light at the intersection of two roads. This example is especially interesting in that it is constructed from several communicating finite-state machines; just as decomposing stick diagrams into cells helped organize layout design, decomposing sequential machines into communicating FSMs helps organize machine design.

Example 5-4: A traffic light controller

We want to control a road using a traffic light:

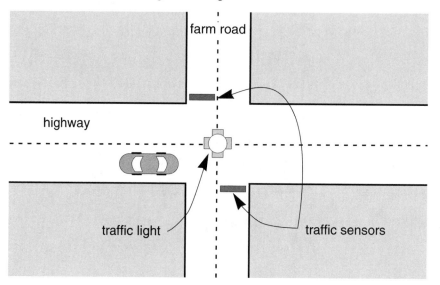

There are many possible schemes to control when the light changes. We could alternate between giving the two roads green lights at regular intervals; that scheme, however, wouldn't give us an interesting sequential machine to study. A slightly more complex and interesting system can be built by taking traffic loads into account. The highway will generally have more traffic, and we want to give it priority; however, we do not want to completely block traffic on the farm road from crossing the highway. To balance these competing concerns, we install traffic sensors on the farm road at the intersection. If there is no traffic waiting on the farm road, the highway always has a green light. When traffic stops at the farm road side of the intersection, the traffic lights are changed to give the farm road a green light as long as there is traffic. But since this simple rule allows the highway light to be green for an interval too short to be safe (consider a second farm road car pulling up just as the highway light has returned to green), we ensure that the highway light (and, for similar reasons, the farm light) will be green for some minimum time.

We must turn this vague, general description of how the light should work into an exact description of the light's behavior. This precise description takes the form of a state transition graph. How do we know that we have correctly captured the English description of the light's behavior as a state transition table? It is very difficult to be absolutely sure, since the English description is necessarily ambiguous, while the state transition table is not. However, we can check the state transition table by mentally executing the machine for several cycles and checking the result given by the state transition table against what we intuitively expect the machine to do. We can also assert several universal claims about the light's behavior: at least one light must be red at any time; lights must always follow a *green → yellow → red* sequence; and a light must remain green for the chosen minimum amount of time.

We will use a pair of communicating sequential machines to control the traffic light:

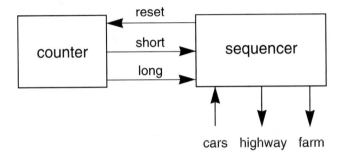

The system consists of a counter and a sequencer. Both are finite-state machines, but each serves a different purpose in the system. The counter counts clock cycles, starting when its reset input is set to 1, and signals two different intervals—the short signal controls the length of the yellow light, while the long signal determines the minimum time a light can be green.

The sequencer controls the behavior of the lights. It takes as inputs the car sensor value and the timer signals; its outputs are the light values, along with the timer reset signals. The sequencer's state transition graph looks like this:

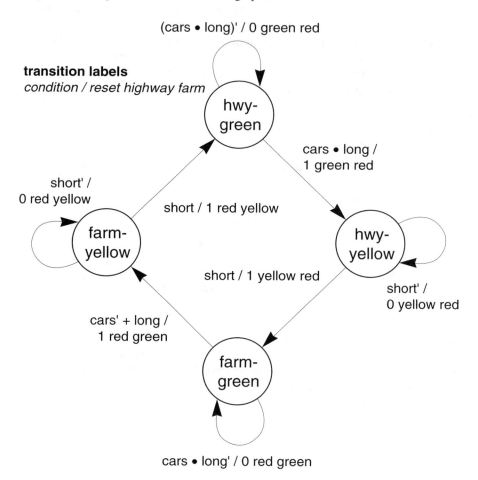

The states are named to describe the value of one of the lights; the complete set of light values, however, is presented at the machine's outputs on every cycle. Tracing through the state transition graph shows that this sequencer satisfies our English specification: the highway light remains green until cars arrive at the farm road (as indicated by the sensor) and the minimum green period (specified by the long timeout) is met. The machine then sets the highway light yellow for the proper amount of time, then sets the highway light to red and the farm light to green. The farm light remains green so long as cars pull up to the intersection, but no longer than the long timeout period. Inspection also shows that the state transition graph satisfies all our assertions:

one light is always red, each light always changes colors in the *green → yellow → red* sequence; and each light, when it turns green, remains green for at least the period specified by the long timer.

We can also write the state transition graph as a table. Some of the transitions in the graph are labeled with OR conditions, such as cars + long'. Since each line in a state transition table can only refer to the AND of input conditions, we must write the OR conditions in multiple lines. For example, one line can specify a transition out of the farm-green state when cars = 1, while another can specify the same next state and outputs when long = 0.

The sequencer and counter work in tandem to control the traffic light's operation. The counter can be viewed as a subroutine of the sequencer—the sequencer calls the counter when it needs to count out an interval of time, after which the counter returns a single value. The traffic light controller could be designed as a single machine, in which the sequencer counts down the long and short time intervals itself, but separating the counter has two advantages. First, we may be able to borrow a suitable counter from a library of pre-designed components, saving us the work of even writing the counter's state transition table. Second, even if we design our own counter, separating the machine states that count time intervals (counter states) from the machine states that make decisions about the light values (sequencer states), clarifies the sequencer design and makes it easier to verify.

We can implement each machine in the traffic light controller just as any other FSM, moving from the state transition table through logic to a final layout. We can either design custom layouts or synthesize standard-cell layouts from optimized logic. However, we have the additional problem of how to connect the two machines. We have three choices. The least palatable is to write a combined state transition table for the sequencer and counter, then synthesize it as a single FSM. Since the states in the combined machine are the Cartesian product of the states in the two component machines, that machine is unacceptably large. The simplest solution is to design each machine separately, then wire them together by hand. This option requires us to intervene after the FSM synthesis task is done, which we may not want to do. The third alternative is to interrupt the FSM synthesis process after logic design, splice together the net lists for the two machines, and give the combined net list to standard cell placement and routing.

5.4.3 State Assignment

State assignment is the design step most closely associated with FSMs. (Input and output signals may also be specified as symbolic values and encoded, but the state variable typically has the most coding freedom because it is not used outside the FSM.) State assignment can have a profound effect on the size of the next state logic, as shown in the next example.

Example 5-5: Encoding a shift register

Here is the state transition table for a two-bit shift register, which echoes its input bit two cycles later:

input	present state	next state	output
0	s00	s00	0
1	s00	s10	0
0	s01	s00	1
1	s01	s10	1
0	s10	s01	0
1	s10	s11	0
0	s11	s01	1
1	s11	s11	1

The state names are, of course, a hint at the optimal encoding. But let's first try another code: s00 = 00, s01= 01, s10 = 11, *s11* = 10. We'll name the present state bits $S_1 S_0$, the next state bits $N_1 N_0$, and the input i. The next state and output equations for this encoding are:

$$output = S_1 \overline{S_0} + \overline{S_1} S_0$$

$$N_1 = i$$

$$N_0 = i \overline{S_1} + i S_1$$

Both the output and next state functions require logic. Now consider the shift register's natural encoding—the history of the last two input bits. The encoding is $s00 = 00$, $s10 = 10$, $s01 = 01$, $s11 = 11$. Plugging these code values into the symbolic state transition table shows that this encoding requires no next state or output logic:

$$output = S_0$$
$$N_1 = i$$
$$N_0 = S_1$$

This example may seem contrived because the shift register function is regular. But changes to the state codes can significantly change both the area and delay of sequencers with more complex state transition graphs. State codes can be chosen to produce logic which can be swept into a common factor during logic optimization—the common factors are found during logic optimization, but exist only because the proper logic was created during state assignment.

State assignment creates two types of common factors: factors in logic that compute functions of the present state; and factors in the next state logic. Input encoding can best be seen as the search for common factors in the symbolic state transition table. Consider this state machine fragment:

input	present state	next state	output
0	s1	s3	1
0	s2	s3	1

If we allow combinations of the present state variable, we can simplify the state transition table as:

input	present state	next state	output
0	$s1 \vee s2$	s3	1

How can we take advantage of the OR by encoding? We want to find the smallest logic which tests for $s1 \vee s2$. For example, if we assume that the state code for the complete machine requires two bits and we encode the state variables as $s1 = 00$, $s2 = 11$, the present state logic is $\overline{S_1}\overline{S_0} + S_1 S_0$. The smallest logic is produced by putting the state codes as close together as possible—that is, minimizing the number of bits in which the two codes differ. If we choose $s1 = 00$, $s2 = 01$, the present state logic reduces to $\overline{S_1}$.

As shown in Figure 5-31, we can interpret the search for symbolic present state factors in the state transition table as a forward search for common next states in the state transition graph [Dev88]. If two states go to the same next state on the same input, the source states should be coded as close together as possible. If the transitions have similar but not identical input conditions, it may still be worthwhile to encode the source states together.

Figure 5-31:
Common next states.

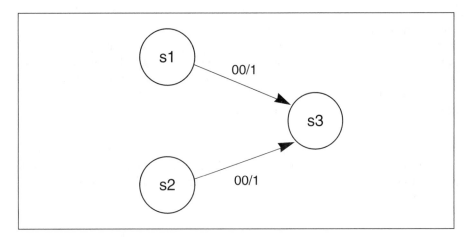

Figure 5-32 illustrates the relationship between bit differences and distance between codes. We can embed a three-bit code in a three-dimensional space: one axis per code bit, where each axis includes the values 0 and 1. Changing one code bit between 0 and 1 moves one unit through the space. The distance between 000 and 111 is three because we have to change three bits to move between the two codes. Putting two codes close together puts them in the same subspace: we can put two codes in the 00-subspace and four in the 1– subspace. We can generate many coding constraints by searching the complete state transition graph; the encoding problem is to determine which constraints are most important.

We can also search backward from several states to find common present states. As shown in Figure 5-34, one state may go to two different states on two different input

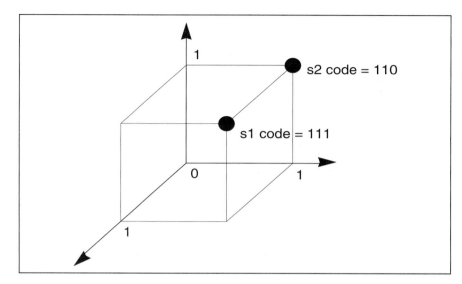

Figure 5-32: State
codes embedded in
a three-dimensional
space.

values. In this case, we can minimize the amount of logic required to compute the next state by making the sink states' codes as close as possible to the source state's code. Consider this example:

input	present state	next state	output
0	s0	s1	1
1	s0	s2	1

We can make use of the input bit to compute the next state with the minimum amount of logic: if $s0 = 00$, we can use the input bit as one bit of the codes for s1 and s2: $s1 = 10$, $s2 = 11$. One bit of the next state can be computed independently of the input condition. Once again, we have encoded $s1$ and $s2$ close together so that we need the smallest amount of logic to compute which next state is our destination.

So far, we have looked at codes that minimize the area of the next state logic and the number of registers. State assignment can also influence the delay through the next state logic; reducing delay often requires adding state bits. Figure 5-33 shows the structure of a typical operation performed in either the next-state or the output logic. Some function $f()$ of the inputs is computed. This value will usually control a conditional operation: either a conditional output or a conditional change in state. Some test of the present state is made to see if it is one of several states. Then those two results are combined to determine the proper output or next state. We can't do much

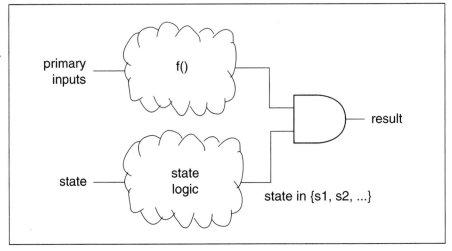

Figure 5-33: *An FSM computes new values from the primary inputs and state.*

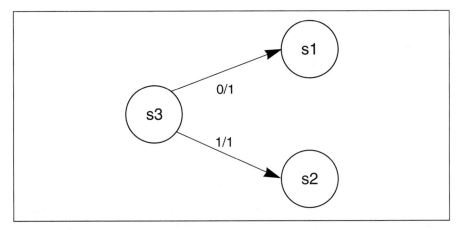

Figure 5-34: *Common present states.*

about the delay through $f()$, but we can choose the state codes so that the important steps on the state are easy to compute. Furthermore, the FSM probably computes several $f()$s for different operations, which in general don't have the same delay. If we can't make all computations on the state equally fast, we can choose the codes so that the fastest state computations are performed on the FSM's critical path.

As shown in Figure 5-35, state codes can add delay both on the output and next state sides. On the output logic side, the machine may need to compute whether the present state is a member of the set which enables a certain output—in the example, the output is enabled on an input condition and when the present state is either s2 or s4. The delay through the logic which computes the state subset depends on whether the state codes were chosen to make that test obvious.

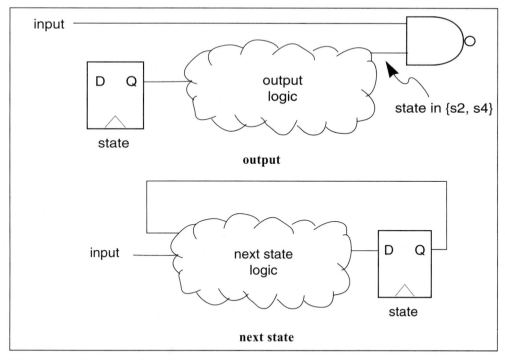

Figure 5-35: *How state codes affect delay.*

A **one-hot code** uses n bits to encode n states; the ith bit is 1 when the machine is in the ith state. We can use such a code to easily compute state subset membership, by simply computing the OR of the state bits in the subset. But this solution has two problems. First, it requires a lot of memory elements for the present state: a machine with 64 states requires at least six memory elements for arbitrary codes, but 64 states for a one-hot encoding. Second, one-hot encoding doesn't help if an output depends on more than one state. It's best to examine the machine for time-critical outputs which depend on the present state and to construct codes which efficiently represent the time-critical state combinations, then use area-minimizing coding for the rest of the states.

On the next state side, the machine needs to compute the next state from the inputs and the present state. The delay to compute the next state depends on the complexity of the next-state function. The fastest next-state logic uses the result of the test of the primary inputs to independently change bits in the state code. For example, setting bit 0 of the next state to 1 and bit 2 to 0 is relatively fast. Computing a new value for bit 0, then setting bit 2 to the complement of bit 1 is slower.

5.5 Power Optimization

As was described in Section 4.4, eliminating glitching is one of the most important techniques for power reduction in CMOS logic. Glitch reduction can often be applied more effectively in sequential systems than is possible in combinational logic. Sequential machines can use registers to stop the propagation of glitches, independent of the logic function being implemented.

Figure 5-36: *Flip -flops stop glitch propagation.*

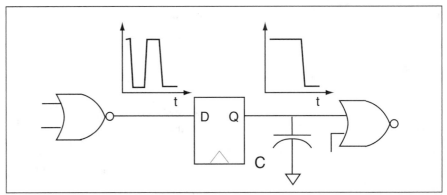

Many sequential timing optimizations can be thought of as retiming [Mon93]. Figure 5-36 illustrates how flip-flops can be used to reduce power consumption by blocking glitches from propagating to high capacitance nodes. (The flip-flop and its clock connection do, of course, consume some power of their own.) A well-placed flip-flop will be positioned after the logic with high signal transition probabilities and before high capacitance nodes on the same path.

Beyond retiming, we can also add extra levels of registers to keep glitches from propagating. Adding registers can be useful when there are more glitch-producing segments of logic than there are ranks of flip-flops to catch the glitches. Such changes, however, will change the number of cycles required to compute the machine's outputs and must be compatible with the rest of the system.

Proper state assignment may help reduce power consumption. For example, a one-hot encoding requires only two signal transitions per cycle—on the old state and new state signals. However, one-hot encoding requires a large number of memory elements. The power consumption of the logic that computes the required current-state and next-state functions must also be taken into account.

5.6 Design Validation

A sequential machine is a chunk of logic large enough to demand its own validation strategy. You can verify functionality both from the top down—checking that your logic matches the machine's description—and from the bottom up—extracting the circuit from the layout and comparing the results of its simulation with the simulation results from the logic network you specified. You must also make sure that the system runs at the required rate; one of the advantages of building sequential systems according to a clocking methodology is that we can verify performance without simulation.

You may have access to true verification tools, which can automatically compare a combinational logic or sequential machine description against the implementation, using tautology or FSM equivalence algorithms. You are more likely to use simulation to validate your design. You can simulate a single description of your machine, such as the register-transfer description, to be sure you designed what you wanted; you can also compare the results of two different simulations, such as the logic and register-transfer designs, to ensure that the two are equivalent. You may need to use several simulators to verify the design, depending on your available tools:

- A register-transfer simulator exhibits the correct cycle-by-cycle behavior at its inputs and outputs, but the internal implementation of the simulator may have nothing to do with the logic implementation. Several specialized languages for hardware description and simulation have been developed. Hardware simulation languages, such as VHDL and Verilog, provide primitives which model the parallelism of logic gate evaluation, delays, *etc.*, so that a structural description like a net list automatically provides accurate simulation. In a pinch, a C program makes a passable register-transfer simulator: the component is modeled as a procedure, which takes inputs for one cycle and generates the outputs for that cycle. However, hardware modeling in C or other general-purpose programming languages requires more attention to the mechanics of simulation.

- A logic simulator accepts a net list whose components are logic gates. The simulator evaluates the output of each logic gate based on the values presented at the gate's inputs. You can trace though the network to find logic bugs, comparing the actual value of a wire to what you think the value should be. Verilog and VHDL can be used for logic simulation: a library provides simulation models for the logic gates; a net list tells the simulation system how the components are wired together.

- A switch simulator models the entire system—both combinational logic gates and memory elements—as a network of switches. Like a logic simulator, the simulator evaluates individual nets, but the simulation is performed at a lower level of abstraction. A switch simulator can find some types of charge sharing bugs, as well. You must use a switch simulator if your circuit contains mixed switch and gate logic; a switch simulator is most convenient for a circuit extracted from a complete layout, since the circuit extractor generates a net list of transistors.

Your should simulate your sequential machine specification—register-transfer description, state transition graph, etc.—before designing the logic to implement the machine. If you specify the wrong function and don't discover the error before implementation, you will waste a lot of logic and layout design before you discover your mistake. This step ensures that your formal description of behavior matches your informal requirements.

To verify your implementation, you should check your logic design against the register-transfer/sequential machine description. Once again, catching any errors before layout saves time and effort. That is definitely true if you design the logic yourself; if the logic was designed by a CAD tool, the results are probably correct, though the more paranoid designers among you may want to perform some simulation to make sure the logic optimizer didn't make a mistake.

You should also extract the circuit from your completed layout, simulate it using the same inputs you used to simulate your logic, and compare the results. Switch or circuit simulation not only check the correctness of the layout, they also identify charge-sharing bugs which can be found only in a switch-level design. Simulation tests which are comprehensive enough to ensure that your original logic design was correct should also spot differences between the logic and the layout. If you do not have a logic simulator available, but you do have layout synthesis, one way to simulate the logic is to generate a layout, then extract a switch-level circuit and simulate it.

If you specify a schematic or net list of the logic before layout, a net list comparison program can check the layout against that schematic. The net list extracted from the layout will use n-type and p-type transistors as its components. If the schematic was designed in terms of logic gates, it can be expanded to a transistor-level schematic. A net list comparison program tries to match up the components and nets in the two schematics to produce a one-to-one correspondence between the two net lists. Such programs usually require that only a few major signals—V_{DD}, V_{SS}, clocks, and the primary inputs and outputs—be identified. If the program can't match up the two net lists, it will try to identify a small part of each circuit which contains the error.

Performance verification—making sure the system runs fast enough—can be separated from functionality if the system is properly designed. If we have obeyed a clocking methodology, we know that the system will work if values arrive at the memory elements within prescribed times. Timing analysis algorithms like those described in Section 4.4 are the best way to ensure that the chip runs at the required rate. Circuit or timing simulation should be used to optimize paths which are expected to be critical. However, unexpected critical paths may have crept into the design. Timing analysis is the guardian which ensures that the paths you optimized are in fact the critical paths.

5.7 Sequential Testing

In the last chapter we studied manufacturing faults in combinational networks and how to test them. Now we are prepared to study the testing of sequential systems, which is made much harder by the inaccessibility of the memory elements to the tester.

A suite of test vectors for a chip is generated using a combination of CAD programs called **automatic test pattern generators** (ATPG) and expert human help. Test generation for combinational networks can be done entirely automatically. Automated methods for sequential circuits are improving, but manual intervention is still required in many cases to provide full test coverage. The designer or test expert may be able to find a test for a fault that a program cannot. Often, however, it is better to redesign a hard-to-test chip to make it easier to find test vectors. Not only does **design for testability** let automatic test generation programs do more of the work, it reduces the risk that the chip will be abandoned in frustration with low test coverage.

Testing a sequential system is much harder than testing a combinational network because you don't have access to all of the inputs and outputs of the machine's combinational logic. Figure 5-37 shows a sequential machine. We want to test the NAND gate for a stuck-at-1 fault at its output, which requires applying 1 to both its inputs. Setting one input to 1 is easy, since the gate's input is tied to one of the machine's primary inputs. (If there were combinational logic between i_1 and the NAND gate's input, finding the proper stimulus would be hard, but we could still apply the value directly, assuming the logic is not redundant.) The other input is more difficult, because it is fed only by logic tied to the machine's state registers. Setting the NAND gate's lower input to 1 requires driving the NOR gate's output to 0; this can be done only when the machine is in a state which has a 1 for either ps_0 or ps_1. Although there may be several states which meet this criterion, getting the machine to a proper state

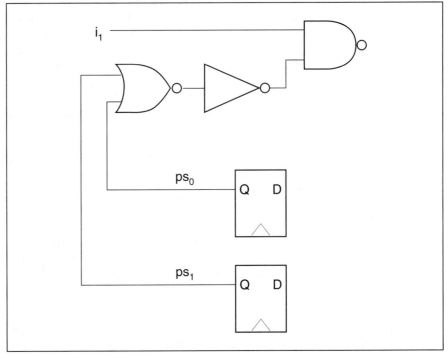

Figure 5-37:
Testing a sequential machine.

may take several cycles. Testing a single fault in the combinational logic may take many, many cycles, meaning that testing all the faults becomes much more expensive than for purely combinational logic.

The state transition graph and state encoding for the machine of Figure 5-37 are given in Figure 5-39. Examining the state transition graph helps us understand how hard it can be to test for a fault in a sequential system. When we start the test sequence, we may not know the machine's present state. (Even if the machine is reset at power-up time, the previous test may have left it in one of several different states.) In that case, we have to find a sequence of inputs to the FSM that drive it to the desired state independent of its starting state. Since this machine has a reset input that lets us get to s_0 from any state in one cycle, we can get to s_3 in three cycles by the sequence $* \rightarrow s_0 \rightarrow s_1 \rightarrow s_3$, where $*$ stands for any state.

At this point, we can apply $i_1 = 0$, run the machine for one more cycle, and perform the test. Of course, some of the combinational logic's primary outputs may be connected only to the next state lines in a way that the result of the test is not visible at the primary outputs. In this case, we must run the machine for several cycles until we observe the test's outcome at the machine's primary outputs.

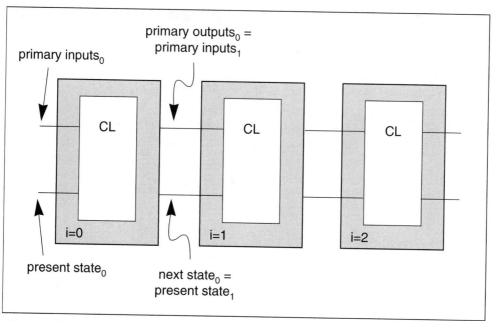

Figure 5-38: *Time-frame expansion of a sequential test.*

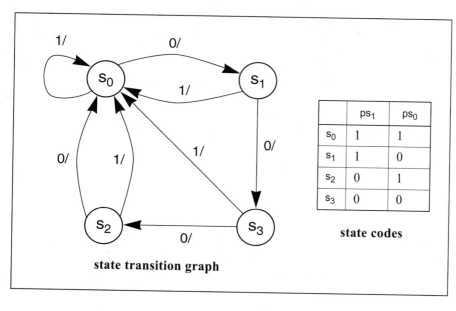

Figure 5-39: *A state transition graph to test.*

	ps_1	ps_0
s_0	1	1
s_1	1	0
s_2	0	1
s_3	0	0

state codes

state transition graph

State assignment may make it impossible to justify the required values in the machine's combinational logic. The next example illustrates the problem.

Example 5-6: Unreachable states

We are given this state transition graph to implement:

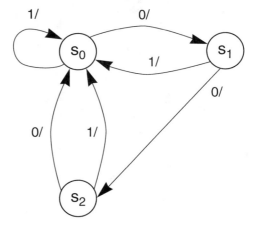

This machine has three states. Let's choose the state assignment $s_0 = 00$, $s_1 = 01$, $s_2 = 10$. Since this state code has two bits, our implementation will actually have *four* states. However, let's ignore this complication for the moment and design the machine's combinational logic taking into account only the states specified in the state transition graph. The truth table for the encoded machine is:

i	$S_1 S_0$	$N_1 N_0$
0	0 0	0 1
1	0 0	0 0
0	0 1	1 0
1	0 1	0 0
0	1 0	0 0
1	1 0	0 0

The equations for the next-state logic are $N_1 = i'S_1'S_0$, $N_0 = i'S_1'S_0'$. This next-state logic creates transitions for the remaining state code, 11. When we use this combinational logic, the state transition graph of the machine we actually implement is this:

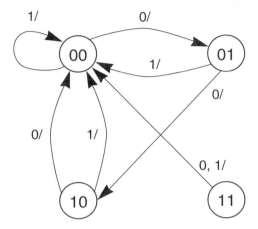

If the machine happens to start in state 11 at power-up, we can apply an input to get it to one of the specified states, but there is no way to get the machine from any of our specified states to the 11 state. If any part of the machine, such as the output logic, requires the present state to be 11 to test for a fault, we can't test for that fault. A **strongly connected** state transition graph has a path from any state to any other state. A reset signal goes a long way to making a state transition graph strongly connected and more easily testable.

To make sequential testing even more difficult, a single fault in the logic can mimic a multiple fault. **Time-frame expansion**, illustrated in Figure 5-38, helps us understand this phenomenon. A sequential test can be analyzed by unrolling the hardware over time: one copy of the hardware is made for each cycle; the copies are connected so that the next state outputs at time t are fed to the present state inputs of the time $t+1$ frame. Time-frame expansion helps us visualize the justification and propagation of the fault over several cycles.

Copying the combinational logic clearly illustrates how a single fault mimics multiple-fault behavior in a sequential test. Each time-frame will have its own copy of the fault. Over several cycles, the faulty gate can block its own detection or observation. Any test sequence must work around the fault under test on every cycle. Test generation programs can help create test vector suites for a machine. But, given the inherent difficulty of testing, we cannot expect miracles. Proper design is the only way to ensure that tests can be found. The chip must be designed so that all logic is made accessible enough that faults can be exercised and the results of a combinational test can be propagated to the pins. Many design-for-testability techniques take advantage of particularities of the component or system being designed; others impose a structure on the system.

Figure 5-40: A level-sensitive scan design (LSSD) system.

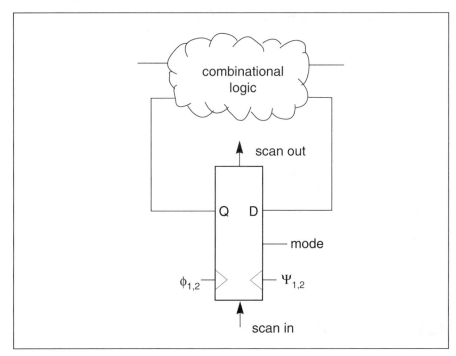

Scan design turns sequential system into combinational testing by making the present state inputs and next state outputs directly accessible. **LSSD** (level-sensitive scan design) was invented at IBM; another scan-path methodology was developed independently at NEC. An LSSD system uses special latches for memory elements, and so runs in two phases. As shown in Figure 5-40, the system has non-scan and scan modes. In non-scan mode, the latches are clocked by ϕ_1 and ϕ_2 and the system operates as any other two-phase system. In scan mode, the latches are clocked by the ψ_1 and ψ_2 clocks, and the latches work as a shift register. The latches are connected in a chain so that all the present state can be shifted out of the chip and the new state can be shifted in.

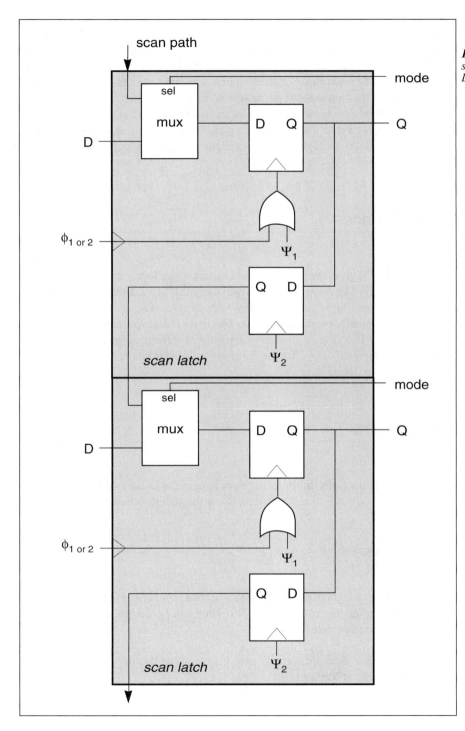

Figure 5-41: *The structure of an LSSD latch.*

Figure 5-41 illustrates the logic design of an LSSD latch [Wil83]. Each LSSD latch can function as a regular latch in non-scan mode and includes latches for both scan-mode clock phases. The memory elements used as components in the scan latch are normal, non-LSSD latches. The latch closest to the D input is shared by the normal and scan paths; the multiplexer at its input determines whether the value clocked into the latch is from the data input or the scan path. The second latch is used only for scan mode. The two latches together form a strict two-phase system in scan mode. This LSSD latch operates as a normal latch in non-scan mode and as a pair of latches in scan mode. The delay through this latch is slightly longer thanks to the multiplexer at the input.

5.8 References

Dietmeyer [Die78] gives the most comprehensive and well-structured survey of memory elements. The Mississippi State University (MSU) Cell Library [MSU89] contains combinational logic, latch, and flip-flop designs. The traffic light controller machine was originally presented in books by Unger [Ung69] and Mead and Conway [Mea80]. The "01"-string recognizer machine was developed at Stanford by Prof. Robert Mathews. The two-phase clocking discipline for latches was introduced in David Noice *et al.* [Noi82].

5.9 Problems

5-1. Animate the transistor-by-transistor operation of the D-latch of Figure 5-2: show which transistors are active and how data flows through the circuit first for CK = 1, then for CK = 0.

5-2. Draw a stick diagram for a two-input multiplexed latch. Place the two transmission gates side-by-side.

5-3. Draw a stick diagram for a latch with one data input and one reset input, where the reset input resets the output to 0. This latch can be designed as a multiplexed latch with one constant data input.

5-4. Draw a circuit diagram for a D-type master-slave flip-flop with a clear input based on the circuit of Figure 5-11.

5-5. Draw a circuit diagram for a T-type (toggle) master-slave flip-flop based on the circuit of Figure 5-12.

5-6. Redesign the hierarchical stick diagram of the counter cell to use the resettable latch.

5-7. Do the two phases of a two-phase, non-overlapping clock have to be of equal length?

5-8. What is the maximum allowable skew as predicted by the Hatamian and Cash constraints for these parameter values: $T = 10$ ns, $t_{pr} = 1$ ns, $t_{sr} = 1$ ns, $t_{sl} = 1$ ns, $t_{pl} = 5$ ns, $t_{pi} = 5$ ns. What is the minimum allowable clock period under that value of skew?

5-9. Design the system block diagram and one-bit cell for a conditional counter, which counts only when its input count = 1.

5-10. Write the state transition table for an eight-bit conditional counter. It has two inputs, count and reset (which returns the counter to the 0 count), and a three-bit binary output of the current count.

5-11. Determine the present state, next state, and output of the "01"-string recognizer machine for the input 10010101110. Assume the machine starts in state bit1.

5-12. How can you determine which wire in the standard-cell layout of the "10"-string recognizer is V_{DD} and which is V_{SS} simply by looking at the layout?

5-13. Write a state transition table for the counter of the traffic light controller, assuming that the short timeout occurs four clock cycles after the counter is reset and that the long timeout occurs several clock cycles after reset.

5-14. Consider a two-phase sequential system in which all the combinational logic is connected between the outputs of the ϕ_1 latches and the inputs of the ϕ_2 latches.

    a) Draw a two-colored block diagram of such a machine.

    b) Draw two such systems, connected so that the outputs of one feed the inputs of the next. Does this system satisfy the two-phase clocking discipline requirements? Explain.

5-15. Draw a transistor schematic for a master-slave flip-flop built from clocked inverters.

5-16. Develop a sequence of tests for the "01"-string recognizer which tests every combinational gate for both stuck-at-0 and stuck-at-1 faults.

5-17. Modify the "01"-string recognizer, adding one primary input and appropriate changes to the logic, to significantly shorten the test sequence for the machine.

5-18. Generate a set of combinational tests for the "01"-string recognizer that test for all stuck-at-0/1 faults.

5-19. Generate a set of sequential tests for the "01"-string recognizer which test for all stuck-at-0/1 faults, assuming you don't know the machine's initial state.

5-20. Design a stick diagram for the LSSD latch of Figure 5-41.

6

Subsystem Design

Highlights:

Pipelines and data paths.

Adders.

Multipliers.

Memory.
PLAs.

6.1 Introduction

Most chips are built from a collection of subsystems: adders, register files, state machines, etc. Of course, to do a good job of designing a chip, we must be able to properly design each of the major components. Studying individual subsystems is also a useful prelude to the study of complete chips because a single component is a focused design problem. When designing a complete chip, such as the HP 7200 CPU shown in Figure 6-1, we often have to perform several different types of computation, each with different cost constraints. A single component, on the other hand, performs a single task; as a result, the design choices are much more clear.

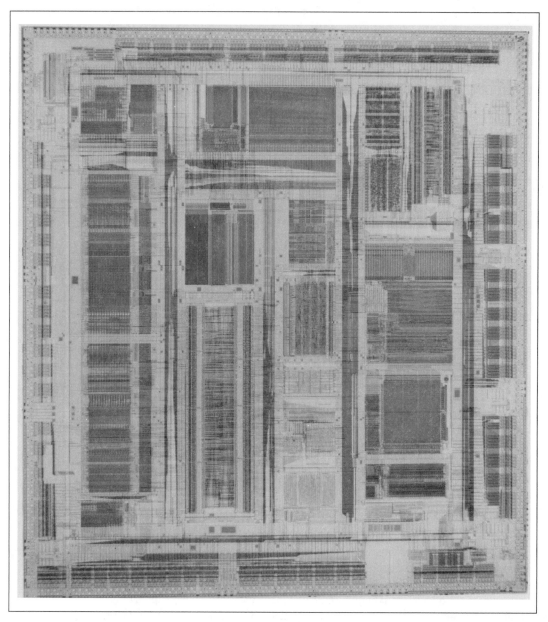

Figure 6-1: *The HP 7200 CPU. (© Copyright 1996 Hewlett-Packard Company. Reproduced with permission.)*

As always, the cost of a design is measured in area, delay, and power. For most components, we have a family of designs that all perform the same basic function, but with different area/delay trade-offs. Having access to a variety of ways to implement a function gives us architectural freedom during chip design. Area and delay costs can be reduced by optimization at each level of abstraction:

- **Layout**. We can make microscopic changes to the layout to reduce parasitics: moving wires to pack the layout or to reduce source/drain capacitance, adding vias to reduce resistance, etc. We can also make macroscopic changes by changing the placement of gates, which may reduce wire parasitics, reduce routing congestion, or both.

- **Circuit**. Transistor sizing is the first line of defense against circuits that inherently require long wires. Advanced logic circuits, such as precharged gates, may help reduce the delay within logic gates.

- **Logic**. As we saw in Chapter 3, redesigning the logic to reduce the gate depth from input to output can greatly reduce delay, though usually at the cost of area.

- **Register-transfer and above**. Proper placement of memory elements makes maximum use of the available clock period. Proper encoding of signals allows clever logic designs that minimize logic delay. Pipelining provides trade-offs between clock period and latency.

While it may be tempting to optimize a design by hacking on the layout, that is actually the least effective way to achieve your desired area/delay design point. If you choose to make a circuit faster by tweaking the layout and you fail, all that layout work must be thrown out when you try a new circuit or logic design. On the other hand, changing the register-transfer design does not preclude further work to fine-tune area or performance. The gains you can achieve by modifications within a level of abstraction grow as you move up the abstraction hierarchy: layout modifications typically give 10%–20% performance improvement; logic redesign can cut delay by more than 20%; we will see how the register-transfer modifications that transform a standard multiplier into a Booth multiplier can cut the multiplier's delay in half. Furthermore, layout improvements take the most work—speeding up a circuit usually requires modifying many rectangles, while the architecture can be easily changed by redrawing the block diagram. You are best off pursuing the largest gains first, at the highest levels of abstraction, and doing fine-tuning at lower levels of abstraction only when necessary.

Logic and circuit design are at the core of subsystem design. Many important components, foremost being the adder, have been so extensively studied that specific optimizations and trade-offs are well understood. However, there are some general principles which can be applied to subsystems. We will first survey some important design concepts, then delve into the details of several common types of subsystem-level components.

6.2 Subsystem Design Principles

6.2.1 Pipelining

Pipelining is a relatively simple method for reducing the clock period of long combinational operations. Complex combinational components can pose serious constraints on system design if their delay is much longer than the delay of the other components. If the propagation time through that combinational element determines the clock period, logic in the rest of the chip may sit idle for most of the clock cycle while the critical element finishes. Pipelining allows large combinational functions to be broken up into pieces whose delays are in balance with the rest of the system components.

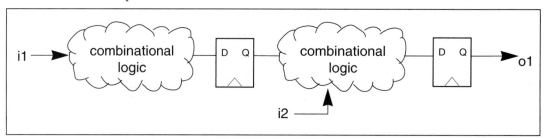

Figure 6-2: *Structure of a pipelined system.*

As illustrated in Figure 6-2, pipelining entails introducing a rank of memory elements along a cut through the combinational logic. We usually want all the outputs to appear on the same clock cycle, which implies that the cut must divide the inputs and outputs into disjoint sets. The pipelined system still computes the same combinational function but that function requires several cycles to be computed.

The number of cycles between the presentation of an input value and the appearance of its associated output is the **latency** of the pipeline. Figure 6-3 shows how clock period and latency vary with the number of stages in a pipeline. Moderate amounts of pipelining cause a great reduction in clock period, but heavy pipelining provides only

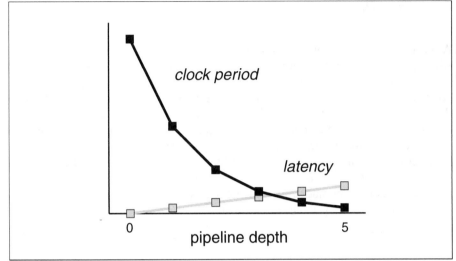

Figure 6-3: *Clock period and latency as a function of pipeline depth.*

modest, and decreasing gains. Meanwhile, latency increases linearly. The total delay through the pipelined system is slightly longer than the combinational delay thanks to the setup and hold requirements of the memory elements. But the clock period can be substantially reduced by pipelining, balancing the pipelined system's critical delay path lengths with those in the rest of the system.

When pipelining, memory elements can be placed along any cut, the best placement balances delays through the combinational logic. As shown in Figure 6-4, if memory elements are placed so that some delay paths are much longer than others, you have recreated in miniature the same conditions that caused you to pipeline the logic in the first place. Perfect pipelining balances the delays between ranks of memory elements, using the principles described in Section 5.4.

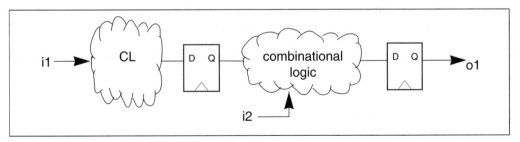

Figure 6-4: *A pipeline with unbalanced stage delays.*

6.2.2 Data Paths

A **data path** is both a logical and a physical structure: it is built from components which perform typical data operations, such as addition and it has a layout structure which takes advantage of the regular logical design of the data operators. Data paths typically include several types of components: registers (memory elements) store data; adders and ALUs perform arithmetic; shifters perform bit operations; counters may be used for program counters. A data path may include point-to-point connections between pairs of components, but the typical data path has too many connections for every component to be connected to every other component. Data is often passed between components on one or more **busses**, or common connections; the number of busses determines the maximum number of data transfers on a clock cycle and is a primary design parameter of data paths.

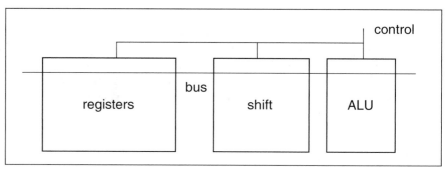

Figure 6-5:
Structure of a typi-
cal bit-slice data
path.

Most data operations are regular—we saw in the previous sections that adders, ALUs, shifters, and other operators can be constructed from arrays of smaller components. The cleanest way to take advantage of this regularity in most cases is to design the layout as a **bit-slice**, as shown in Figure 6-5. A bit-slice, as the name implies, is a one-bit version of the complete data path, and the *n*-bit data path is constructed by replicating the bit-slice. Typically, data flows horizontally through the bit-slice along point-to-point connections or busses, while control signals (which provide read and write signals to registers, opcodes for ALUs, *etc.*) flow vertically.

Bit-slice layout design requires careful, simultaneous design of the cells that comprise the data path. Since the bit-slice must be stacked vertically, any signals that pass through the cells must be tilable—the signals must be aligned at top and bottom. Horizontal constraints are often harder to satisfy. The V_{DD} and V_{SS} lines must run horizontally through the cells, as must busses. Signals between adjacent cells must also be aligned. While the vertical wires usually distribute signals, the horizontal wires are often interrupted by logic gates. The transistors in the cells impose constraints on the

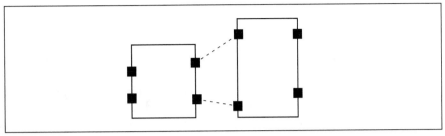

Figure 6-6: Abutting cells may require moving pins or stretching.

layout that may make it hard to place horizontal connections at the required positions. As shown in Figure 6-6, cells often need to be stretched beyond their natural heights to make connections with other cells.

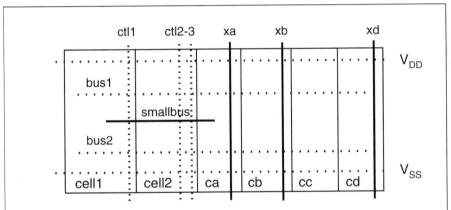

Figure 6-7: A simple wiring plan.

The data path's layout design also requires careful consideration of layer assignments. With a process that provides two levels of metal, metal 1 is typically used for horizontal wires and metal 2 for vertical wires. A **wiring plan** helps organize your thoughts on the wires required, their positions, and the best layers to use for each wire. A black-and-white wiring plan is shown in Figure 6-7; you should draw your wiring plans in color to emphasize layer choices.

Two circuit design problems unique to data path design are registers and busses. The circuit chosen for the register depends on the number of registers required. If only a few registers are needed, a standard latch or flip-flop from Section 5.2 is a perfectly reasonable choice. If many registers are required, area can be saved by using an *n*-port static RAM, which includes one row enable and one pair of bit lines for each port. Although an individual bit/word can allow only one read or write operation at a time, the RAM array can support the simultaneous, independent reading or writing of two words simply by setting the select lines of the two words high at the same time. One SRAM port is required for each bus in the data path.

A bit-slice with connections between all pairs of components would be much too large: not only would it be much too tall because of the large number of horizontal wires needed for the connections, but it would also be made longer by the many signals required to control the connections. Data paths are almost always made with busses that provide fewer connections at much less cost in area. Since the system probably doesn't need all data path components to talk to each other simultaneously, connections can be shared. But these shared connections often require special circuits that take up a small amount of space while providing adequate speed of communication. While a multiplexer provides the logical function required to control access to the bus, a mux built from static complementary gates would be much too large.

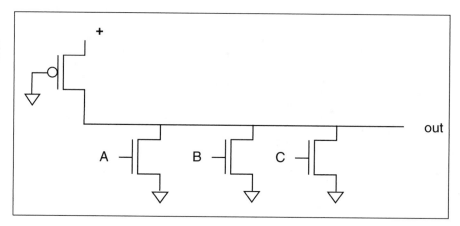

Figure 6-8: A pseudo-nMOS bus circuit.

A more clever circuit design for a bus is as a distributed NOR gate: the common wire forms the NOR gate's output, while pulldowns at the sources select the source and set the NOR gate's output. (All devices connected to the bus can read it in this scheme.) The circuit choices for busses are much like those for the advanced gate circuits of Section 3.5: pseudo-nMOS, shown in Figure 6-8, and precharged, shown in Figure 6-9. The trade-offs are also similar: the pseudo-nMOS bus is slow but does not require a separate precharge phase.

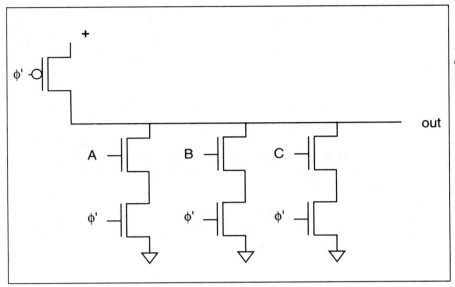

Figure 6-9: A precharged bus circuit.

6.3 Combinational Shifters

A **shifter** is most useful for arithmetic operations since shifting is equivalent to multiplication by powers of two. Shifting is necessary, for example, during floating-point arithmetic. The simplest shifter is the shift register, which can shift by one position per clock cycle. However, that machine isn't very useful for most arithmetic operations—we generally need to shift several bits in one cycle and to vary the length of the shifts.

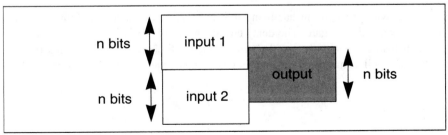

Figure 6-10: How a barrel shifter performs shifts and rotates.

A **barrel shifter** [Mea80] can perform n-bit shifts in a single combinational function, and it has a very efficient layout. It can rotate and extend signs as well. Its architecture is shown in Figure 6-10. The barrel shifter accepts $2n$ data bits and n control signals and produces n output bits. It shifts by transmitting an n-bit slice of the 2n data bits to the outputs. The position of the transmitted slice is determined by the control bits; the exact operation is determined by the values placed at the data inputs. Consider two examples:

- Send a data word d into the top input and a word of all zeroes into the bottom input. The output is a right shift (imagine standing at the output looking into the barrel shifter) with zero fill. Setting the control bits to select the top-most n bits is a shift of zero, while selecting the bottom-most n bits is an n-bit shift that pushes the entire word out of the shifter. We can shift with a ones fill by sending an all-ones word to the bottom input.

- Send the same data word into both the top and bottom inputs. The result is a rotate operation—shifting out the top bits of a word causes those bits to reappear at the bottom of the output.

How can we build a circuit that can select an arbitrary n bits and how do we do it in a reasonably sized layout? A barrel shifter with n output bits is built from a $2n$ vertical by n horizontal array of cells, each of which has a single transistor and a few wires. The schematic for a small group of contiguous cells is shown in Figure 6-11. The core of the cell is a transmission gate built from a single n-type transistor; a complementary transmission gate would require too much area for the tubs. The control lines run vertically; the input data run diagonally upward through the system; the output data run horizontally. The control line values are set so that exactly one is 1, which turns on all the transmission gates in a single column. The transmission gates connect the diagonal input wires to the horizontal output wires; when a column is turned on, all the inputs are shunted to the outputs. The length of the shift is determined by the position of the selected column—the farther to the right it is, the greater the distance the input bits have travelled upward before being shunted to the output.

Note that, while this circuit has many transmission gates, each signal must traverse only one transmission gate. The delay cost of the barrel shifter is largely determined by the parasitic capacitances on the wires, which is the reason for squeezing the size of the basic cell as much as possible. In this case, area and delay savings go hand-in-hand.

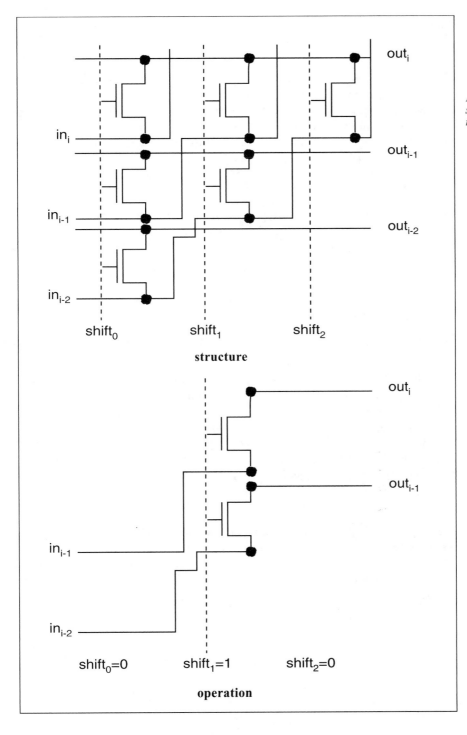

Figure 6-11: A section of the barrel shifter.

6.4 Adders

The adder is probably the most studied digital circuit. There are a great many ways to perform binary addition, each with its own area/delay trade-offs. A great many tricks have been used to speed up addition: encoding, replication of common factors, and precharging are just some of them. The origins of some of these methods are lost in the mists of antiquity. Since advanced circuits are used in conjunction with advanced logic, we need to study some higher-level addition methods before covering circuits for addition.

The basic adder is known as a **full adder**. It computes a one-bit sum and carry from two addends and a carry-in. The equations for the full adder's functions are simple:

$$s_i = a_i \oplus b_i \oplus c_i$$
$$c_{i+1} = a_i b_i + a_i c_i + b_i c_i$$

(EQ 6-1)

In these formulas, s_i is the sum at the ith stage and c_{i+1} is the carry out of the ith stage.

The n-bit adder built from n one-bit full adders is known as a **ripple-carry adder** because of the way the carry is computed. The addition is not complete until the $n\text{-}1^{th}$ adder has computed its s_{n-1} output; that result depends on c_i input, and so on down the line, so the critical delay path goes from the 0-bit inputs up through the c_i's to the n-1 bit. (We can find the critical path through the n-bit adder without knowing the exact logic in the full adder because the delay through the n-bit carry chain is so much longer than the delay from a and b to s.) The ripple-carry adder is area efficient and easy to design but is slow when n is large.

Speeding up the adder requires speeding up the carry chain. The **carry-lookahead adder** is one way to speed up the carry computation. The carry-lookahead adder breaks the carry computation into two steps, starting with the computation of two intermediate values. The adder inputs are once again the a_i's and b_i's; from these inputs, P (propagate) and G (generate) are computed:

$$P_i = a_i + b_i$$
$$G_i = a_i \cdot b_i$$

(EQ 6-2)

If $G_i = 1$, there is definitely a carry out of the ith bit of the sum—a carry is generated. If $P_i = 1$, then the carry from the i-1th bit is propagated to the next bit. The sum and carry equation for the full adder can be rewritten in terms of P and G:

$$s_i = c_i \oplus P_i \oplus G_i$$
$$c_{i+1} = G_i + P_i c_i$$

(EQ 6-3)

The carry formula is smaller when written in terms of P and G, and therefore easier to recursively expand:

$$
\begin{aligned}
c_{i+1} &= G_i + P_i \cdot (G_{i-1} + P_{i-1} \cdot c_{i-1}) \\
&= G_i + P_i G_{i-1} + P_i P_{i-1} \cdot (G_{i-2} + P_{i-2} \cdot c_{i-2}) \\
&= G_i + P_i G_{i-1} + P_i P_{i-1} G_{i-2} + P_i P_{i-1} P_{i-2} c_{i-2}
\end{aligned}
$$

(EQ 6-4)

The c_{i+1} formula of Equation 6-4 depends on c_{i-2}, but not c_i or c_{i-1}. After rewriting the formula to eliminate c_i and c_{i-1}, we used the speedup trick of Section 4.4—we eliminated parentheses, which substitutes larger gates for long chains of gates. There is a limit beyond which the larger gates are slower than chains of smaller gates; typically, four levels of carry can be usefully expanded.

A depth-4 carry-lookahead unit is shown in Figure 6-12. The unit takes the P and G values from its four associated adders and computes four carry values. Each carry output is computed by its own logic. The logic for c_{i+3} is slower than that for c_i, but the flattened c_{i+3} logic is faster than the equivalent ripple-carry logic.

There are two ways to hook together depth-b carry-lookahead units to build an n-bit adder. The carry-lookahead units can be recursively connected to form a tree: each unit generates its own P and G values, which are used to feed the carry-lookahead unit at the next level of the tree. A simpler scheme is to connect the carry-ins and carry-outs of the units in a ripple chain. This approach is most common in chip design because the wiring for the carry-lookahead tree is hard to design and area-consuming.

The **carry-skip adder** [Leh61] looks for cases in which the carry out of a set of bits is the same as the carry in to those bits. This adder makes a different use of the carry-propagate relationship. A carry-skip adder is typically organized into m-bit *groups*; if the carry is propagated for every bit in the stage, then a bypass gate sends the stage's carry input directly to the carry output. The structure of the carry chain for a carry-

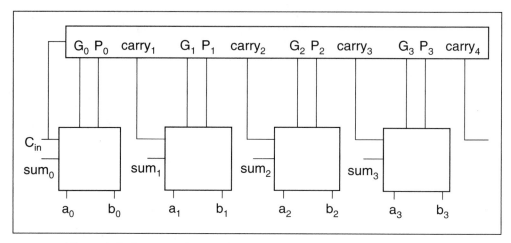

Figure 6-12: *Structure of a carry lookahead adder.*

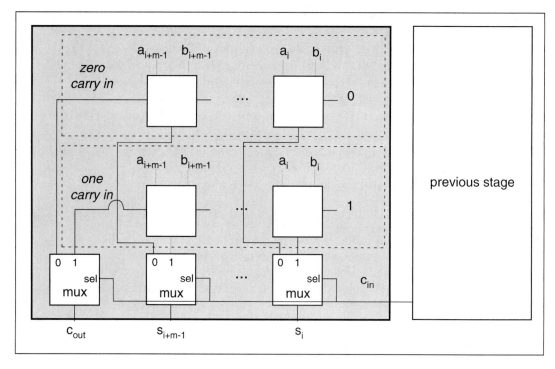

Figure 6-13: *The carry-select adder.*

skip adder divided into groups of bits is shown in Figure 6-13. A true carry into the group and true propagate condition P at every bit in the group is needed to cause the carry to skip. It is possible to determine the optimum number of bits in a group [Kor93].The worst case for the carry signal occurs when there is a carry propagated through every bit, but in this case P_i will be true at every bit. Therefore, the longest path for the carry begins when the carry is generated at the bottom bit of the bottom group (rippling through the remainder of the group), is skipped past the intermediate groups, and ripples through the last group; the carry must necessity ripple through the first and last groups to compute the sum. Using some simple assumptions about the relative delays for a ripple through a group and skip, Koren estimates the optimum group size for an n-bit word as

$$k_{opt} = \sqrt{n/2} .$$
(EQ 6-5)

Since the carry must ripple through the first and last stages, the adder can be further speeded up by making the first and last groups shorter than this length and by lengthening the middle groups.

The **carry-select adder** computes two versions of the addition with different carry-ins, then selects the right one. Its structure is shown in Figure 6-13. As with the carry-skip adder, the carry-select adder is typically divided into m-bit stages. The second stage computes two values: one assuming that the carry-in is 0 and another assuming that it is 1. Each of these candidate results can be computed by your favorite adder structure. The carry-out of the previous stage is used to select which version is correct: multiplexers controlled by the previous stage's carry-out choose the correct sum and carry-out. This scheme speeds up the addition because the ith stage can be computing the two versions of the sum in parallel with the i-1th's computation of its carry. Once the carry-out is available, the ith stage's delay is limited to that of a two-input multiplexer.

No matter what style of circuit is used for the carry chain, we can size transistors to reduce the carry chain delay. By increasing widths of transistors on critical delay paths, we can steadily reduce the carry delay to the ring oscillator limit [Fis92]. In a carry-lookahead adder, the carry-in signal is of medium complexity; depending on the exact logic implementation, either P or G will be most critical, with the other signal being less critical than carry-in. At the adder's high-order bits, the non-carry outputs of the carry chain also become critical, requiring larger transistors; otherwise, the stages of an optimally-sized adder are identical.

Precharged circuits are an obvious way to speed up the carry chain. One of the most interesting precharged adders is the **Manchester carry chain** [Mea80], which computes the sum from P and G. Two bits of a Manchester carry chain are shown in Figure 6-14. The storage node, which holds the complement of the carry (c_i'), is charged to 1 during the precharge phase. If $G_i = 1$ during the evaluate phase, the storage node is discharged, producing a carry into the next stage. If $P_i = 1$, then the ith storage node is connected to the $i\text{-}1^{th}$ storage node; in this case, the ith storage node can be discharged by the P_{i-1} pulldown or, if the C_{i-1} transmission gate is on, by a preceding P pulldown. In the worst case, the $n\text{-}1^{th}$ storage node will be discharged through the 0th pulldown and through n transmission gates, but in the typical case, the pulldown path for a storage node is much shorter. The widest transistors should be at the least-signif-

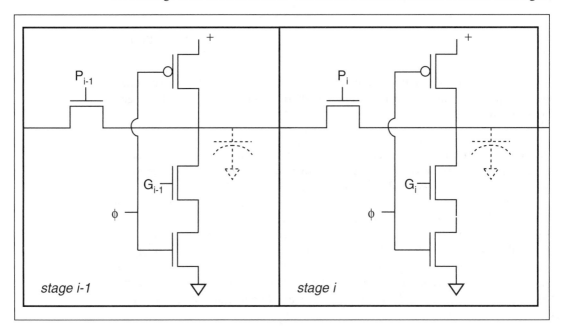

Figure 6-14: *A Manchester carry chain.*

icant bit stage since they see the largest load. The next example describes the use of a Manchester carry chain in a high-performance microprocessor.

Example 6-1: *Carry chain of the DEC Alpha 21064 microprocessor*

The first DEC Alpha microprocessor, the 21064, used a Manchester carry chain [Dob92]. That processor had a 64-bit word with the carry chain organized into groups of 8 bits to enable byte-wise operation. The carry circuit at each stage was made entirely of n-type devices:

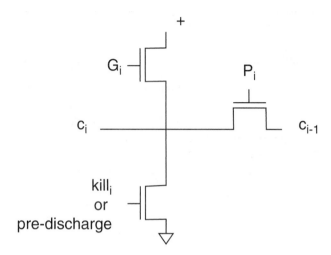

The carry chain uses carry propagate, generate, and kill signals. The chain is pre-discharged and then selectively charged using the n-type pullups. Pre-discharging was used to avoid the threshold drop on the carry signal introduced by precharging through an n-type device; precharging to V_{DD}-V_t was deemed to provide unacceptable noise margins. The pass transistors for the carry were sized with the largest transistors at the least-significant bit.

Each group of eight bits was organized into a 32-bit carry-select, and a logarithmic carry-select technique was used to generate the 64-bit sum. As a result, there were two carry chains in each 8-bit group: one for the zero-carry-in case and another for the one-carry-in case.

Serial adders present an entirely different approach to high-speed arithmetic—they require many clock cycles to add two n-bit numbers, but with a very short cycle time. Serial adders can work on nybbles (4-bit words) or bytes, but the bit-serial adder [Den85], shown in Figure 6-15 is the most extreme form. The data stream consists of three signals: the two numbers to be added and an LSB signal that is high when the current data bits are the least significant bits of the addends. The addends appear LSB first and can be of arbitrary length—the end of a pair of numbers is signaled by the

LSB bit for the next pair. The adder itself is simply a full adder with a memory element for the carry. The LSB signal clears the carry register. Subsequently, the two input bits are added with the carry-out of the last bit. The serial adder is small and has a cycle time equal to that of a single full adder.

*Figure 6-15: A
bit-serial adder.*

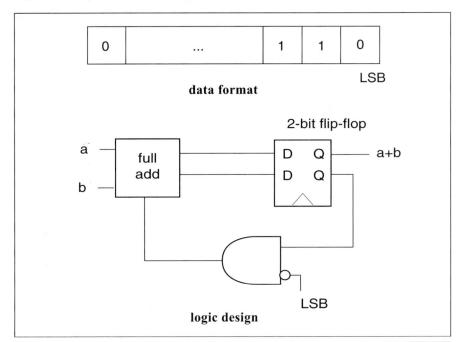

Table 6-1 Average
power consumption
of 16-bit adders
calculated from
circuit simulation.
*(After Callahan and
Schwartzlander.
Copyright © 1996
Kluwer Academic
Publishers. Used
with permission.)*

adder type	power (mW)
ripple-carry	0.117
constant width carry-skip	0.109
carry-lookahead	0.171
carry-select	0.216

Callaway and Schwartzlander [Cal96] evaluated the power consumption of several types of parallel adders. Their results from circuit simulation of 16-bit adders are summarized in Table 6-1. In general, slower adders consume less power. The exception is the carry-skip adder, which uses less current than the ripple-carry adder because, although it consumes a higher peak current, the current falls to zero very quickly. They measured the average power consumption of fabricated circuits and showed that these estimates were well-correlated with the behavior of actual circuits.

6.5 ALUs

The **arithmetic logic unit**, or **ALU**, is a modified adder. While an ALU can perform both arithmetic and bit-wise logical operations, the arithmetic operations' requirements dominate the design.

A basic ALU takes two data inputs and a set of control signals, also called an **opcode**. The opcode, together with the ALU's carry-in, determine the ALU's function. For example, if the ALU is set to add, then $c_0 = 0$ produces $a+b$ while $c_0 = 1$ produces $a+b+1$.

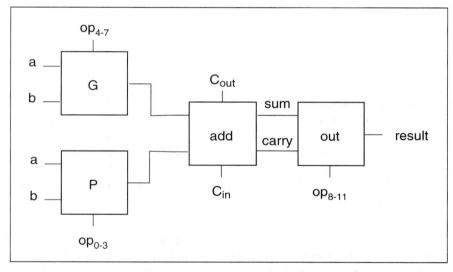

Figure 6-16: A three function block ALU.

The logic to compute all possible ALU functions can be large unless it is carefully designed. The ALU presents an ideal opportunity to use transmission gate logic by building a three function block ALU [Mea80], shown in Figure 6-16. The function block is shown in Figure 6-17; it takes two data inputs and their complements along with four control signals and can compute all 16 possible functions of the two data inputs.

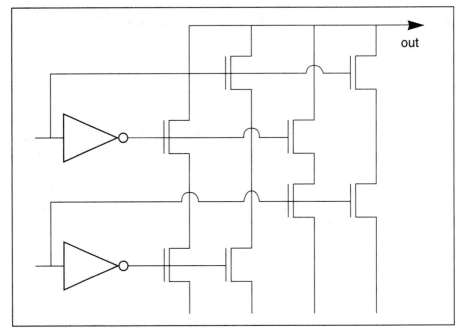

Figure 6-17: A two-input function block.

The ALU is built around an adder. The three function blocks require a total of 12 opcode bits to determine the ALU's function. The two function blocks at the inputs compute signals used to drive the adder's P and G inputs, but those signals are not necessarily the propagate and generate functions defined for carry-lookahead addition—they are signals required to drive the adder block to compute the required values. The ALU's output is computed by the final function block from the adder's carry and the propagate signal.

Not all ALUs must implement the full set of logical functions. If the ALU need implement only a few functions, the function block scheme may be overkill. An ALU which, for example, must implement only addition, subtraction, and one or two bitwise functions can usually be implemented using static gates.

6.6 Multipliers

Multiplier design starts with the elementary school algorithm for multiplication. Consider the simple example of Figure 6-18. At each step, we multiply one digit of the multiplier by the full multiplicand; we add the result, shifted by the proper number of bits, to the partial product. When we run out of multiplier digits, we are done. Single-digit multiplication is easy for binary numbers—binary multiplication of two bits is performed by the AND function. The computation of partial products and their accu-

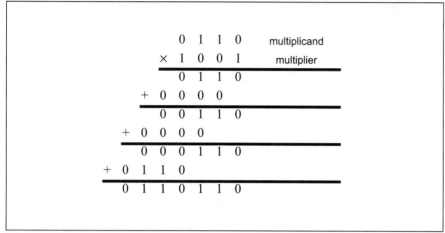

Figure 6-18:
Multiplication using the elementary school algorithm.

mulation into the complete product can be optimized in many ways, but an understanding of the basic steps in multiplication is important to a full appreciation of those improvements.

One simple, small way to implement the basic multiplication algorithm is the **serial-parallel** multiplier of Figure 6-19, so called because the n-bit multiplier is fed in serially while the m-bit multiplicand is held in parallel during the course of the multiplication. The multiplier is fed in least-significant bit first and is followed by at least m zeroes. The result appears serially at the end of the multiplier chain. A one-bit multiplier is simply an AND gate. The sum units in the multiplier include a combinational full adder and a register to hold the carry. The chain of summation units and registers performs the shift-and-add operation—the partial product is held in the shift register chain, while the multiplicand is successively added into the partial product.

One important complication in the development of efficient multiplier implementations is the multiplication of two's-complement signed numbers. The **Baugh-Wooley multiplier** [Bau73] is the best-known algorithm for signed multiplication because it maximizes the regularity of the multiplier logic and allows all the partial products to have positive sign bits. The multiplier X can be written in binary as

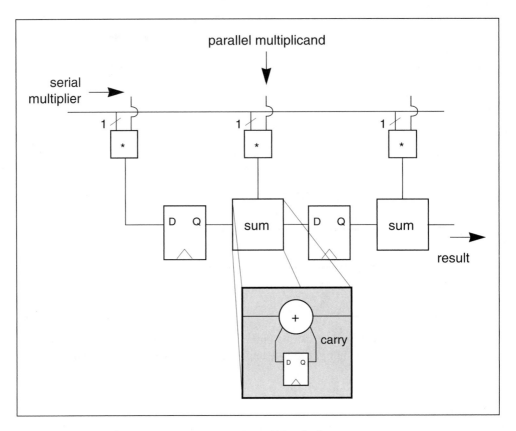

Figure 6-19: *Basic structure of a serial-parallel multiplier.*

$$X = x_{n-1}2^{n-1} + \sum_{i=0}^{n-2} x_i 2^i, \qquad \text{(EQ 6-6)}$$

where n is the number of bits in the representation. The multiplicand Y can be written similarly. The product P can be written as

$$P = p_{2n-1}2^{2n-2} + \sum_{i=0}^{2n-2} p_i 2^i. \qquad \text{(EQ 6-7)}$$

When this formula is expanded to show the partial products, it can be seen that some of the partial products have negative signs:

$$P =$$ (EQ 6-8)

$$\left[x_{n-1}y_{n-1}2^{2n-2} + \sum_{i}^{n-2} \sum_{j}^{n-2} x_i y_j 2^{i+j} \right] - \left[\sum^{n-2} (x_{n-1}y_i + y_{n-1}x_i)2^{n-1+i} \right]$$

The formula can be further rewritten, however, to move the negative-signed partial products to the last steps and to add the negation of the partial product rather than subtract. Further rewriting gives this final form:

$$P = 2^{n-1}\left(-2^n + 2^{n-1} + \bar{x}_{n-1}2^{n-1} + x_{n-1} + \sum_{i=0}^{n-2} x_{n-1}\bar{y}_i 2^i \right).$$ (EQ 6-9)

Each partial product is formed with AND functions and the partial products are all added together. The result is to push the irregularities to the end of the multiplication process and allow the early steps in the multiplication to be performed by identical stages of logic.

The elementary school multiplication algorithm (and the Baugh-Wooley variations for signed multiplication) suggest a logic and layout structure for a multiplier which is surprisingly well-suited to VLSI implementation—the **array multiplier**. The structure of an array multiplier for unsigned numbers is shown in Figure 6-20. The logic structure is shown in parallelogram form both to simplify the drawing of wires between stages and also to emphasize the relationship between the array and the basic multiplication steps shown in Figure 6-18. As when multiplying by hand, partial products are formed in rows and accumulated in columns, with partial products shifted by the appropriate amount. In layout, however, the y bits generally would be distributed with horizontal wires since each row uses exactly one y bit.

Notice that only the last adder in the array multiplier has a carry chain. The earlier additions are performed by full adders which are used to reduce three one-bit inputs to two one-bit outputs. Only in the last stage are all the values accumulated with carries. As a result, relatively simple adders can be used for the early stages, with a faster (and presumably larger and more power-hungry) adder reserved for the last stage. As a result, the critical delay path for the array multiplier follows the trajectory shown in Figure 6-21.

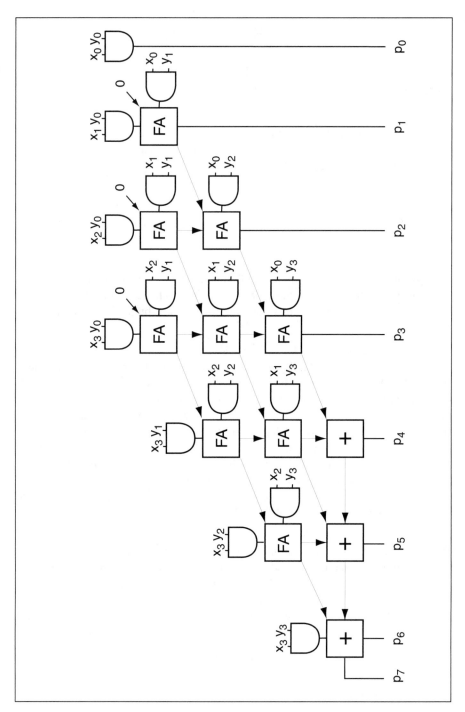

Figure 6-20:
*Structure of an
unsigned array
multiplier.*

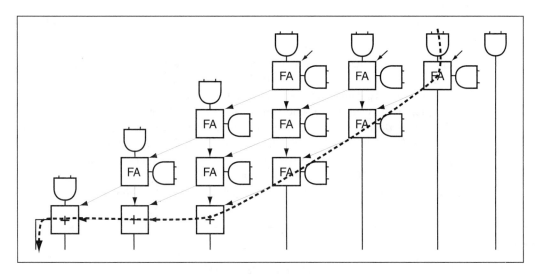

Figure 6-21: *The critical delay path in the array multiplier.*

Pipelining reduces cycle time but doesn't reduce the total time required for multiplication. One way to speed up multiplication is **Booth encoding** [Boo51], which performs several steps of the multiplication at once. Booth's algorithm takes advantage of the fact that an adder-subtractor is nearly as fast and small as a simple adder. In the elementary school algorithm, we shift the multiplicand x, then use one bit of the multiplier y if that shifted value is to be added into the partial product. The most common form of Booth's algorithm looks at three bits of the multiplier at a time to perform two stages of the multiplication.

Consider once again the two's-complement representation of the multiplier y:

$$y = (-2)^n y_n + 2^{n-1} y_{n-1} + 2^{n-2} y_{n-2} + \dots \qquad \textbf{(EQ 6-10)}$$

We can take advantage of the fact that $2^a = 2^{a+1} - 2^a$ to rewrite this as

$$y = 2^n (y_{n-1} - y_n) + 2^{n-1} (y_{n-2} - y_{n-1}) + 2^{n-2} (y_{n-3} - y_{n-2}) + \dots . \qquad \textbf{(EQ 6-11)}$$

Now, extract the first two terms:

$$2^n(y_{n-1} - y_n) + 2^{n-1}(y_{n-2} - y_{n-1}).$$ **(EQ 6-12)**

Each term contributes to one step of the elementary-school algorithm: the right-hand term can be used to add x to the partial product, while the left-hand term can add $2x$. (In fact, since y_{n-2} also appears in another term, no pair of terms exactly corresponds to a step in the elementary school algorithm. But, if we assume that the y bits to the right of the decimal point are 0, all the required terms are included in the multiplication.) If, for example, $y_{n-1} = y_n$, the left-hand term does not contribute to the partial product. By picking three bits of y at a time, we can determine whether to add or subtract x or $2x$ (shifted by the proper amount, two bits per step) to the partial product. Each three-bit value overlaps with its neighbors by one bit. Table 6-2 shows the contributing term for each three-bit code from y.

Table 6-2 *Action
s during Booth
multiplication.*

$y_i\ y_{i-1}\ y_{i-2}$	increment
0 0 0	0
0 0 1	x
0 1 0	x
0 1 1	2x
1 0 0	-2x
1 0 1	-x
1 1 0	-x
1 1 1	0

Let's try an example to see how this works: $x = 011001$ (25_{10}), $y = 101110$ (-18_{10}). Call the ith partial product P_i. At the start, $P_0 = 00000000000$ (two six-bit numbers give an 11-bit result):

1. $y_1 y_0 y_{-1} = 100$, so $P_1 = P_0 - (10 \cdot 011001) = 11111001110$.
2. $y_3 y_2 y_1 = 111$, so $P_2 = P_1 + 0 = 11111001110$.
3. $y_5 y_4 y_3 = 101$, so $P_3 = P_2 - 0110010000 = 11000111110$.

In decimal, $y_1 y_0 y_{-1}$ contribute $-2x \cdot 1$, $y_3 y_2 y_1$ contribute $0 \cdot 4$, and $y_5 y_4 y_3$ contribute $-x \cdot 16$, giving a total of $-18x$. Since the multiplier is -18, the result is correct.

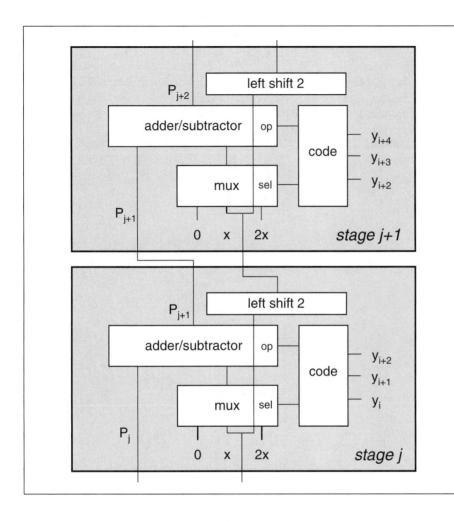

Figure 6-22:
Structure of a Booth multiplier.

Figure 6-22 shows the detailed structure of a Booth multiplier. The multiplier bits control a multiplexer which determines what is added to or subtracted from the partial product. Booth's algorithm can be implemented in an array multiplier since the accumulation of partial products still forms the basic trapezoidal structure. In this case, a column of control bit generators on one side of the array analyzes the triplets of y bits to determine the operation in that row.

Another way to speed up multiplication is to use more adders to speed the accumulation of partial products. The best-known method for speeding up the accumulation is the **Wallace tree** [Wal64], which is an adder tree built from **carry-save adders**, which is simply an array of full adders whose carry signals are not connected, as in the early stages of the array multiplier. A carry save adder, given three n-bit numbers a, b, c, computes two new numbers y, z such that $y + z = a + b + c$. The Wallace tree performs the three-to-two reductions; at each level of the tree, i numbers are combined to form $\lceil 2i/3 \rceil$ sums. When only two values are left, they are added with a high-speed adder. Figure 6-23 shows the structure of a Wallace tree. The partial products are introduced at the bottom of the tree. Each of the z outputs is shifted left by one bit since it represents the carry out.

Figure 6-23: Structure of a Wallace tree.

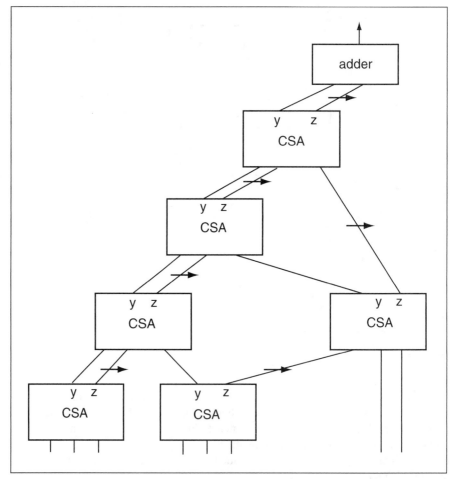

A Wallace tree multiplier is considerably faster than a simple array multiplier because its height is logarithmic in the word size, not linear. However, in addition to the larger number of adders required, the Wallace tree's wiring is much less regular and more complicated. As a result, Wallace trees are often avoided by designers who do not have extreme demands for multiplication speed and for whom design complexity is a consideration.

Callaway and Schwartzlander [Cal96] also evaluated the power consumption of multipliers. They compared an array multiplier and a Wallace tree multiplier (both without Booth encoding) and found that the Wallace tree multiplier used significantly less power for bit widths between 8 and 32, with the advantage of the Wallace tree growing as word length increased.

6.7 High-Density Memory

So far, we have built memory elements out of circuits that exhibit mostly digital behavior. By taking advantage of analog design methods, we can build memories that are both smaller and faster. Memory design is usually best left to expert circuit designers; however, understanding how these memories work will help you learn how to use them in system design. On-chip memory is becoming increasingly important as levels of integration increase to allow both processors and useful amounts of memory to be integrated on a single chip [Kog95].

Read-only memory (ROM), as the name implies, can be read but not written. It is used to store data or program values that will not change, because it is the densest form of memory available. An increasing number of digital logic processes support **flash memory**, which is the dominant form of electrically erasable PROM memory. There are two types of read-write **random access memories**: **static** (SRAM) and **dynamic** (DRAM). SRAM and DRAM use different circuits, each of which has its own advantages: SRAM is faster but uses more power and is larger; DRAM has a smaller layout and uses less power. DRAM cells are also somewhat slower and require the dynamically stored values to be periodically refreshed, just as in a dynamic latch.

Some types of memory are available for integration on a chip, while others that require special processes are generally used as separate chips. Commodity DRAMs are based on a one-transistor memory cell. That cell requires specialized structures, such as poly-poly capacitors, which are built using special processing steps not usually included in ASIC processes. A design that requires high-density ROM or RAM is usually partitioned into several chips, using commodity memory parts. Medium density memory, on the order of one K bytes, can often be put on the same chip with the logic that uses it, giving faster access times, as well as greater integration.

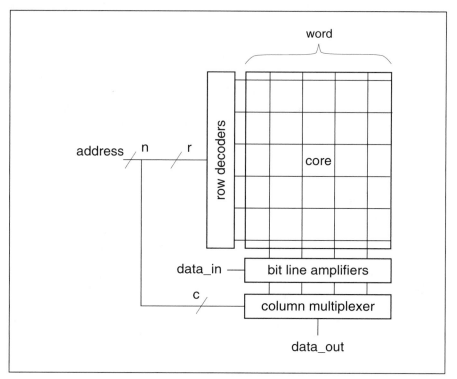

A RAM or ROM is used by presenting it with an address and receiving the value stored at that address some time later. Details differ, of course: large memories often divide the address into row and column sections, which must be sent to the memory separately, for example. The simplest and safest way to use memory in a system is to treat it as a strict sequential component: send the address to the memory on one cycle and read the value on the next cycle.

The architecture of a generic RAM/ROM system is shown in Figure 6-24. Think of the data stored in the memory core as being organized into n bit-wide words. The address decoders (also known as row decoders) translate the binary address into a unary address—exactly one word in the core is selected. A read is performed by having each cell in the selected word set the bit and bit' lines to the proper values: bit = 1, bit' = 0 if a 1 is stored in the array, for example. The bit lines are typically precharged, so the cell discharges one of the lines. The bit lines are read by circuits that sense the value on the line, amplify it to speed it up, and restore the signal to the proper voltage levels. A write is performed by setting the bit lines to the desired values and driving that value from the bit lines into the cell. If the core word width is narrower than the final word width (for example, a one-bit wide RAM typically has a core much wider than one bit to make a more compact layout), a multiplexer uses the bottom few bits of the address to select the desired bits out of the word.

The row decoders are not very complex: they typically use NOR gates to decode the address, followed by a chain of buffers to allow the circuit to drive the large capacitance of the word line. There are two major choices for circuits to implement the NOR function: pseudo-nMOS and precharged. Pseudo-nMOS circuits are adequate for small memories, but precharged circuits offer better performance (at the expense of control circuitry) for larger memory arrays. Figure 6-25 shows a precharged row decoder; the true and complement forms of the address lines can be distributed vertically through the decoders and connected to the NOR pulldowns as appropriate for the address to be decoded at that row.

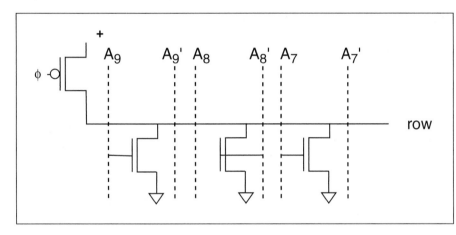

Figure 6-25: A precharged row decoder.

The column decoders are typically implemented as pass transistors on the bit lines. As shown in Figure 6-26, each output bit will be selected from several columns. The multiplexer control signals can be generated at one end of the string of multiplexers and distributed to all the mux cells.

6.7.1 ROM

A read-only memory is programmed with transistors to supply the desired values. A common circuit is the NOR array shown in Figure 6-27. It uses a pseudo-nMOS NOR gate: a transistor is placed at the word-bit line intersection for which bit' = 0. Programmable ROMs (PROMs) are also available. The simplest PROM uses a high voltage to blow selected fuses in the array; this type of programming cannot be undone.

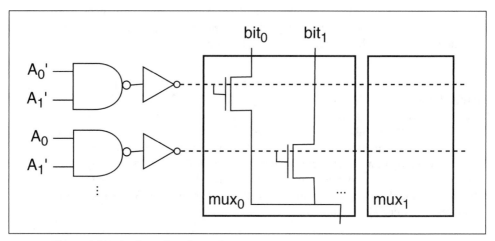

Figure 6-26: *A column decoding scheme.*

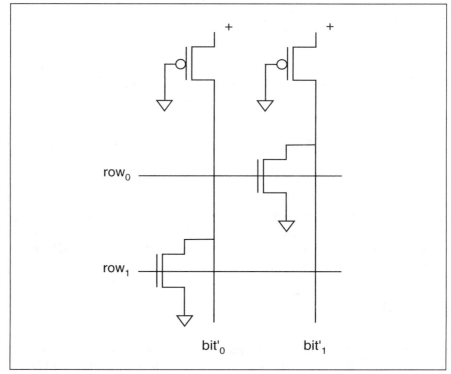

Figure 6-27:
Design of a ROM core.

Electrically or UV-erasable ROMs allow the ROM to be erased, then reprogrammed many times. However, both programmable and erasable memories usually require special processing steps which are not often available to the average VLSI designer.

6.7.2 Static RAM

Basic static RAM circuits can be viewed as variations on the designs used for latches and flip-flops; more aggressive static RAMs make use of design tricks originally developed for dynamic RAMs to speed up the system.

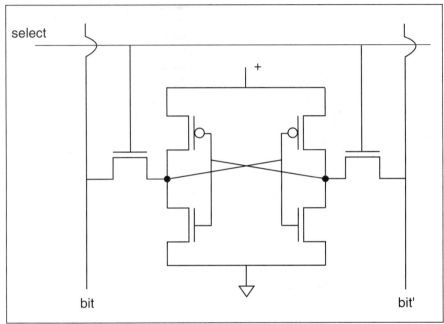

Figure 6-28:
*Design of an
SRAM core cell.*

The SRAM core circuit is shown in Figure 6-28. The value is stored in the middle four transistors, which form a pair of inverters connected in a loop (try drawing a gate-level version of this schematic). The other two transistors control access to the memory cell by the bit lines. When **select** = 0, the inverters reinforce each other to store the value. A read or write is performed when the cell is selected:

- To read, bit and bit' are precharged to V_{DD} before the **select** line is allowed to go high. One of the cell's inverters will have its output at 1, and the other at 0; which inverter is 1 depends on the value stored. If, for example, the right-hand inverter's output is 0, the bit' line will be drained to V_{SS} through that inverter's pulldown and the bit line will remain high. If the opposite value is stored in the cell, the bit line will be pulled low while bit' remains high.

- To write, the bit and bit' lines are set to the desired values, then **select** is set to 1. Charge sharing forces the inverters to switch values, if necessary, to store the desired value. The bit lines have much higher capacitance than the inverters, so the charge on the bit lines is enough to overwhelm the inverter pair and cause it to flip state.

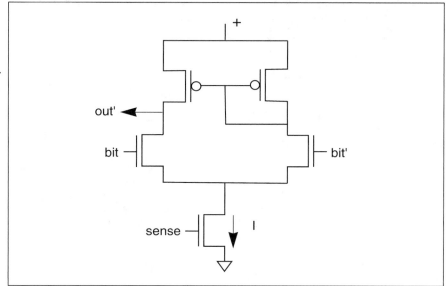

Figure 6-29: A differential pair sense amplifier for an SRAM.

The layout of a pair of SRAM cells in the SCMOS rules is shown in Figure 6-31.

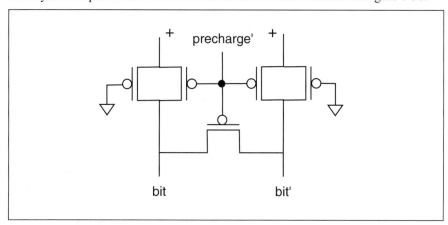

Figure 6-30: An SRAM pre-charge circuit.

A **sense amplifier**, shown in Figure 6-29, makes a reasonable bit line receiver for modest-size SRAMs. The n-type transistor at the bottom acts as a switchable current source—when turned on by the *sense* input, the transistor pulls a fixed current *I* through the sense amp's two arms. Kirchoff's current law tells us that the currents through the two branches must sum to *I*. When one of the bit lines goes low, the current through that leg of the amplifier goes low, increasing the current in the other leg. P-type transistors are used as loads. For an output of the opposite polarity, both the output and the pullup bias connection must be switched to the opposite sides of the circuit. More complex circuits can determine the bit line value more quickly [Gla85].

A precharging circuit for the bit lines is shown in Figure 6-30. Precharging is controlled by a single line. The major novelty of this circuit is the transistor between the bit and bit' lines, which is used to equalize the charge on the two lines.

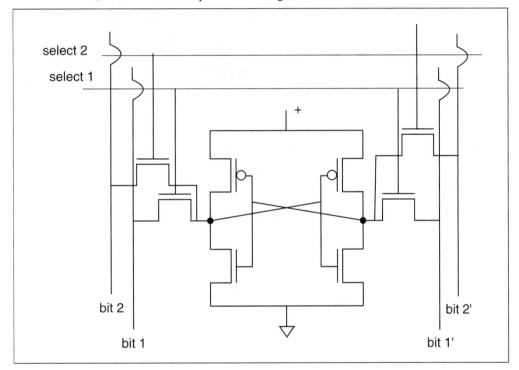

Figure 6-32: *A dual-ported SRAM core cell.*

Many designs require multi-ported RAMs. For example, a register file is often implemented as a multi-port SRAM. Each port consists of address input, data outputs, and select and read/write lines. When select is asserted on the ith port, the ith address is used to read or write the addressed cells using the ith set of data lines. Reads and writes on separate ports are independent, although the effect of simultaneous writes to

Figure 6-31:
Layout of a pair of SRAM core cells.

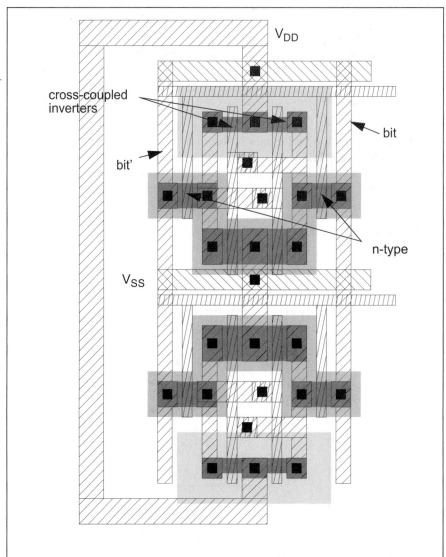

a port are undefined. The circuit schematic for a two-port SRAM core cell is shown in Figure 6-32. Each port has its own pair of access transistors. The transistors in the cross-coupled inverters must be resized moderately to ensure that multiple port activations do not affect the stored value, but the circuit and layout designs do not differ radically from the single-ported cell.

6.7.3 The Three-Transistor Dynamic RAM

The simplest dynamic RAM cell uses a three-transistor circuit [Reg70]. This circuit is fairly large and slow. It is sometimes used in ASICs because it is denser than SRAM and, unlike one-transistor DRAM, does not require special processing steps.

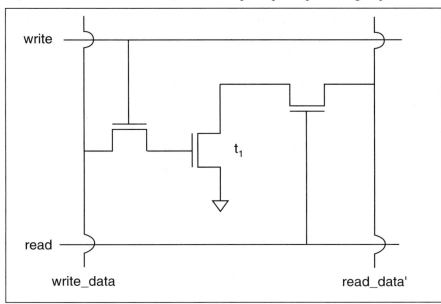

Figure 6-33:
Design of a three-transistor DRAM core cell.

The three-transistor DRAM circuit is shown in Figure 6-33. The value is stored on the gate capacitance of t_1; the other two transistors are used to control access to that value:

- To read, read_data' is precharged to V_{DD}. We then set read to 1 and write to 0. If t_1's gate has a stored charge, then t_1 will pull down the read_data' signal, else read_data' will remain charged. Read_data', therefore, carries the complement of the value stored on t_1.

- To write, the value to be written is set on write_data, write is set to 1, and read to 0. Charge sharing between write_data and t_1's gate capacitance forces t_1 to the desired value.

Substrate leakage will cause the value in this cell to decay. The value must be **refreshed** periodically—a refresh interval of 1 ms is consistent with the approximate leakage rate of typical processes. The value is refreshed by rewriting it into the cell, being careful of course to rewrite the original value.

6.7.4 The One-Transistor Dynamic RAM

The one-transistor DRAM circuit quickly supplanted the three-transistor circuit because it could be packed more densely, particularly when advanced processing techniques are used. The term one-transistor is somewhat of a misnomer—a more accurate description would be one-transistor/one-capacitor DRAM, since the charge is stored on a pure capacitor rather than on the gate capacitance of a transistor. The design of one-transistor DRAMs is an art beyond the scope of this book. But since embedded DRAM is becoming more popular, it is increasingly likely that designers will build chips with one-transistor DRAM subsystems, so it is useful to understand the basics of this memory circuit.

Figure 6-34:
Circuit dia-
gram for a one-
transistor
DRAM core
cell.

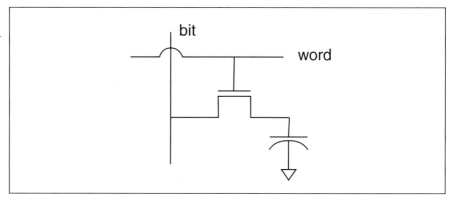

Figure 6-34 shows the circuit diagram of a one-transistor DRAM core cell. The cell has two external connections: a bit line and a word line. The value is stored on a capacitor guarded by a single transistor. Setting the word line high connects the capacitor to the bit line. To write a new value, the bit line is set accordingly and the capacitor is forced to the proper value. When reading the value, the bit line is first precharged before the word line is activated. If the storage capacitor is discharged, then charge will flow from the bit line to the capacitor, lowering the voltage on the bit line. A sense amp can be used to detect the dip in voltage; since the bit line provides only a single-ended input to the bit line, a reference voltage may be used as the sense amp's other input. One common way to generate the reference voltage is to introduce dummy cells which are precharged but not read or written. This read is destructive—the zero on the capacitor has been replaced by a one during reading. As a result, additional circuitry must be placed on the bit lines to pull the bit line and storage capacitor to zero when a low voltage is detected on the bit line. This cell's value must also be

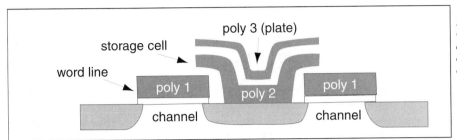

Figure 6-35:
*Cross-section of
a pair of stacked-
capacitor DRAM
cells.*

refreshed periodically, but it can be refreshed by reading the cell.

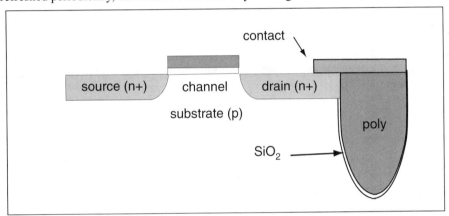

Figure 6-36:
*Cross section of
a one-transis-
tor DRAM cell
built with a
trench capaci-
tor.*

Modern DRAMs are designed with three-dimensional structures to minimize the size of the storage cell. The two major techniques for DRAM fabrication are the **stacked capacitor** and the **trench capacitor**. The cross-section of a pair of stacked capacitor cells is shown in Figure 6-35 [Tak85]. The cell uses three layers of polysilicon and one level of metal: the word line is fabricated in poly 1, the bottom of the capacitor in poly 2, and the top plate of the capacitor in poly 3. The bit line is run in metal above the capacitor structures. The capacitor actually wraps around the access transistor, packing a larger parallel plate area in a smaller surface area. The bottom edge of the bottom plate makes the contact with the access transistor, saving additional area. The trench capacitor cell cross-section is shown in Figure 6-36 [Sun84]. A trench is etched into the chip, oxide is formed, and the trench is filled with polysilicon. This structure automatically connects the bottom plate to the grounded substrate; a contact is used to directly connect the polysilicon plate to the access transistor.

One should not expect one-transistor DRAMs that can be fabricated on a logic process to be equivalent to commodity DRAMs. The processing steps required to create a dense array of capacitors are not ideal for efficient logic transistors, so high-density DRAMs generally have lower-quality transistors. Since transistors make up a relatively small fraction of the circuitry in a commodity DRAM, those chips are optimized for the capacitors. However, in a process designed to implement large amounts of logic with some embedded DRAM, processing optimizations will generally be made in favor of logic, resulting in less-dense DRAM circuitry. In addition, the sorts of manufacturing optimizations possible in commodity parts are also not generally possible in logic-and-DRAM processes, since more distinct parts will be manufactured, making it more difficult to measure the process. As a result, embedded DRAM will generally be larger and slower than what one would expect from evaluating commodity DRAM in a same-generation process. However, substantial performance gains can be obtained simply by eliminating chip-to-chip connections, as shown in the next example.

Example 6-2: A processor-in-memory system

The EXECUBE processor [Kog95] is known as a processor-in-memory system since it integrates CPUs on a dynamic RAM chip. The prototype EXECUBE has eight processing elements (PEs). Each PE includes a 16-bit CPU and 32 kilobytes of dynamic RAM. Since each CPU has its own block of DRAM, all the processors can perform simultaneous independent memory accesses. Here is a photomicrograph of an EXECUBE chip, showing the eight PE and memory subsystems:

Photo © 1996 IEEE.

Integrating the CPU and memory has three main advantages. First, much higher memory bandwidth is allowed, since the width of a memory access is not limited by the available pins. Second, because pins are not limited, the wide memory can be divided into multiple addresses. Typical microprocessors read a single sequence of bytes to reduce the number of address bits required, while the EXECUBE can read seven uncorrelated addresses simultaneously. Third, power consumption is greatly reduced—a 50 MHz EXECUBE processor fabricated in 0.8 µm technology consumes only 2.7 Watts.

6.8 Field-Programmable Gate Arrays

A **field-programmable gate array** (FPGA) is a block of programmable logic that can implement multi-level logic functions. FPGAs are most commonly used as separate commodity chips that can be programmed to implement larger functions than can be fit into a programmable logic device (PLD). However, small blocks of FPGA logic can be useful components on-chip to allow the user of the chip to customize part of the chip's logical function.

An FPGA block must implement both combinational logic functions and interconnect to be able to construct multi-level logic functions. There are several different technologies for programming FPGAs, but most logic processes are unlikely to implement anti-fuses or similar hard programming technologies, so we will concentrate on SRAM-programmed FPGAs. A programmable interconnect point can be constructed by using a pass transistor to connect two wires and controlling its gate by the value stored in an SRAM cell. When the SRAM cell stores a logic 1, the pass transistor will connect the two wires. A combinational logic block must be able to implement any combinational function of n bits, depending on its programming. This functionality is equivalent to a truth table. A block of SRAM can be used to implement the truth table—the address determines the row of the truth table required by the current value of the combinational inputs, while the value stored at that address is the function's output. Combinational logic blocks typically include a flip-flop which can be selectively connected between the combinational output itself and the programmable wiring connected to the block; adding the flip-flop simplifies programming in many cases.

6.9 Programmable Logic Arrays

The **programmable logic array** (**PLA**) is a specialized circuit and layout design for two-level logic. While the PLA is not as commonly used in CMOS technology as in nMOS, due to the different gate circuits used in the two technologies, CMOS PLAs can efficiently implement certain types of logic functions.

The architecture of a PLA, as shown in Figure 6-37, is very simple: it uses two levels of logic, one implementing the ANDs (called **product terms**) and another implementing the ORs. One of the best features of the PLA is that it can compute several functions at once, which can share product terms. Two-level functions are built from both the true and complement forms of the variables, so the inputs supply both forms, usually with a pair of inverters on the true form as buffers. Some sort of buffer is usually placed at the output. The architecture also suggests the attractiveness of the layout: the inputs to the AND plane flow vertically, while the outputs flow horizontally and emerge at the right side. The OR plane is simply a $90°$ rotation of the AND plane.

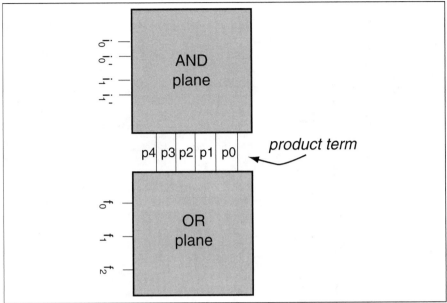

Figure 6-37:
Organization of a
PLA.

Such a layout can be very compact, but we need to find a gate structure compatible with this layout organization.

We clearly cannot use fully complementary gates for this layout style—wiring both the pullups and pulldowns would be too complex. The most common form of the CMOS PLA uses precharged gates for both AND and OR planes [Sho88]. Using a non-complementary gate lets us use very regular layouts for the wires: input signals are evenly spaced in one direction and output signals are also evenly spaced in the perpendicular direction. Figure 6-38 shows a logic diagram for a doubly-precharged PLA drawn to correspond with the layout. The product term lines p1, etc. flow from the AND plane to the OR plane. The circuits in the AND and OR planes are identical, except for programming transistors, which determine the PLA's **personality**. In the AND plane, each precharged gate's output node runs vertically through the cell. Pull-downs run between the output node and V_{SS} lines which run parallel to the output lines; inputs run horizontally to be attached to the pulldowns' gates.

The AND/OR plane is very simple to generate. We first lay down a grid of wires for input signals, V_{SS}, and output signals. We can create a pulldown transistor at the intersection of input and output signals by adding small amounts of diffusion and poly along with a via. The pulldown can be added to the wiring by superimposing one cell on another; the pulldown cell is called a **programming tab** thanks to the shape of the transistor and via. The same cell can be used for both planes since the OR plane

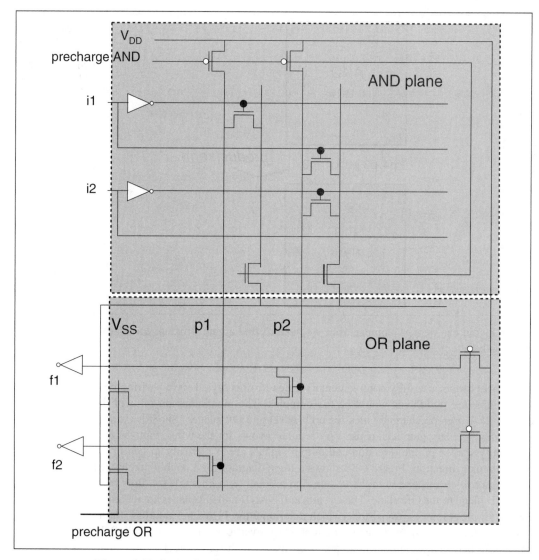

Figure 6-38: *Internal architecture of a PLA. (Masakazu Shoji, CMOS Digital Circuit Technology © 1988, p. 220. Adapted by permission of Prentice Hall, Englewood Cliffs, New Jersey.)*

is simply a 90° rotation of the AND plane. Figure 6-39 shows a section of an AND/OR plane with four inputs running vertically in poly and two outputs running horizontally in metal1. Each pair of input lines shares a ground line running in n-diffusion; pairs also share open space for a programming tab's via, so that one via can be used for two pulldown transistors on opposite sides of the via.

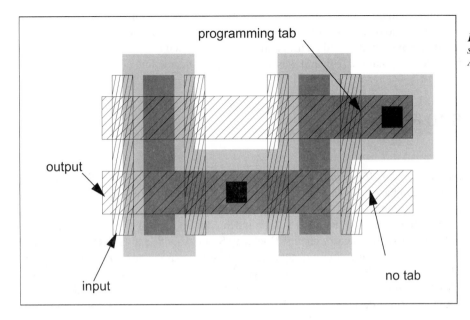

Figure 6-39: A section of a PLA AND/OR plane.

The delay through the PLA is determined largely by the load introduced by the vertical and horizontal wires. The neat layout of the PLA keeps us from using large transistors in the pulldowns to speed up the gates—a large pulldown would not only add blank space down the entire row but would also lengthen the perpendicular wires. This doubly-precharged structure also complicates clocking.

Which functions are best implemented as PLAs? Those functions which are true for about half their input vectors are well-suited. If the PLA has very few programming tabs, we are wasting most of its area and are probably better off using fully complementary gates. If the PLA is nearly full, we can complement the functions to produce a nearly empty PLA. PLAs are also good for implementing several functions that share many common product terms, since the AND-OR structure makes it easy to send one product term to many different ORs. CPU microcode often has these characteristics.

Standard two-level minimization algorithms can be used to optimize the PLA personality. An optimization unique to PLAs is **folding**. If one region of the PLA is empty of programming tabs, it may be possible to remove that section and fold another section of the PLA into the newly freed space. Folding can leave the PLA with inputs and outputs on all four sides. While folding can dramatically reduce the size of the PLA, it also makes the PLA's layout very sensitive to changes. A small change in the logic may require a use of a single programming tab in the region removed for folding, enough to undo the entire folding operation. Folded PLAs may, with a small logic change, unfold themselves like origami pieces in the middle of a chip design, destroying floorplans which relied on the small size of the folded PLA. As a result, PLA folding is less popular today than it was when PLAs first came into common use.

6.10 References

Introduction to Algorithms by Cormen, Leiserson, and Rivest [Cor90] includes a detailed comparison of the computational complexities of various addition and multiplication schemes. As they point out, asymptotically efficient algorithms aren't always the best choice for small n. I am indebted to Jack Fishburn for an explanation of transistor sizing in carry chains. Books by Shoji [Sho88] and Glasser and Dobberpuhl [Gla85] describe circuit designs for interesting components; *The MIPS-X RISC Microprocessor* [Cho89] provides a good survey of the components used for a CMOS microprocessor. Hodges and Jackson [Hod83] give a good introduction to RAM and ROM design; more detailed discussions can be found in Glasser and Dobberpuhl and Shoji. The static RAM layout of Figure 6-31 was designed by Kirk Nolan of Princeton University. Keitel-Schulz and Wehn [Kei01] discuss embedded DRAM technologies.

6.11 Problems

6-1. Produce a complete table of the barrel shifter's possible operations. List all possible combinations of top and bottom inputs and the resulting operation.

6-2. Identify the critical delay path through the barrel shifter.

6-3. Design a logic gate network for the full adder:

> a) using two-level logic;
>
> b) using multi-level logic.

6-4. What combination of input values produces the longest delay through a Manchester carry chain?

6-5. The output of an adder feeds one input to a second adder; the second adder's other input comes directly from a primary input. Sketch the critical delay path through this system of two adders.

6-6. What is the false delay path in the carry-skip adder?

6-7. Design a stick diagram for a serial adder cell.

6-8. Design a one-row layout of a four-bit carry lookahead unit. You can use inverters and two- and three-input NANDs and NORs in the design.

    a) Show the desired locations of primary inputs and outputs around the edge of the cell.

    b) Show your decomposition of the functions into gates.

    c) Show a good placement of the gates in the row and explain your criteria for judging the layout.

    d) Sketch the design of the wiring channel underneath the gates, showing where wires enter and exit and their assignment to tracks in the channel.

6-9. Consider an ALU design:

    a) Enumerate all the possible functions of a two-input ALU.

    b) For each possible function, list the control inputs to the three function blocks of the three function block ALU.

6-10. Design the logic for an ALU that can perform these functions: addition, subtraction, AND, OR, NOT.

6-11. Develop stick diagrams for an unsigned array multiplier:

    a) Draw stick diagrams for the full adder used in the core cell and for a carry-skip adder to be used in the final stage.

    b) Draw a hierarchical stick diagram for the full array showing the layer assignments and organization of wires and cells.

6-12. Draw a block diagram for an array multiplier that uses Booth encoding. Use the adder-subtractor as a basic building block of the array.

6-13. Design components of a Booth multiplier:

    a) Design the logic for one bit of the adder-subtracter.

    b) Design a stick diagram for your adder-subtractor.

6-14. Design and analyze an 8×8 Wallace tree multiplier:

a) Draw the block diagram for an eight-bit carry save adder.

b) Draw the complete block diagram for the multiplier.

c) Draw a basic floorplan for the multiplier including the partial product generators and Wallace tree.

d) Identify the critical delay path through the multiplier.

6-15. Show how a bit-serial adder adds the two numbers $a = 0101$ and $b = 0110$ (these numbers are written here with the MSB first). Show the adder's inputs and outputs for every clock cycle until the addition is complete.

6-16. How many groups must node capacitances be split into to properly simulate an SRAM at the switch level?

6-17. What is the minimum-delay encoding for a modulo-8 counter?

6-18. What is the relationship between a truth table and the pattern of programming tabs in a PLA's AND and OR planes?

7

Floorplanning

Highlights:

Floorplaning styles and methodology.

Global routing.

Clock distribution.

Power distribution.

Packaging and pads.

7.1 Introduction

In the last chapter we built architectures from fairly abstract components. This chapter looks at the chip in more detail. We will assume that the block diagram is fixed; now we will study chip-level layout and circuit design. The size of the design problem requires us to develop different methods than we used to design the layout for a single NAND gate. But the basic objectives—area, delay, power—are still the same.

7.2 Floorplanning Methods

Floorplanning is chip-level layout design. When designing a leaf cell, we used transistors and vias as our basic components; floorplanning uses the adders, registers, and FSMs as the building blocks. The fundamental difference between floorplanning and leaf-cell design is that floorplanning works with components that are much larger than the wires connecting them. This great size mismatch forces us to analyze the layout differently and to make different trade-offs during design.

Figure 7-1: A typical layout, built from a variety of styles.

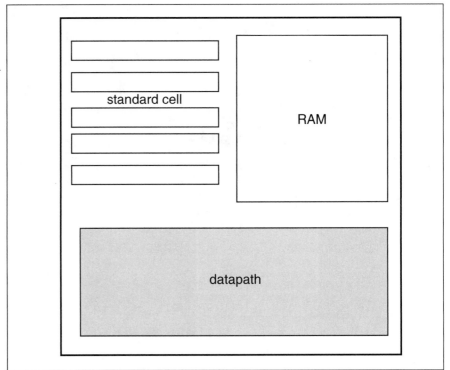

Many chips are composed from cells of a variety of shapes and sizes, as shown in Figure 7-1. We call the layout cells **blocks** during floorplanning because we use them like building blocks to construct the floorplan. In **bricks-and-mortar** style layout, the cells may have radically different sizes and shapes. The layout program must place the components on the chip by position and orientation, leaving sufficient space between the components for the necessary wires. Blocks may be redesigned to change their aspect ratio in order to improve the floorplan. As we will see, the more complex traffic pattern of wiring areas makes routing wires in a bricks-and-mortar layout much harder than in a standard cell layout. (Some people use the term *standard cell* for any layout, including brick-and-mortar, which is built from pre-designed components. Since standard cell is a much abused term, be sure you understand its meaning in the context in which it is used.)

The next example shows the floorplan for a large chip.

Example 7-1: *Floorplan of the IBM Power 2 Super Chip*

The Power 2 Super Chip (P2SC) is a large microprocessor. It has over 15 million transistors (5.7 million logic, 9.3 million cache) on an $18.2 \times 18.4\text{mm}^2$ die. The chip is fabricated in a $0.27\mu\text{m}$, 5-level-metal process. The chip comes in 120 and 135 MHz versions.

The chip photomicrograph has been overlaid below with the floorplan showing the major functional units:

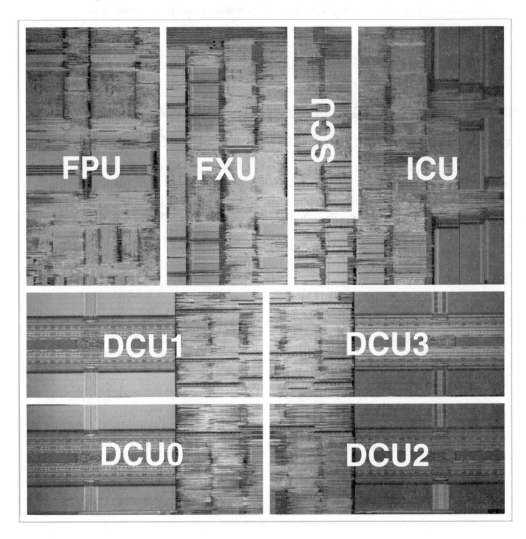

Photo courtesy of IBM.

This chip is large enough that each of the units in the chip-level floorplan has its own internal floorplan. The DCU units contain memory arrays as well as driver and control logic. The ICU unit contains several data paths along with the necessary control logic.

Floorplanning is divided into three phases: block placement, global routing, and detailed routing. These three phases successively refine the design until the layout is complete. Block placement, as the name implies, places the blocks on the chip.

Floorplanning occurs throughout the design process:

- Early in the design process, a floorplan is designed using estimates of the sizes of the blocks and of the number of wires between those blocks. The area required for wiring is estimated during floorplanning. This initial floorplan serves as a budget for the design—if the sizes of components or of wires actually implemented are significantly different from those in the initial floorplan, the floorplan needs to be rethought. Having a budget for blocks and wires encourages the designers of those sections to live within their allocated areas.

- The design of the initial floorplan defines the interface requirements for the blocks. Once those blocks are designed, the chip layout can be assembled from the blocks. Blocks may need to be modified due to errors in estimating the properties of the blocks during floorplanning.

Even with layout design divided into placement and routing, the design of the complete chip layout is a daunting task. Chip-level wiring design is usually divided into two phases: **global routing** assigns wires to routing channels between the blocks; **detailed routing** designs the layouts for the wiring. Placement and global routing divide the routing region into smaller sections that can be designed independently, greatly simplifying the detailed routing of those sections.

Interactive floorplan editors and global placement-and-routing tools are a big help in floorplan design. The size of the floorplan and the disparity in scales between large blocks and individual wires make it difficult to manage a manually-designed floorplan. Floorplanning tools may allow you to enter blocks with pinouts, plot rat's nests to evaluate routability, define routing channels, and perform global routing. Global layout tools perform detailed block placement, global routing, and detailed routing, making sure that the results of the various switchbox and channel routing tools are assembled into a complete layout.

Global and detailed routing design layouts for signals. We use specialized methods to route power/ground nets and clocks because each has specialized requirements. V_{DD} and V_{SS} must be supplied to the logic gates with minimal voltage drop and must be adequately wide to carry the required current. Clocks must be distributed throughout the chip to minimize skew. Special care must be taken on both power/ground and clock nets in large chips.

7.2.1 Block Placement and Channel Definition

Figure 7-2: A floorplan sketch.

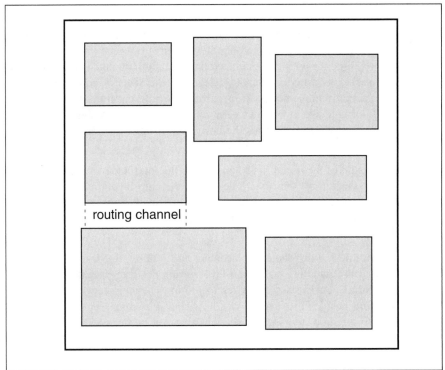

A first cut at floorplanning is to arrange the blocks to minimize wasted space. As illustrated in Figure 7-2, a good way to manually experiment with floorplans is to draw the blocks on graph paper, cut them out, and arrange them on another block of graph paper. A block is characterized by its area and its **aspect ratio** (the ratio of its width to its height). The wiring between the blocks (such as a rat's nest plot) can be used to adjust the positions of the blocks.

Don't forget to try different rotations and reflections of the blocks. The more similar the blocks, the easier they are to interchange. Interchangeability makes wiring optimization easier. You will probably not be able to avoid a design with a few large blocks and a few small ones, but it may be worthwhile to combine or split some of the blocks, especially blocks built from standard cells, to equalize block sizes as much as possible.

Experimentation with floorplans may suggest that a change to the shape or size of a block would make the floorplanner's job much easier. Changing the shape of a block or splitting a block in two may make use of white space that cannot be removed any other way. If area is important, floorplanning before block design starts can help guide the design of blocks. However, blocks are not infinitely malleable. The internal reorganization of wires and components necessary to change a block's aspect ratio may make it larger, particularly if the block is a tightly designed array of tiled cells; the larger cell may not fit in the desired hole, even with its new shape. Redesigning a block may also unacceptably increase internal delays by making critical wires too long. Floorplanning and block design should go hand-in-hand for best results, but neither can dominate the other.

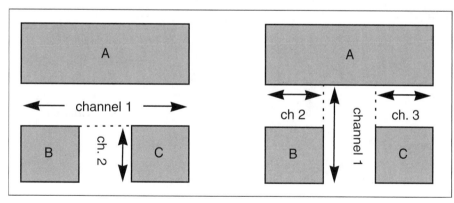

Figure 7-3:
Alternative channel definitions.

Too much floorplanning without consideration of routing is dangerous. In fact, it is hard to talk about placement without global routing, because most placement decisions are determined by the space needed for wires, not block shape. We have already seen routing channels, which have pins defined on two opposite sides and so can be stretched in one dimension to accommodate more wires. A **switchbox** is a routing area with connections anywhere along its four sides, which means that it cannot grow in either direction. Switchboxes are useful in connecting abutting channels. To take wiring into account during placement, we must define routing channels and switch-boxes, then assign nets to paths through those channels/switchboxes. (The job of designing each routing channel/switchbox is left until later.)

The first job is to define the routing channels and switchboxes. We want to break up the space between the blocks into rectangular regions for simplicity during detailed routing; this step is known as **channel definition**, though it defines both two-sided channels and four-sided switchboxes. A natural idea is to use the blocks' edges to define the routing channels; while that idea is useful, it doesn't uniquely define a set of channels for a floorplan. Figure 7-3 shows several examples of channel definition. In each case, we use the blocks to define the ends of rectangular routing regions—think of each block casting a shadow over the space left for routing, with each shadow defining its own routing region.

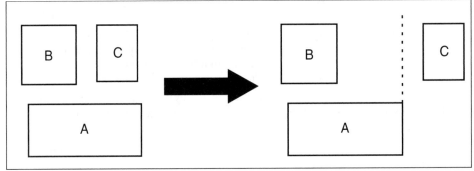

Figure 7-4: Channel definition changes with block spacing.

Channel definition has no single solution for two different reasons. First, by moving the lighting to change the shadows cast, we can change the channel definition. There is no way to choose the optimum division of the chip into routing regions, though it is probably best to use fewer, larger channels than to break the chip into many small regions. Second, we can change the way shadows are cast by changing the distance between the blocks, as shown in Figure 7-4.

We don't know the space required between two blocks until we know the number of wires routed through the channel; the best solution is to base the spacing on a rough guess of the required wiring capacity. In both cases, small changes to the problem can give us very different configurations of routing channels and switchboxes with which to work. However, it is very difficult to tell until detailed routing is complete which is best. Often, any of several choices will give roughly equivalent results.

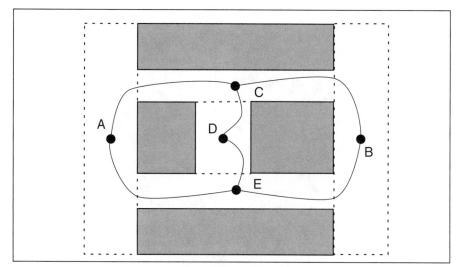

Figure 7-5: *The channel graph.*

The full geometric description of the channels and switchboxes is cumbersome for global routing—all we really need is the topology of the paths between blocks. The **channel graph** reduces the floorplan to a description of the routes between blocks. As shown in Figure 7-5, each channel is represented by a node. An edge is added between two nodes if those channels abut each other. The paths through the graph correspond to global routes—paths from channel to channel. For example, two distinct paths from channel A to channel B are (AC, CB) and (AC, CD, DE, EB). Which path is better depends on the locations of the pins to be connected and the congestion of the channels. At this point, the exact layout of the wire within each channel doesn't matter; that is to be determined by detailed routing.

If two channels intersect at a T, we can directly connect them. However, the channels in the channel graph cannot be routed in arbitrary order, and that fact influences our choice of floorplan designs. Figure 7-6 illustrates the problem. The connections to a channel are made at its top and bottom; we have complete control over the placement of pins along those edges, but the track to which a wire is assigned is determined by the routing algorithm or human designer. When two channels meet, the top and bottom pins of one are determined, in part, by the tracks of another. We don't know where a wire through the two channels will enter the vertical channel until it is assigned a track in the horizontal channel. Therefore, we must route the horizontal channel first. Any nets in the horizontal channel that connect to the vertical channel must be extended to the far end of the channel. The extended net defines an **end pin** of the first channel on the second channel.

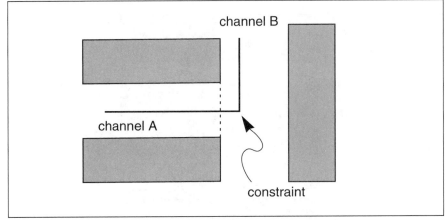

Figure 7-6:
Channels must be routed in order.

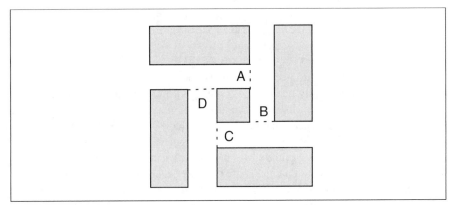

Figure 7-7:
Windmill structures introduce irresolvable constraints on routing order.

Channel ordering is a problem for placement and global routing because some channel graphs don't have any feasible routing order. Figure 7-7 shows a **windmill** [Pre79]. Careful examination shows that each channel depends on the one to its left: B depends on A, C on B, D on C, and D on A. As a result, there is no channel we can route first and guarantee that the complete structure can be successfully routed.

The best solution to windmills is to avoid them completely. If the floorplan is a **slicing structure** [Ott80], as shown in Figure 7-8, it has no windmills. A slicing structure can be recursively sliced down to its blocks—a slice is a straight cut through the routing region that separates the chip into two sections. Each section forms a smaller floorplan that can be cut again. Note that there need be only one slice through the whole chip—successive slices can cut the pieces cut by the original slice. (A standard cell layout has several parallel slices through the complete floorplan.) Since a windmill cannot be sliced, any floorplan that is slicable is guaranteed to have no wind-mills. As a result, its channels can be routed in order such that all the pins on every channel are well-defined at the time it is routed.

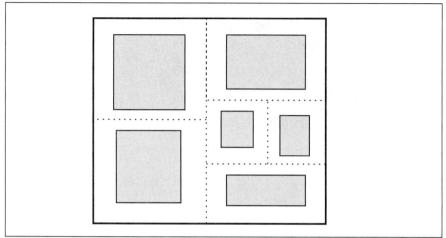

Figure 7-8: A slicable floor-plan.

7.2.2 Global Routing

Figure 7-9:
Line-probe
routing.

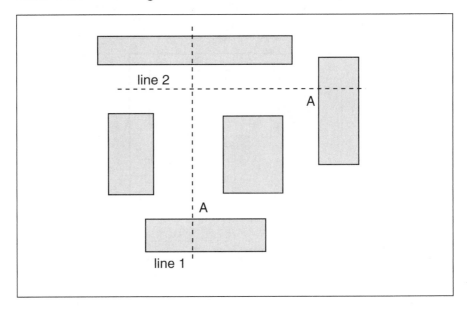

We will discuss global routing algorithms in Chapter 10. A good algorithm for hand routing is the **line-probe** method introduced by Mikami and Tabuchi [Mik68] and by Hightower [Hig69]. Line-probe routing can work on arbitrary-shaped routing regions; for the moment, we will restrict ourselves to the rectangular switchbox area, with each wire segment already routed as an additional obstruction in the switchbox. As shown in Figure 7-9, the line-probe algorithm starts at one pin on the net and constructs a series of lines along which the other pin may lie. The first probe line is perpendicular to the face which holds the pin and extends to the first obstacle encountered. If the other pin can be reached from this probe line, we are done. If not, we move to the far end of the probe line and construct a new, perpendicular line. A probe line stops when it hits the switchbox edge or an existing wire segment. The search stops when a probe line runs past the other pin or when there is nowhere left to probe. The route between the pins follows the probe lines. This algorithm may not find the shortest route for the wire—it is not even guaranteed to find an existing path. But it often works in practice and is very fast.

We want to choose a global routing that gives the best detailed routing for all the channels and switchboxes. However, it is difficult to estimate the exact results of detailed routing at this stage. At this level of abstraction, a good goal is to equalize **channel utilization**—the number of wires that start, end, or flow through a channel.

Density gives a good estimate of utilization without routing the channels. Your goal is to assign wires to paths such that all channels are about equally full. A few rules help:

- The first nets to route are those whose delays are critical. These wires need to be as short as possible, so they should get priority for the channels that give the shortest paths.

- Some wires may stay entirely within one channel or require a short trip between a few channels with one obvious choice. Route these wires early to get them out of the way.

- Don't be afraid to rip up and reroute wires. You must route the wires in some order. If you find that an earlier decision was bad, remove the wires that are in the way, route the new wires, then reroute the ripped-up wires.

It may be necessary to go through the placement-global routing cycle several times. Some wiring problems may not be fixable by ripping up other wires. If necessary, change the floorplan to get around serious problems. Remember, however, that floorplans are complex—a change that improves one wire's route may make another important wire's route much worse. Repeatedly designing floorplans and global routes is also tedious; you are much more likely to try several floorplans and converge on a good one if you have a CAD tool that can do the detail work for you.

7.2.3 Switchbox Routing

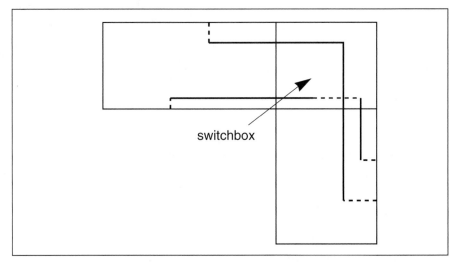

Figure 7-10: *A switchbox formed at the intersection of two channels.*

Switchbox routing is harder than channel routing because we can't expand the switchbox to make room for more wires. As shown in Figure 7-10, a switchbox may be used to route wires between intersecting channels. The track assignments at the ends of the channels define the pins for the switchbox. It is often possible to define channels without requiring switchboxes, but switchboxes may sometimes be necessary in floorplanning.

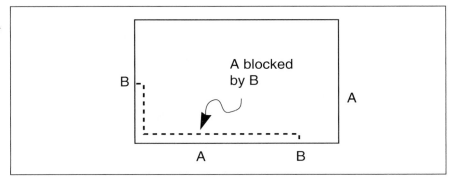

Figure 7-11: *Net ordering can be critical when routing switch-boxes.*

Since a switchbox has fixed pins on all four sides, there are no obvious preferred directions to suggest layer assignments. It is tempting to use the same layer to route both horizontal and vertical segments of a wire, but that can create problems, as shown in Figure 7-11. If the A and B pins are on the same layer, routing B as shown completely blocks A—there is no room to insert a via for the A net.

A common strategy for switchbox routing is to arbitrarily pick layers for vertical and horizontal problems, then to treat the switchbox as a routing problem with fixed pins at the ends of the channel. While a channel routing algorithm may sometimes fail due to the added constraints, channel routing algorithms can often give reasonably good results in such circumstances.

7.2.4 Power Distribution

Power distribution presents two problems: sizing wires to ensure they can deliver the required power without being destroyed; and designing a global power distribution network that runs both V_{DD} and V_{SS} entirely in metal.

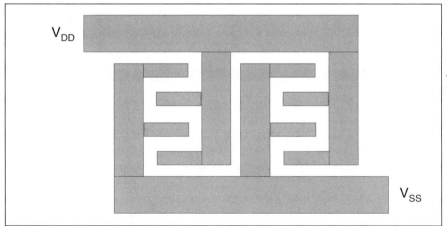

Figure 7-12: *Interdigitated power and ground trees.*

We encountered the metal migration limit in Section 2.4—if too much current is carried through a wire, the wire quickly disintegrates. Power lines are usually routed as trees, with the power supply at the root and the logic gates connected to the twigs. As seen in Figure 7-12, each branch must be wide enough to carry the currents in all of its branches. If your logic gates use only a few standard transistor sizes, computing power line width is easy: compute the power consumed by the gates on each twig, then move up the tree, adding together the power requirements of branches at one level to get the required sizes for the branches at the next level.

When designing large chips, using simple rules to choose power supply widths may not be sufficient. Metal migration is not the only problem in power supply distribution. Large currents that do not cause metal migration may still cause **power supply noise** due to IR drops in the power supply network. Circuit simulations of the power supply network, using high-level models for major components which models their current requirements over time, can be used to analyze the power network.

Decoupling capacitors—capacitors across the power supply pins—are traditionally used in printed-circuit board design to reduce power supply noise. In large chips, decoupling capacitors may be used on-chip as well. One DEC Alpha microprocessor [Gie97] uses 250 nF of on-chip capacitance to decouple the power supply. This microprocessor also illustrates a more radical approach to power supply decoupling: a separate chip is bonded directly on top of the microprocessor to add 1 µf of decoupling capacitance. The chip capacitor uses a 2 cm^2 pMOS device to implement the large capacitance. The decoupling capacitor is connected to the microprocessor with approximately 160 wirebond pairs. An alternative solution to power supply noise is to modify the sequential design of the machine. The next example illustrates potential problems in the design of large power supply networks.

Example 7-2: Power distribution in the BELLMAC-32A

The BELLMAC-32A [Sho82] was an early 32-bit single-chip microprocessor. The basic layout of the power bus is shown below:

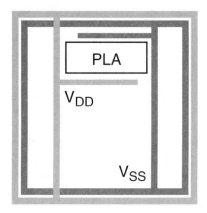

The layout was designed with power pads at both ends, which reduced power bus noise by a factor of four, since both impedance and switching current were cut in half.

The designers conducted circuit simulations to determine the impedance-induced noise on the power bus. An early design resulted in a 2 V swing on the power lines. The large voltage spike was due to the precharging of PLAs, which occurred on the same clock phase as the precharging of a large decoder. Adding pads was insufficient to reduce the power bus noise to an acceptable level. The problem was solved by changing the timing of the PLA precharge phase—precharging was moved to phase different than that used to precharge the decoder.

This example shows how major events with many correlated transistor actions, such as precharging, can be responsible for power noise. Initialization also causes problems in modern microprocessors. Most large microprocessors no longer use single-cycle resets. They instead reset the machine over several cycles to avoid large amounts of activity that can cause current spikes. This, naturally, complicates the sequential design of the machine.

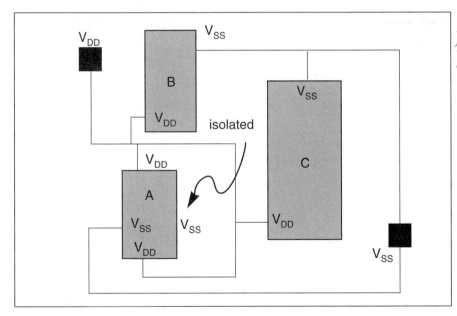

Figure 7-13: A floorplan which isolates a ground pin.

Design of a planar power network requires attention to cell design even before routing begins. If we orient all the cells so that their V_{DD} pins are all on the same side of a dividing line through the cell, we are guaranteed that a planar routing exists

[Sye82]. Conversely, if we don't satisfy the pin placement condition, we are guaranteed that a planar routing does not exist. We can ensure consistent power/ground pin placement either by reordering power/ground pins or by internally routing V_{DD} and V_{SS} connections—if each cell has only one V_{DD} and one V_{SS} pin, the condition is guaranteed to be satisfied. An example of an ill-conceived floorplan is shown in Figure 7-13: one of A's V_{SS} pins becomes surrounded by the V_{DD} net.

7.2.5 Clock Distribution

The problems caused by clock skew were discussed in Section 5.4. The main job of clock routing design is to control clock skew from the clock pad to all the memory elements. The major obstacle to clock distribution is capacitance, with resistance playing a secondary but important role. Gate capacitance cannot be avoided on a clock line even on modest-size chips. Consider, for example, a chip with 25,000 logic gates. Having 5,000 transistors on that chip whose gates are connected to the clock is not unreasonable. Those transistors have a total gate capacitance of 4.5 pF in a typical 0.5 μm process; driving that load at the same rate as one minimum-size inverter drives another would require a transistor 5,000 times wider than minimum size. Clock distribution is made more challenging by the fact that the large capacitive load must be driven to produce a very sharp transition. The slope of the gate signal affects switching speed. A slow-rising clock edge will cause serious performance problems.

Figure 7-14:
Clock delay
vs. position.

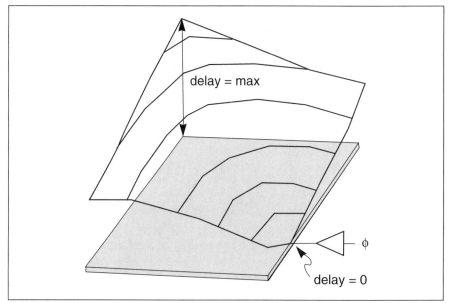

Clocking is a floorplanning problem because clock delay varies with position on the chip. As a result, clock delay must be taken into account both in the placement of logic blocks and in the design of the clocking network. Figure 7-14 shows a typical map of clock delay *vs*. position, where height on the surface above the chip gives the clock delay at that point. Memory elements that are logically related should be connected to the clock signal tapped at roughly the same position, implying that those memory elements and the combinational logic between them should all be placed close together in the layout. The designers of the Alpha processor built such a delay map of the clock signal to determine when the clock arrived at latches throughout the chip [Dob92].

There are two complementary ways to improve clock distribution:

- **Physical design.** The layout can be designed to make clock delays more even, or at least more predictable.

- **Circuit design.** The circuits driving the clock distribution network can be designed to minimize delays using several stages of drivers.

In general, both techniques must be used to distribute a clock signal with adequate characteristics.

Let us first consider the physical design of clock distribution networks. The two most common styles of physical clocking networks are the **H tree** and the **balanced tree**. The H tree is a very regular structure which allows predictable delay. The balanced tree takes the opposite approach of synthesizing a layout based on the characteristics of the circuit to be clocked.

An H tree is shown in Figure 7-15. It is a recursive construction of Hs—given one level of H structure, four smaller H structures can be added at the four endpoints of the H bars. The H tree structure can be recursively refined to any level of required detail. The widths of the wires in the H tree can be adjusted to account for variations in load capacitance to equalize skew throughout the H tree. Buffers can also be added into the H tree network to increase drive capability. An H tree network can be thought of as a top-down clock distribution methodology since the floorplan of the H tree determines the floorplan of the logic to which it is connected. Since skew increases with physical distance in the H tree, memory elements must be grouped together to make use of the same or nearby distribution points in the H tree network.

A balanced tree clock network, illustrated in Figure 7-16, is generated by placement and routing. Memory elements are clustered into groups. The clustering is used to

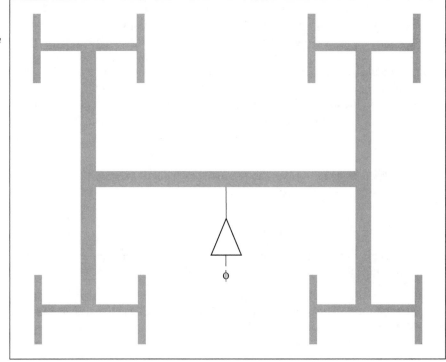

Figure 7-15: *An H-tree.*

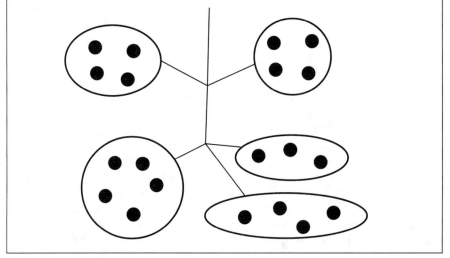

Figure 7-16: *A balanced clock tree.*

guide placement and a clocking tree is then synthesized based on the skew information generated during clustering. The tree is irregular in shape but has been balanced

during design to minimize skew. Once again, wire widths can be varied in the tree and buffers can be added. Several tools exist for generating balanced clock trees.

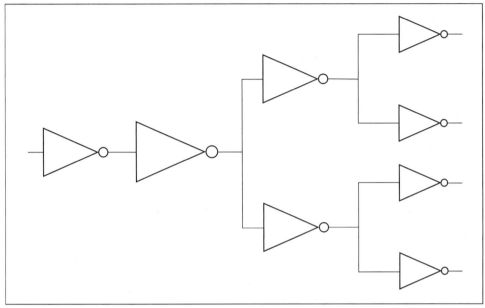

Figure 7-17: *A clock distribution tree.*

Two strategies can be used to distribute the clock: using a driver chain, as shown in Section 3.3.8, to drive the entire load from a single point; or distributing drivers through the clock wiring, forming a hierarchical clock distribution system [Fri86]. If a hierarchical system like the one in Figure 7-17 is used, the resistance and capacitance of the clock wiring needs to be analyzed to determine the points at which buffers should be inserted. Of course, since the drivers will probably be inverting, care must be taken to use an even number of driver stages to avoid delivering an inverted clock signal to the memory elements.

The next example describes the design of a clocking network which uses a regular physical design with a sophisticated clock driving network.

Example 7-3: Clock distribution in the DEC Alpha 21164

The DEC Alpha 21164 [Bow95] contains 16.5 million transistors on a 16.5 mm × 18.1 mm die. The basic floorplan of the clock distribution system is shown below:

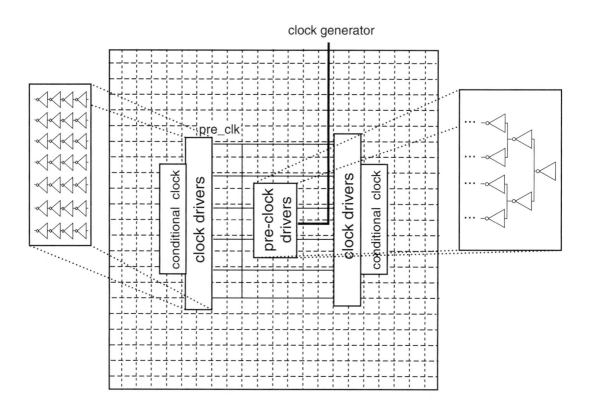

clock generator

The clock generator drives the first stage of clock driver, located on the center of the chip and known as the pre-clock driver. This network consists of a six-level inverter tree and generates 24 outputs on the **PRE_CLK** signal shown in the figure. That signal is fed to two final clock driver systems, one on each side of the chip. Each final clock driver contains 44 drivers with four levels of inverters on each output. The last clock driver inverter level has a total gate width of 58 cm. The system also includes a set of 12 conditional clock drivers on each side. The final clock drivers are connected to a regular clock grid, shown in the figure as dotted lines. The clock grid is laid out in metal 3 and metal 4.

The clock interconnect and gate load present a total capacitive load of 3.75 *pF*. The clocking network provides the clock to instruction and execution units within 65 *ps* of skew.

7.2.6 Floorplanning Tips

Floorplanning a moderate-size chip is necessary but not overwhelming. A few simple rules of thumb help you get to an easy-to-implement, easy-to-change design more quickly.

- **Develop a wiring plan.** You should think about how to use layers to make connections as part of planning your wiring. The horizontal metal 1-vertical metal 2 scheme used by channel routers is an example of a wiring plan. Sketch the plan to help you think about rational, regular schemes for assigning layers to wires. Using different layers for different directions or for different types of nets helps reduce alternatives to a manageable number and make choices clearer. Hand-crafted blocks, such as data paths, may have their own wiring plans.

- **Sweep small components into larger ones.** Block diagrams often have isolated gates or slightly larger components. While these small components help describe system operation, they create lots of problems during floorplanning: they require extra effort for power/ground routing and they disrupt the flow of wires across the chip. Put these small components into an existing larger block or create a glue logic block to contain all the miscellaneous elements.

- **Design wiring that looks simple.** If your sketch of the block diagram or floorplan looks like a plate of spaghetti, it will be hard to route. More importantly, it will be harder to change the design when you need to make logic changes or redesign to reduce delays. Move blocks, then move pin locations to simplify routing topology.

- **Design planar wiring.** A set of nets is planar if all the nets can all be routed in the plane without crossing. While most interesting chips don't have planar wiring, a subset of the wires may be planar. It may help organize your thinking to first design a floorplan on which the most important signals have a planar routing, then add the less-critical signals later.

- **Draw separate wiring plans for power and clock signals.** You may want to include these signals in your floorplan sketch, but they may be hard to distinguish from the maze of signals on the chip. A separate chart of power and clock routing will help you convince yourself that your design is good for signals, power, and clock.

7.2.7 Design Validation

Chip assembly is when your earlier efforts at design verification are judged—if you did a good job of checking the pieces, the chip should work relatively quickly. As with subsystems, design validation breaks down into checking the structure and performance.

In both cases, you should check each block in the floorplan individually, then check the complete chip after assembly. Each block should be extracted, then simulated and checked with a timing verifier. Checking blocks before assembling the layout can save lots of work: the size, shape, and pinout of the block may change after a bug fix, especially if the layout was created by a synthesis tool. But the fact that each block works doesn't imply that the chip will work. Besides wiring errors, the most common chip-level bugs are interface errors, such as one block emitting an active-low signal and the receiving block expecting an active-high signal. Delay problems may also explode at the chip level, either due to unanticipated long wires or very long chains of logic that were not recognized earlier.

The amount of design-rule checking required depends on the CAD tools used to build the chip and your willingness to catch layout errors after fabrication. Layouts designed using editing systems which don't provide design-rule checking should definitely be checked by a separate system—the probability of a person designing a large layout with no errors is close to zero. Running a final design-rule check on the complete layout is still standard practice at many companies, even though correct-by-construction CAD tools are in common use. The cost of an error in both money and time is large enough that a final check is prudent.

Beyond checking for layout errors, you should check the assumptions on which your delay calculations were made. Extract parasitics, check their values for reasonableness, then rerun timing verifications and simulations to be sure the circuits are fast enough when driving the actual parasitics.

7.3 Off-Chip Connections

A chip isn't very useful if you can't connect it to the outside world. You rarely see chips themselves because they are encased in **packages**. A complete discussion of packaging and its engineering problems fills several books [Bak90] [Ser89]. The packaging problem most directly relevant to the designer is the pads which connect the chip's internals to the package and surrounding circuitry—the chip designer is responsible for designing the pad assembly. But first, we will briefly review packages and some of the system and electrical problems they introduce.

7.3.1 Packages

Chips are much too fragile to be given to customers in the buff. The package serves a variety of important needs: its pins provide manageable solder connections; it gives the chip mechanical support; it conducts heat away from the chip to the environment; ceramic packages in particular protect the chip from chemical damage.

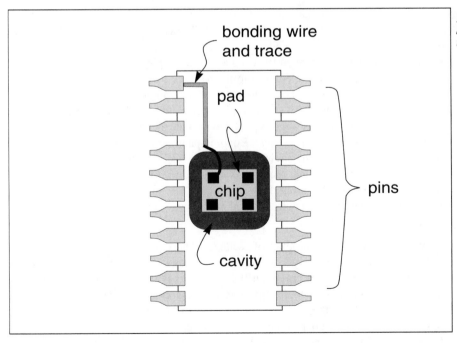

Figure 7-18:
Structure of a typical package.

Figure 7-19: An
empty package
showing the sub-
strate contact and
bonding areas.

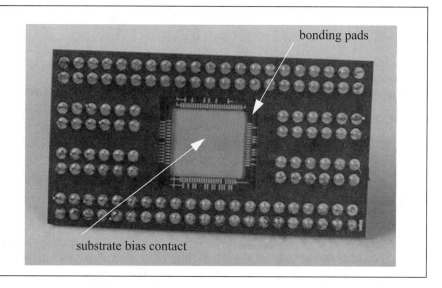

Figure 7-19: An empty package showing the substrate contact and bonding areas.

Figure 7-18 shows a schematic of a generic package (though high-density packaging technologies often have very different designs). The chip sits in a cavity. The circuit board connects to the pins at the edge (or sometimes the bottom) of the package. Wiring built into the package (called traces) goes from the pins to the edge of the cavity; very fine **bonding wires** that connect to the package's leads are connected by robot machines to the chip's pads. The **pads** are metal rectangles large enough to be soldered to the leads. Figure 7-19 shows a photograph of a package before a chip has been bonded to it. The cavity is gold-plated to provide a connection to the chip's substrate for application of a bias voltage. Bonding pads connected to the package's pins surround the four sides of the cavity. In a ceramic package, the cavity is sealed by a lid; to make a plastic package, the chip is soldered to a bare wiring frame, then the plastic is injection- molded around the chip-frame assembly. Ceramic packages offer better heat conductivity and environmental protection.

Figure 7-20 shows several varieties of packages. These packages vary in cost and the number of available pins; as you would expect, packages with more pins cost more money. The **dual in-line package** (DIP) is the cheapest and has the fewest number of pads, usually no more than 40. The **plastic leadless chip carrier** (PLCC) has pins around its four edges; these leads are designed to be connected to printed circuit boards without through-board holes. PLCCs typically have in the neighborhood of 128 pins. The **pin grid array** (PGA) has pins all over its bottom and can accommodate about 256 pins. The ball grid array (BGA) uses solder balls to connect to the package across the entire bottom of the package. The **plastic quad flat pack**

PGA

DIP

PLCC

BGA (courtesy IBM)

Figure 7-20: Common package types.

(PQFP), which is not shown, resembles the PLCC in some respects, but has a different pin geometry.

Packages introduce system complications because of their limited pinout. Although 256 may seem like a lot of pins, consider a 32-bit microprocessor—add together 32-bit data and address ports for both instructions and data, I/O bus signals, and miscellaneous control signals, and those pins disappear quickly. Cost-sensitive chips may not be able to afford the more expensive packages and so may not get as many pins as they could use. Off-chip bandwidth is one of the most precious commodities in high-performance designs. It may be necessary to modify the chip's internal architecture to perform more work on-chip and conserve communication. It may also be necessary to communicate in units smaller than the natural data size, adding rate conversion circuitry to the chip and possibly slowing it down.

Packages also introduce electrical problems. The most common problems are caused by the inductance of the pins and the printed circuit board attached to them—on-chip wires have negligible inductance, but package inductance can introduce significant voltage fluctuations.

<hr/>

Example 7-4: Power line inductance

Inductance causes the most problems on the power line because the largest current swings occur on that pin. (Inductance can also cause problem for signals in very high-frequency parts.) Package inductance caused headaches for early VLSI designers, who didn't expect the package to introduce such serious problems. However, these problems can be easily fixed once they are identified.

The system's complete power circuit looks like this:

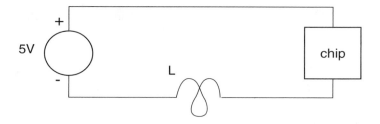

An off-chip power supply is connected to the chip. The inductance of the package and the printed circuit board trace is in series with the chip. (The chip also contributes capacitance to ground on the V_{DD} and V_{SS} lines which we are ignoring here.) The voltage across the inductance is

$$v_L = L\frac{di_L}{dt}.$$

In steady state there is no voltage drop across the inductance. But, if the current supplied by the power supply changes suddenly, v_L momentarily increases, and since the inductance and chip are in series, the power supply voltage seen at the chip decreases by the same amount. How much will the voltage supplied to the chip drop? Assume that the power supply current changes by 1 A in 1 ns, a large but not impossible value. A typical value for the package and printed circuit board total inductance is 0.5 nH [Ser89]. That gives a peak voltage drop of $v_L = 0.5{\times}10^{-9}H \cdot 1A/1{\times}10^{-9}s = 0.5V$, which may easily be large enough to cause dynamic circuits to malfunction. We can avoid this problem by introducing multiple power and ground pins. Running current through several pins in parallel reduces di/dt in each pin, reducing the total voltage drop. The first-generation Intel Pentium™ package, for example, has 497 V_{CC} pins and an equal number of V_{SS} pins [Int94].

7.3.2 The I/O Architecture

Pads are distributed around the edge of the chip. (Advanced, high-density packaging schemes devote a layer of metal to pads and distribute them across the entire chip face.) Each pad must be large enough to have a wire (or a solder bump) soldered to it; it must also include input or output circuitry, as appropriate.

A typical **pad frame** is shown in Figure 7-21. Each pad is built to a standard width and height, for simplicity. Each pad has large V_{DD} and V_{SS} lines running through it. A pad includes a large piece of metal to which the external wire is soldered. If the pad requires external circuitry, it is usually put on the side of the pad closest to the chip core. The **chip core** fits in the middle of the pad ring. If the pad ring is not completely filled with pads, spacers are added to keep the power lines connected. The placement of pads around the ring is usually determined by the required order of pins on the package—the wires to the package cannot be crossed without danger of shorting, so if the package pins are required in a certain order, the pads must be arranged in that order. The order of pins on the package determines routability of the board and electrical noise among other things; the order of pins on a package has been known to determine which candidate design wins a design contest.

Figure 7-21:
Architecture of a
pad frame.

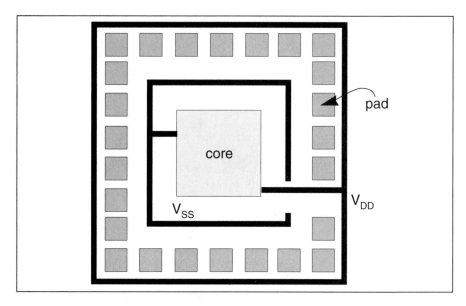

V_{DD} and V_{SS} pads are the easiest pads to design because they require no circuitry—each is a blob of metal connected to the appropriate ring. How much current can be supplied by one of these pads? The pad is much larger than any single wire connected to it, so the current in each direction is limited by the outgoing wire. If we want to use multiple power pins to limit inductive voltage drop, we can use several V_{DD} and V_{SS} pads around the ring.

7.3.3 Pad Design

Pads used for input and output signals require different supporting circuitry. A pad used for both input and output, sometimes known by its trade name of Tri-state pin[1], combines elements of both inputs and output pads.

The input pad may include circuitry to shift voltage levels, *etc.* The main job of an input pad is to protect the chip core from static electricity. People or equipment can easily deliver a large static voltage to a pin when handling a chip. MOS circuits are particularly sensitive to static discharge because they use thin oxides—the transistor photomicrograph in Chapter 2 shows how small the gate oxide is in comparison with the submicron-length channel. The gate oxide, which may be a few hundred Ang-

---

1. Tri-state is a trademark of National Semiconductor.

stroms thick, can be totally destroyed by a single static jolt, shorting the transistor's gate to its channel.

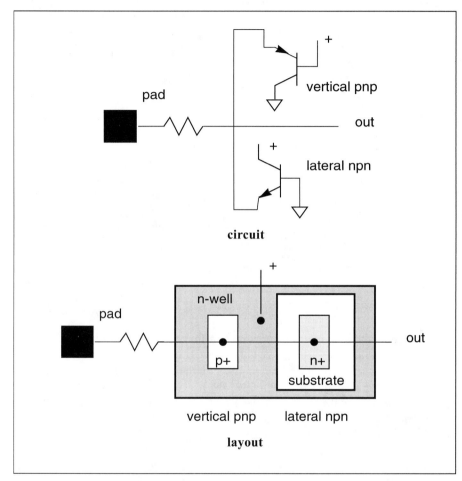

Figure 7-22: *Electrostatic discharge protection using parasitic bipolar transistors. (After Glasser and Dobberpuhl [Gla85], The Design and Analysis of VLSI Circuits, © 1985 by Addison-Wesley Publishing Company, Inc. Reprinted with permission of the publisher.)*

An input pad puts protective circuitry between the pad itself and the transistor gates in the chip core. Electrostatic discharge (ESD) can cause two types of problems: dielectric rupture and charge injection [Vin98]. When the dielectric ruptures, chip structures can short together. The charge injected by an ESD event can be sufficient to damage the small on-chip devices.

A commonly used **electrostatic discharge** (ESD) protection circuit is shown in Figure 7-22 [Gla85]. The resistor, which is usually made of a long diffusion run between the pad and the protection circuitry, helps limit the current caused by a voltage spike. Parasitic bipolar transistors are used as diodes to draw excess current from the output node. The npn transistor limits the negative-going voltage swing to 0.7 V below V_{SS}, while the pnp transistor limits the positive-going swing to 0.7 V above V_{DD}. The standard masks can be used to create both the pnp and npn transistors, but the layout must be carefully designed to minimize the chance of latch-up.

Figure 7-23: An output pad circuit.

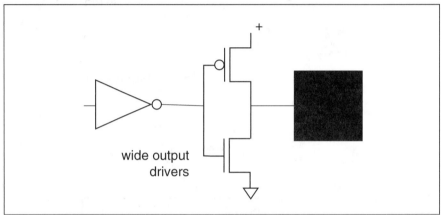

Electrostatic discharge protection is not needed for an output pad because the pad is not connected to any transistor gate. The main job of an output pad is to drive the large capacitances seen on the output pin. Within the chip, λ scaling ensures that we can use smaller and smaller transistors, but the real world doesn't shrink along with our chip's channel length. The output pad's circuitry, shown in Figure 7-23, includes a chain of inverters to drive the large off-chip load. Figure 7-25 shows the layout of an SCMOS output pad from MOSIS. Because the final stages use very large transistors, creative layout techniques are used to reduce the pad's size. Transistors are often folded to reduce the pad's height; the transistors may also be wrapped around the extra space surrounding the pad.

Three-state pads, used for both input and output, help solve the pin count crunch—if we don't need to use all combinations of inputs and outputs simultaneously, we can switch some pins between input and output. The pad cannot, of course, be used as an input and output simultaneously—the chip core is responsible for switching between modes. The pad requires electrostatic discharge protection for when the pad is used as an input, an output driver for when it is used as an output, plus circuitry to switch the pad between input and output modes. The circuit of Figure 7-24 can be used for mode switching. The n-type and p-type transistors are used to drive the pad when it is used as an output—the logic gates are arranged so that the output signal turns on exactly

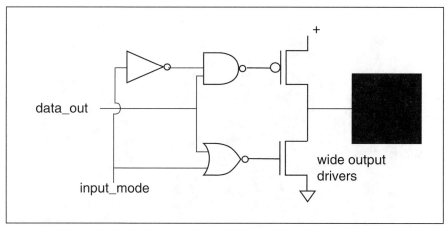

Figure 7-24: *A three-state pad circuit.*

one of the two transistors when input_mode is 0. To use the pad as an input, input_mode is set to 1: the NOR gate emits a 0, turning off the pulldown, and the NAND gate emits a 1, turning off the pullup. Since both driver transistors are disabled, the pad can be used as an input. (The required ESD circuitry is not shown in the schematic.)

Pads may also include circuitry to support **boundary scan** [Par92], which configures the chip's pins as an LSSD chain. Chips that support boundary scan can be chained to form a single scan path for all the chips on a printed circuit board. Boundary scan makes the printed circuit board much easier to test because it makes the chips separately observable and controllable. Boundary scan support requires some circuitry on the pins corresponding to the chip's primary inputs and outputs, along with a small controller and a few pins dedicated to boundary scan configuration.

7.4 References

Physical Design Automation of VLSI Systems [Pre88] includes chapters on placement and routing, which cover a variety of algorithms. Glasser and Dobberpuhl [Gla85] describe pad buffer design and electrostatic discharge protection. Friedman's edited volume of collected papers on clock distribution [Fri95] provides a valuable overview of the electrical problems of clock distribution.

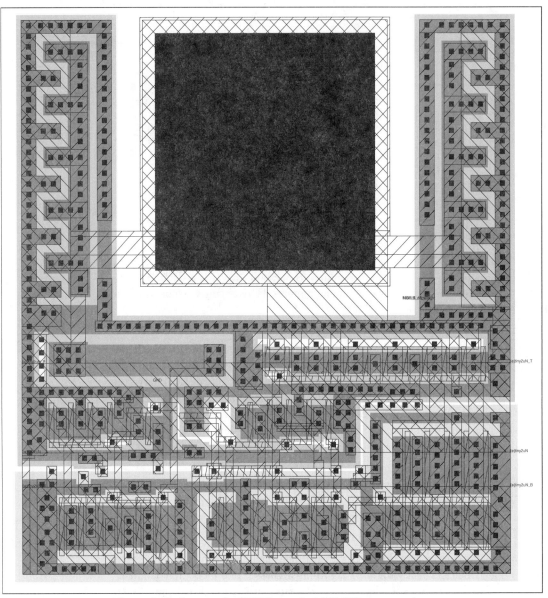

Figure 7-25: *Layout of a MOSIS-supplied SCMOS output pad.*

7.5 Problems

7-1. Pins with the same name in the floorplan below are on the same net. Design a global route, equalizing channel utilization and wire lengths.

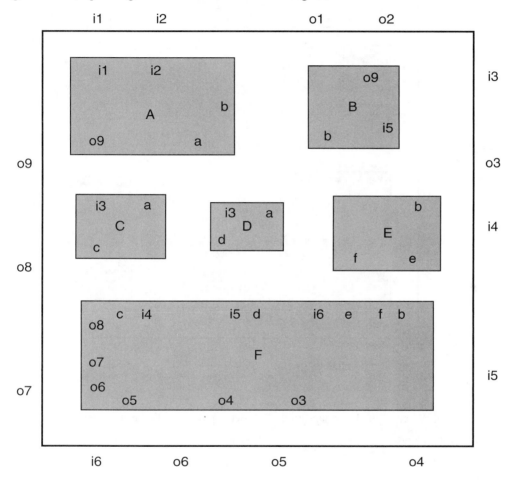

7-2. For the floorplan and channel decomposition below:

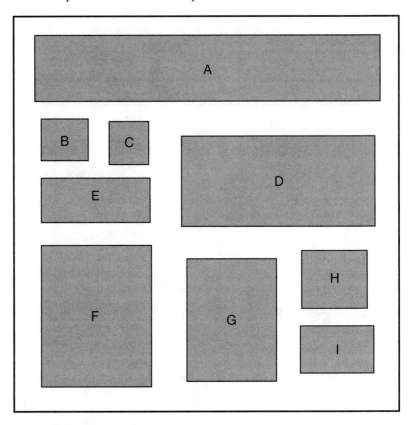

a) Draw a channel graph.

b) Give a feasible order for routing the channels.

7-3. Which of the floorplans on the following page is a slicing structure? Explain.

a)

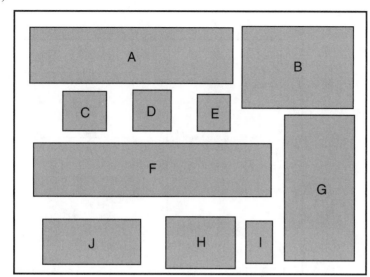

b)

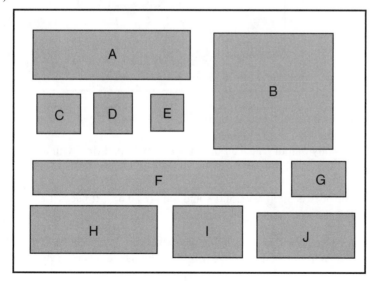

7-4. Can the floorplan below be routed with planar power and ground nets? If so, give a routing. If not, explain why.

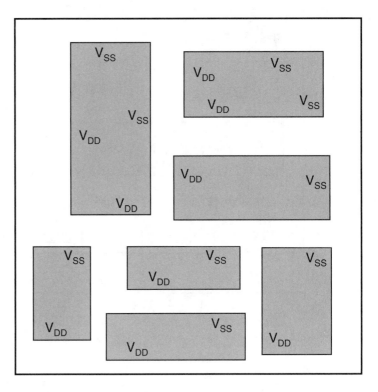

7-5. How much current can be supplied by a 10 λ wide power line in a pad ring?

7-6. How many output pads can be supported by a pad ring with a 10 λ wide power line?

7-7. A chip core is 3000 λ × 2500 λ and requires 0.8 A. It needs 18 signal input pads and 19 signal output pads.

    a) How many V_{DD} and V_{SS} pads are required assuming a 12 λ power ring?

    b) Will the total chip size be limited by the chip core or by the pad ring?

7-8. Why doesn't an output pad require electrostatic discharge protection circuitry?

7-9. A standard TTL load is 50 pF.

a) If each stage of an output pad is 1.4 times larger than the previous stage, how large are the transistors in the pad's final stage such that the output pin's rise or fall time is no more than 2 *ns*?

b) How many stages are required to buffer a signal from a minimum-size inverter?

7-10. Assume that a TTL-compatible output pad must produce 8 *mA* of current.

a) Counting only the output-stage current, how wide must the power lines be to support ten such output pads?

b) What is a reasonable estimate of the total current required for each pad, including both intermediate drivers and the output stage? How much does this increase the required width of the power lines?

8

Architecture Design

Highlights:

Hardware description languages (HDLs).

Register-transfer design.

High-level synthesis.

Low-power architectures.

Systems-on-chips and embedded CPUs.

Architecture testing.

8.1 Introduction

A good digital system design is more than a jumble of components. You must design an **architecture** that executes the desired function and that meets area, performance, and testability constraints. Simply executing the specified function is the easy part— there are many candidate architectures that will execute almost any function. What makes chip design challenging is sorting through all the possible designs to find those few which are small and fast enough.

We'll start with a review of hardware description languages (HDLs). HDLs allow us to capture designs at a variety of levels of abstraction, but they frequently used for register-transfer designs. The register-transfer abstraction is the foundation of architecture design. A register-transfer is a complete specification of what the chip will do on every cycle. However, even though the register-transfer system is specified only as Boolean functions, not logic, the burden of specifying the complete sequential behavior can be too great, causing you to miss important architectural opportunities. By studying scheduling and binding, we can understand how to optimize a register-transfer design to improve area, speed, and testability. During the design process, modeling the system as a program is critical to ensuring that what you design is what you intended.

8.2 Hardware Description Languages

8.2.1 Modeling with Hardware Description Languages

Hardware description languages (HDLs) are the most important modern tools used to describe hardware. HDLs become increasingly important as we move to higher levels of abstraction. While schematics can convey some information very clearly, they are generally less dense than textual descriptions of languages. Furthermore, a textual HDL description is much easier for a program to generate than is a schematic with pictorial information such as placement of components and wires.

In this section we will introduce the use of the two most widely used hardware description languages, **Verilog** [Tho98,Smi00] and **VHDL** [IEE93,Bha95], in architectural and logical modeling. Since both these languages are built on the same basic framework of event-driven simulation, we will start with a description of the fundamental concepts underlying the two languages. We will then go on to describe the details of using VHDL and Verilog to model hardware. We don't have room to discuss all the details of Verilog and VHDL modeling, but this brief introduction should be enough to get you started with these languages. We will also briefly consider the use of C as a hardware modeling language.

Both Verilog and VHDL started out as simulation languages—they were designed originally to build efficient simulations of digital systems. Some other hardware description languages, such as **EDIF**, were designed to describe the structure of nets and components used to build a system. Simulation languages, on the other hand, are designed to be executed. Simulation languages bear some resemblance to standard programming languages. But because they are designed to describe the parallel execution of hardware components, simulation languages have some fundamental differences from sequential programming languages.

There are two important differences between simulation and sequential programming languages. First, statements are not executed in sequential order during simulation. When we read a sequential program, such as one written in C, we are used to thinking about the lines of code being executed in the order in which they were written. In contrast, a simulation may describe a series of logic gates all of which may change their outputs simultaneously. If you have experience with a parallel programming language, you may be used to this way of thinking. Second, most simulation languages must support some notion of real time in order to provide useful results. Even parallel programming languages usually do not explicitly support real time. Time may be measured in nanoseconds for more realistic simulation or in some more abstract unit such as gate delays or clock cycles in faster, more abstract simulators. One important job of the simulator is to determine how long it takes to compute a given value. Delay information determines not only clock speed but also proper operation: glitches caused by unbalanced delays may, for example, cause a latch to be improperly clocked. Simulating functional behavior in the absence of time can be relatively easy; however, the simulator must go to a great deal of effort to compute the time at which values are computed by the simulated hardware.

Simulation languages serve as specialized programming languages for the **simulation engines** that execute simulations. Both VHDL and Verilog are built on top of **event-driven simulators**.

Event-driven simulation is a very efficient algorithm for hardware simulation because it takes advantage of the activity levels within the hardware simulation. In a typical hardware design, not all the nets change their values on every clock cycle: having fewer than 50% of the nets in a system keep their value on any given clock cycle is not unusual. The most naive simulation algorithm for a clocked digital system would scan through all the nets in the design for every clock cycle. Event-driven simulation, in contrast, ignores nets that it knows are not active.

An **event** has two parts: a value and a time. The event records the time at which a net takes on a new value. During simulation, a net's value does not change unless an event records the change. Therefore, the simulator can keep track of all the activity in the system simply by recording the events that occur on the nets. This is a sparse representation of the system's activity that both saves memory and allows the system activity to be computed more efficiently.

Figure 8-1: Event-
driven simulation
of a gate.

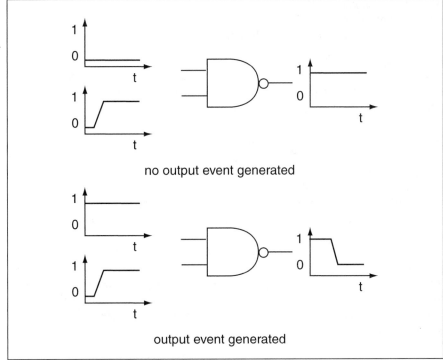

Figure 8-1 illustrates the event-driven simulation of gates; the same principle can be applied to digital logic blocks at other levels of abstraction as well. The top part of the figure shows a NAND gate with two inputs: one input stays at 0 while the other changes from a 0 to a 1. In this case, the NAND gate's output does not change—it remains 1. The simulator determines that the output's value does not change. Although the gate's input had an event, the gate itself does not generate a new event on the net connected to its output. Now consider the case shown on the bottom part of the figure: the top input is 1 and the bottom input changes from 0 to 1. In this case, the NAND gate's output changes from 1 to 0. The activity at the gate's input in this case causes an event at its output.

The event-driven simulator uses a **timewheel** to manage the relationships between components. As shown in Figure 8-2, the timewheel is a list of all the events that have not yet been processed, sorted in time. When an event is generated, it is put in the appropriate point in the timewheel's list. The simulator therefore processes events in the time order in which they occur by pulling events in order from the head of the timewheel's list. Because a component with a large internal delay can generate an event far in the future from the event that caused it, operations during simulation may

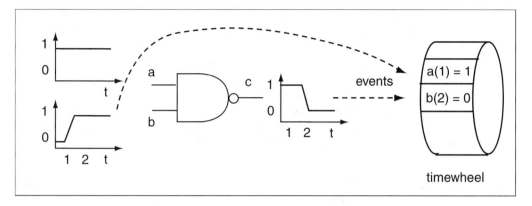

Figure 8-2: *The event-driven timewheel.*

occur in a very different order than is apparent from the order of statements in the HDL program.

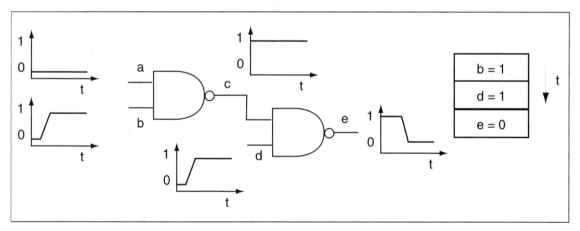

Figure 8-3: *Order of evaluation in event-driven simulation.*

As shown in Figure 8-3, an event caused by the output of one component causes events to appear at the inputs of the components being driven by that net. As events are put into the timewheel, they are ordered properly to ensure causality, so that the simulator events are processed in the order in which they occur in the hardware. In the figure, the event at the input causes a cascade of other events as activity flows through the system.

There are two ways to describe a design for simulation: **structural** or **behavioral modeling**. A structural model for a component is built from smaller components. The structural model specifies the external connections, the internal components, and the nets that connect them. The behavior of the model is determined by the behavior of the components and their connections. A behavioral model is more like a program—it uses functions, assignments, etc. to describe how to compute the component's outputs from its inputs. However, the behavioral model deals with events, not with variables as in a sequential programming language. Simulation languages define special constructs for recognizing events and for generating them.

Whether a component is described structurally or behaviorally, we must define it and use it. As in a programming language, a hardware description language has separate constructs for the type definition of a component and for instantiations of that component in the design of some larger system. In C, the statement struct { int a; char b; } mydef; defines a data structure called mydef. However, that definition does not allocate any instances of the data structure; memory is committed for an instance of the data structure only by declaring a variable of type mydef. Similarly, in order to use a component, we must have a definition of that component available to us. The module that uses the component does not care whether the component is modeled behaviorally or structurally. In fact, we often want to simulate the system with both behavioral and structural models for key components in order to verify the correctness of the behavioral and structural models. Modern hardware description languages provide mechanisms for defining components and for choosing a particular implementation of a component model.

We can use the simulator to exercise a design as well as describe it. Testing a design often requires complex sequences of inputs that must be compared to expected outputs. If you were testing a physical piece of hardware, you would probably wire up a test setup that would supply the necessary input vectors and capture and compare the resulting output vectors. We can do the equivalent during simulation. We build components to generate the inputs and test the outputs; we then wire them together with the component we want to test and simulate the entire system. This sort of simulation setup is known as a **testbench**.

Both VHDL and Verilog were originally designed as simulation languages. However, one of their principal uses today is as a synthesis language. A VHDL or Verilog model can be used to define the functionality of a component for logic synthesis (or for higher levels of abstraction such as behavioral synthesis). The synthesis model can also be simulated to check whether it is correct before going ahead with synthesis. However, not all simulatable models can be synthesized. Synthesis tools define a **synthesis subset** of the language that defines the constructs they know how to handle. The synthesis subset defines a modeling style that can be understood by the synthesis tool and also provides reasonable results during simulation. There may in fact be several different synthesis subsets defined by different synthesis tools, so you should understand the synthesis subset of the tool you plan to use.

The most common mode of synthesis is **register-transfer synthesis**. RT synthesis uses logic synthesis on the combinational logic blocks to optimize their implementations, but the registers are placed at the locations specified by the designer. The result is a sequential machine with optimized combinational logic. The signals in the combinational section of the model may or may not appear in the synthesized implementation, but all the registers appear in the implementation. (Some register-transfer synthesis tools may use state assignment algorithms to assign encodings to symbolic-valued symbols.) Although there are several RT synthesis tools on the market, the most commonly used RT synthesis subset is the one defined for the Synopsys Design Compiler(TM), which is accepted by that tool and several others.

One item to look out for in RT synthesis is the **inferred storage element**. A combinational logic block is defined by a set of assignments to signals. If those signals form a cycle, many synthesis tools will insert a storage element in the loop to break the cycle. While the inferred storage element can be handy if you want it to appear, it can cause confusion if the combinational cycle was caused by a bug in the synthesis model. The synthesis tool will emit a warning message when an inferred storage element is inserted into an implementation; that warning is generally used to warn of an unintended combinational cycle in the design.

8.2.2 VHDL

VHDL is a general-purpose programming language as well as a hardware description language, so it is possible to create VHDL simulation programs ranging in abstraction from gate-level to system. VHDL has a rich and verbose syntax that makes its models appear to be long and verbose. However, VHDL models are relatively easy to understand once you are used to the syntax. We will concentrate here on register-transfer simulation in VHDL, using the sequencer of the traffic light controller of Section 5.4.2 as an example. The details of running the simulator will vary substantially depending on which VHDL simulator you use and your local system configuration. We will concentrate here on basic techniques for coding VHDL simulation models.

Figure 8-4:
Abstract types in
VHDL.

```
package lights is
    -- this package defines constants used by the
    -- traffic light controller light encoding
    subtype light is bit_vector(0 to 1);
    constant red : light := B"00";
    constant green : light := B"01";
    constant yellow : light := B"10";
end lights;
```

VHDL provides extensive type-definition facilities: we can create an abstract data type and create signals of that data type, rather than directly write the simulation model in terms of ones and zeroes. Abstract data types and constants serve the same purposes in hardware modeling that they do in programming: they identify design decisions in the source code and make changing the design easier. Figure 8-4 shows a set of type definitions for the traffic light controller. These data types are defined in a package, which is a set of definitions that can be used by other parts of the VHDL program. Since we encode traffic light values in two bits, we define a data type called light to hold those values. We also define the constants red, green, and yellow and their encoding; the syntax B"00" defines a constant bit vector whose value is 00. If we write our program in terms of these constants, we can change the light encoding simply by changing this package. (VHDL is case-insensitive: yellow and YELLOW describe the same element.) When we want to use this package in another section of VHDL code, the use statement imports the definitions in the package to that VHDL code.

Figure 8-5: A
VHDL entity decla-
ration.

```
-- define the traffic light controller's pins
entity tlc_fsm is
    port( CLOCK: in BIT; -- the machine clock
    reset : in BIT; -- global reset
    cars : in BIT; -- signals cars at farm road
    short, long : in BIT; -- short and long timeouts
    highway_light : out light := green; -- light values
    farm_light : out light := red ;
    start_timer : out BIT -- signal to restart timer
    );
end;
```

VHDL requires an entity declaration for each model; Figure 8-5 shows the entity declaration for the traffic light controller. The entity declaration defines the model's primary inputs and outputs. We can define one or more bodies of code to go with the entity; that allows us to, for example, simulate once with a functional model, then swap in a gate-level model by changing only a few declarations.

```
combin : process(state, hg)
begin
highway_light <= green;
end process combin;
```

Figure 8-6: A process in a VHDL model.

The basic unit of modeling in VHDL is the **process**. A process defines the actions that are taken whenever any input to the process is activated by an event. As shown in Figure 8-6, a process starts with the name of the process and a **sensitivity list**. The sensitivity list declares all the signals to which the process is sensitive: if any of these signals changes, the process should be evaluated to update its outputs. In this case, the process proc1 is sensitive to a, b, and c. Assignment to a signal are defined by the <= symbol. The first assignment is straightforward, assigning the output x to the or of inputs a and b.

```
if (b or c) = '1' then           if (b or c) = '1' then
     y <= '1';                        y <= '1'
else                             else
     y <= '0';                        z <= a or b;

  assignment to y                 assignment to y or z
```

Figure 8-7: Conditional assignment in VHDL.

The second statement in the example defines a conditional assignment to y. The value assigned to y depends on the value of the conditional's test. Figure 8-7 shows the combinational logic that could be used to implement this statement.

What if a signal is not assigned to in some case of a conditional? Consider, for example, the conditional of Figure 8-7. If (b or c) = '1' then y is assigned a value; if not, then z is assigned a value. This statement illustrates some subtle differences between the semantics of simulation and synthesis:

- During simulation, the simulator would test the condition and execute the statements in the selected branch of the conditional. The signal referred to in the branch not taken would retain its value since no event is generated for that signal. This case is somewhat similar to sequential software.

- Synthesis may interpret this statement as don't-care conditions for both y and z: y's value is a don't-care if (b or c) is not '1', while z is a don't care if (b or c) is '1'. However, unlike in software, both y and z are always evaluated. Although a C program with this sort of conditional would assign to either y or z but not both, the logic shown in the figure makes it clear that both y and z both are combinational logic signals.

These differences are minor, but they do highlight the differences between simulation and logic synthesis. A simulation run results in a single execution of the machine; with different inputs, the simulation would have produced different outputs. Don't-care values could be used in simulation, but they can cause problems for later stages of logic that may not know what value to produce. Logic synthesis, in contrast, results in the structure of the machine that can be run to produce desired values. Don't-care values are very useful to logic synthesis during minimization.

Table 8-1 *Some elements of VHDL syntax.*

a and b	Boolean AND
a or b	Boolean OR
not a	Boolean NOT
a <= b	signal assignment, less than or equal to
a = b	equality
a = b	equality
after 5 ns	time

Table 7-1 shows the syntax of a few typical VHDL expressions. VHDL modelers can build complex signals with arrays of signals and bundles of different signals. Signals also need not carry binary values. By defining a series of VHDL functions, one can create a signal definition that works on a variety of logical systems: three-valued logic (0, 1, x); or symbolic logic such as the states of a state machine (s1, s2, s3). VHDL defines a basic bit type that provides two values of logic, '0' and '1'. The library IEEE.std_logic_1164 defines a nine-valued signal type known as std_ulogic.

Here is a complete, simple VHDL model of the traffic light controller:

```
Library IEEE;
use IEEE.std_logic_1164.all;
use work.lights.all; -- use the traffic light controller data types

-- define the traffic light controller's pins
entity tlc_fsm is
    port( CLOCK: in BIT; -- the machine clock
    reset : in BIT; -- global reset
    cars : in BIT; -- signals cars at farm road
    short, long : in BIT; -- short and long timeouts
    highway_light : out light := green; -- light values
    farm_light : out light := red ;
    start_timer : out BIT -- signal to restart timer
        );
end;

-- define the traffic light controller's behavior
architecture register_transfer of tlc_fsm is

        -- internal state of the machine
        -- first define a type for symbolic control states,
        -- then define the state signals
        type ctrl_state_type is (hg,hy,fg,fy);
        signal ctrl_state, ctrl_next : ctrl_state_type := hg;

begin

-- the controller for the traffic lights
ctrl_proc_combin : process(ctrl_state, short, long, cars)
begin
if reset = '1' then
        -- reset the machine
        ctrl_next <= hg;
 else
        case ctrl_state is
        when hg =>
            -- set lights
            highway_light <= green;
            farm_light <= red;
            -- decide what to do next
            if (cars and long) = '1' then
                ctrl_next <= hy;
                start_timer <= '1';
            else -- state doesn't change
                ctrl_next <= hg;
```

```vhdl
                    start_timer <= '0';
                end if;
            when hy =>
                -- set lights
                highway_light <= yellow;
                farm_light <= red;
                -- decide what to do next
                if short = '1' then
                    ctrl_next <= fg;
                    start_timer <= '1';
                else
                    ctrl_next <= hy;
                    start_timer <= '0';
                end if;
            when fg =>
                -- set lights
                highway_light <= red;
                farm_light <= green;
                -- decide what to do next
                if (not cars or long) = '1' then -- sequence to yellow
                    ctrl_next <= fy;
                    start_timer <= '1';
                else
                    ctrl_next <= fg;
                    start_timer <= '0';
                end if;
            when fy =>
                -- set lights
                highway_light <= red;
                farm_light <= yellow;
                -- decide what to do next
                if short = '1' then
                    ctrl_next <= hg;
                    start_timer <= '1';
                else
                    ctrl_next <= fy;
                    start_timer <= '0';
                end if;
        end case; -- main state machine
    end if; -- not a reset
end process ctrl_proc_combin;

-- the sync process updates the present state of the controller
sync: process(CLOCK)
begin
    wait until CLOCK'event and CLOCK = '1';
    ctrl_state <= ctrl_next;
end process sync;

end register_transfer;
```

The description has several parts. The first statements declare the libraries needed by this model. The VHDL simulator or synthesis tool gathers the declarations and other information it needs from these libraries. The next statement is the entity declaration. After the entity declaration, we can have one or more architecture statements. An architecture statement actually describes the component being modeled for simulation or synthesis. We may want to have several architecture description for a component at different levels of abstraction or to have faster simulation models for some purpose. The architecture of this model is named register_transfer; this name has no intrinsic meaning in VHDL and is used only to identify the model. After the architecture declaration proper, we can define signals, required type definitions, etc.

This model has two processes, one for the combinational behavior and another for the sequential behavior. Each process begins with its sensitivity list—the signals that should cause this process to be reevaluated when they change. The **combinational process** first uses an if to check for reset, then uses a case statement to choose the right action based on the machine's current state. In each case, we may examine primary inputs that help determine the proper action in this state, then set outputs and the next state as appropriate. There may be several combinational processes in a register-transfer model, which would correspond to a system partitioned into several communicating machines.

The **sequential process** is written in a particular style that is recognized by synthesis and works properly during simulation. The sequential process is activated by activity on the clock or reset lines. A reset causes the machine's state to be reset. A clock edge is tested for by the condition (CLOCK'event AND CLOCK = '1'), which checks for an event on CLOCK and a '1' value for CLOCK after the event takes place. During simulation, this statement ensures that the machine's state changes only on a positive clock edge. Logic synthesis looks for this statement to identify the sequential process, which tells the synthesis tool what signals need flip-flops to hold the machine's state.

Figure 8-8 shows several useful constructs in VHDL. The output avec is defined as a vector using the std_logic_vector type defined by the std_logic library. This definition defines the bits of the vector from 11 downto 0, rather than from 0 up to 11; this makes a difference if two 12-bit vectors with opposite endianness are connected together. The constant zerovec is a constant value of a vector type; constants may also be scalars. The last construct shows an adder; the + symbol is overloaded by the library to provide the necessary functionality.

Figure 8-8: Several useful VHDL constructs.

```
            avec : out std_logic_vector(11 downto 0);

                              vector

         constant zerovec: std_logic_vector(0 to 7) := "00000000";

                          constant vector

        sum <= a + b;

                              adder
```

Logic synthesizers will generally add inferred memory elements to break combinational cycles. Such inferred latches should be carefully inspected to be sure that they are not the result of errors in the combinational logic description.

We need to execute this process on simulation vectors to be sure it is correct. Your simulator may have a graphical user interface which lets you enter and see waveforms on your screen. It is also possible to write a VHDL model which exercises the simulator. The virtue of this approach is that it captures your simulation vector set, allowing you to run the vectors many times on the design as it evolves and save the exerciser program as documentation of your design.

Here is a testbench for the traffic light controller:

```
Library IEEE;
use IEEE.std_logic_1164.all;
use work.lights.all;
use work.tlc_fsm;

entity tlc_fsm_exerciser is
    -- this entity declaration is purposely empty
end;

architecture stimulus of tlc_fsm_exerciser is

    component tlc_fsm -- tlc_fsm is the circuit under test
        -- this port declaration is a copy of the declaration
        -- in the tlc_fsm register_transfer model
        port( CLOCK: in BIT; -- the machine clock
        reset : in BIT; -- global reset
```

```
        cars : in BIT; -- signals cars at farm road
        short, long : in BIT; -- short and long timeouts
        highway_light : out light := green; -- light values
         farm_light : out light := red ;
        start_timer : out BIT -- signal to restart timer
        );
        end component;

-- the signals which connect to the
-- circuit under test
signal clock, reset, cars, short, long, start_timer : BIT;
signal highway_light, farm_light : light;

begin

-- connect the exerciser's signals to the
-- circuit under test
    tlc_fsm_cut : tlc_fsm port map(clock,reset,cars,short,long,
highway_light,farm_light,start_timer);

-- the tester process generates outputs and checks inputs
tester : process
begin
        -- reset the circuit under test
        reset <= '1';
        clock <= '0'; wait for 5 ns; clock <= '1'; wait for 5 ns; -- tick
        reset <= '0';
        -- check that machine is in HG state
        assert(highway_light = green);
        assert(farm_light = red);
        -- put a car at the farm road
        -- should respond after long timeout
        cars <= '1';
        clock <= '0'; wait for 5 ns; clock <= '1'; wait for 5 ns; -- tick
        assert(highway_light = green);
        assert(farm_light = red);
        long <= '1';
        clock <= '0'; wait for 5 ns;
        assert(start_timer = '1');
        clock <= '1'; wait for 5 ns; -- tick
        assert(highway_light = yellow);
        assert(farm_light = red);
end process tester;

end stimulus;

-- tell VHDL which tlc_fsm to use
configuration stimulate of tlc_fsm_exerciser is
        for stimulus
            for all : tlc_fsm
                use entity work.tlc_fsm(register_transfer);
            end for;
```

```
                    end for;
                end stimulate;
```

This code isn't meant to be a thorough test of the machine, but an example of how to write such exerciser programs. We used the assert statement to test the values emitted by the sequencer: if the condition specified in the assertion isn't true, the simulator stops and flags the error. This testbench assumes that the traffic light controller model was separately compiled and available to the compiler. The testbench includes one architecture declaration that has no inputs or outputs. It defines an instance of the traffic light controller component to be tested, naming the component UUT. (UUT stands for **unit under test**, a testing term for the component being tested.) The port declaration in this architecture provides the local names for the signals wired to tlc_fsm. The testbench has two processes that apply signals to tlc_fsm: one applies a reset signal; the other applies a simple sequence of inputs. Simulators generally let you interactively examine signals so no explicit output is necessary. The final declaration in the testbench is the configuration statement, which binds the instantiated component UUT to a particular entity, namely tlc_fsm, stored in a library.

8.2.3 Verilog

Figure 8-9: A structural Verilog model.

```
// this is a comment
module adder(a,b,cin,sum,cout);
    input a, b, cin;
    output sum, cout;

    // sum
    xor #2
        s(sum,a,b,cin);
    // carry out
    and #1
        c1(x1,a,b);
        c2(x2,a,cin);
        c3(x3,b,cin);
    or #1
        c4(cout,x1,x2,x3);
endmodule
```

Verilog is in many respects a very different language from VHDL. Verilog has much simpler syntax and was designed for efficient simulation (although it now has a synthesis subset). Figure 8-9 gives a simple Verilog structural model of an adder. The module statement and the succeeding input and output statement declare the adder's inputs and outputs. The following statements define the gates in the adder and the connections between them and the adder's pins. The first gate is an XOR; the #2 modifier declares that this gate has a delay of two time units. The XOR's name is s and its parameters follow. In this case, all the XOR's pins are connected directly to other pins. The next statements define the AND and OR gates used to compute the carry out. Each of these gates is defined to have a delay of one time unit. The carry out requires internal wires c1, c2, and c3 to connect the AND gates to the OR gate. These names are not declared in the module so Verilog assumes that they are wires; wires may also be explicitly declared using the wire statement at the beginning of the module.

a & b	Boolean AND
a | b	Boolean OR
~a	Boolean NOT
a = b	assignment
a <= b	concurrent assignment, less than or equal to
a >= b	greater than or equal to
==	equality
2'b00	two-bit binary constant with value 00
#1	time

Table 8-2 *Some elements of Verilog syntax.*

Table 7-2 summarizes some basic syntax for Verilog expressions. We can use the 'define compiler directive (similar to the C #define preprocessor directive) to define a constant:

'define aconst 2'b00

Verilog uses a four-valued logic that includes the value x for unknown and z for high-impedance. Table 7-4 and Table 7-3 show the truth tables for four-valued AND and OR functions. These additional logic values help us better simulate the analog behavior of digital circuits. The high-impedance z captures the behavior of disabled three-

Figure 8-10: A
Verilog testbench.

```
module testbench;
    // this testbench has no inputs or outputs
    wire awire, bwire, cinwire, sumwire, coutwire;

    // declare the adder and its tester
    adder a(awire,bwire,cinwire,sumwire,coutwire);
    adder_teser at(awire,bwire,cinwire,sumwire,coutwire);
endmodule

module adder(a,b,cin,sum,cout);
    input a, b, cin;
    output sum, cout;

    // sum
    xor #2
        s(sum,a,b,cin);
    // carry out
    and #1
        c1(x1,a,b);
        c2(x2,a,cin);
        c3(x3,b,cin);
    or #1
        c4(cout,x1,x2,x3);
endmodule

module adder_tester(a,b,cin,sum,cout);
    input sum, cout;
    output a, b, cin;
    reg a, b, cin;

    initial
        begin
            $monitor($time,,"a=%b, b=%b, cin=%cin, sum=%d,
cout=%d",
                        a,b,cin,sum,cout);
            // waveform to test the adder
            #1 a=0; b=0; cin=0;
            #1  a=1; b=0; cin=0;
            #2  a=1; b=1; cin=1;
            #2 a=1; b=0; cin=1;
        end
endmodule
```

valued gates. The unknown value is a conservative, pessimistic method for dealing with unknown values and is particularly helpful in simulating initial conditions. A circuit often needs a particular set of initial conditions to behave properly, but nodes may come up in unknown states without initialization. If we assumed that certain nodes were 0 or 1 during simulation, we may optimistically assume that the circuit works

Table 8-3 Truth table for OR in four-valued logic.

	0	1	x	z
0	0	1	x	x
1	1	1	1	1
x	x	1	x	x
z	x	1	x	x

Table 8-4 Truth table for AND in four-valued logic.

	0	1	x	z
0	0	0	x	x
1	0	1	x	x
x	x	x	x	x
z	x	x	x	x

when, in fact, it fails to operate in some initial conditions. The unknown x is an absorbing node value, as illustrated by a comparison of the four-valued AND and OR functions. The AND function's output is unknown if either of the inputs is x (or z). That is because the AND's value is 1 only when both inputs are 1; when we do not know one of the inputs, we cannot know whether the output is 0 or 1. The OR function, in contrast, has an x output only when one of the inputs is 0 and the other is x; if one input is 1, the output is known to be 1 independent of the other input's value.

Figure 8-10 shows a testbench for the adder. The testbench includes three modules. The first is the testbench itself. The testbench wires together the adder and the adder tester; it has no external inputs or outputs since none are needed. The second module is the adder, which is unchanged from Figure 8-9. The third module is the adder's tester. adder_test generates a series of inputs for the adder and prints the results. In order to hold the adder's inputs for easier observation of its behavior, adder_test's outputs are declared as registers with the reg statement. The initial statement allows us to define the initial behavior of the module, which in this case is to apply a stimulus to the adder and print the results. The $monitor statement is similar to a C printf statement. The succeeding statements define the changes to signals at different times. The first #1 line occurs one time unit after simulation starts; the second statement occurs one time unit later; the last two each occur two time units apart. At each time point, the $monitor statement is used to print the desired information.

Here is a synthesizable register-transfer Verilog model for the sequencer of the traffic light controller of Section 5.4.2 as an example.

```
module tlc_fsm(clock,reset,cars,short,long,
               highway_light,farm_light,start_timer);
    input clock, reset, cars, short, long;
    output [1:0] highway_light, farm_light;
    output start_timer;

    reg [1:0] highway_light, farm_light;
    reg start_timer;

    reg [1:0] current_state, next_state;

    // light encoding: 11 = green, 00 = red, 10 = yellow
'define GREEN '2b11
'define RED '2b00
'define YELLOW '2b10

    // state encoding: 00 = hwy-green, 01 = hwy-yellow,
    // 10 = farm-green, 11 = farm-yellow
'define HG '2b00
'define HY '2b01
'define FG '2b10
'define FY '2b11

    // combinational portion
    always @(ctrl_state or short or long or cars) begin
    case (ctrl_state)
        when HG: begin // state hwy-green
            // set lights
            highway_light = GREEN;
            farm_light = RED;
            // decide what to do next
            if (cars and long) then
                begin ctrl_next = HY; start_timer = 1; end
            else begin ctrl_next = HG; start_timer = 0; end
        end
    when HY: begin // state highway-yellow
        // set lights
        highway_light = YELLOW;
        farm_light = RED;
```

```
        // decide what to do next
          if (short) then begin ctrl_next = FG; start_timer = 1; end
          else begin ctrl_next = HY; start_timer = 0; end
    end
    when FG: // state farm-green
        // set lights
        highway_light = RED;
        farm_light = green;
        // decide what to do next
          if (~cars or long) then
              begin ctrl_next = FY; start_timer = 1; end
          else begin ctrl_next = FG; start_timer = 0; end
    end
    when FY: // state farm-yellow
        // set lights
        highway_light = RED;
        farm_light = YELLOW;
        // decide what to do next
          if (short) then begin ctrl_next = HG; start_timer = 1; end
          else begin ctrl_next = FY; start_timer = 0; end
        end
    endcase
    end

    // sequential portion
    always @(posedge clock or negedge reset) begin
        if (~reset)
            ctrl_state <= 0;
        else
            ctrl_state <= ctrl_next;
    end
    endmodule
```

The model defines internal multi-bit signals to hold the machine's current and next state; those signals are registered to maintain the machine's state. The model has two sections, one for the combinational logic and another to hold the machine's state. Each section is defined as a process using the **always** statement. The synthesis tool relies on this syntactic structure to determine the various components of the synthesizable model.

Each section is guarded by an always statement that defines when each is evaluated. The combinational portion is executed when any of the combinational logic's inputs (including the machine's current state) changes as specified by the @() expression that guards the combinational portion. All the combinational inputs need to be included in the guard statement and each output should be assigned to in every case of the combinational block. The combinational section uses a case statement to extract out the various possible states, which in this case are given a specific binary encoding.

The sequential portion is executed on either a positive clock edge or a downgoing reset signal. The reset condition should be specified first, with the clock edge behavior written in the final else clause. The sequential portion uses a new form of assignment, the <= for non-blocking assignment. This form of assignment ensures that all the bits of the FSM's state are updated concurrently. The sequential portion changes little from synthesizable model to synthesizable model, unlike the combinational section, which defines the machine's unique logic.

Figure 8-11: A synthesizable Verilog description written with continuous assigns.

```
module adder2(a,b,cin,sum,cout);
    input a, b, cin;
    output sum, cout;

    // use continuous assign
    assign sum = a ^ b ^ cin; ^ is xor
    assign cout = (a and b) or (a and cin) or (b and cin);
endmodule;
```

Figure 8-11 shows an alternative synthesizable description of the adder. This model uses the assign statement, also known as the **continuous assign**. A continuous assign always drives values onto the net, independent of whether the assign's inputs have changed. The function is not implemented with logic gate models which would generate events only when their outputs changed.

Figure 8-12: A loop in synthesizable Verilog.

```
for (i=0; i<N; i=i+1)
    x[i] = a [i] & b[i];
```

We can use for loops in synthesizable Verilog to perform similar functions over arrays of signals. Figure 8-12 shows a for loop that is used to assign the value to each member of the array x[].

We can use the x value to specify don't-cares for logic synthesis. Assigning an x to a signal specifies a don't-care condition that can be used during logic synthesis.

Logic synthesis will create inferred storage elements when an output is not assigned to in a combinational section. Inferred storage elements should be carefully examined to ensure that they are wanted and not the result of a mistake in the combinational logic description.

8.2.4 C as a Hardware Description Language

Even if you can't use your program as the source for synthesis, the experience of writing and executing a register-transfer simulator is invaluable.

```
sync: process
begin
    wait until CLOCK'event and
CLOCK='1';
    state <= state_next;
    x <= x_next;
    y <= y_next;
end process sync;

combin: process
begin
    case state is
        when S0 =>
            x_next <= a + b;
            state_next <= S1;
        when S1 =>
            y_next <= a - c;
            if i1 = '1' state_next <= S3;
            else state_next <= S2; end if;
        ...
    end case;
end process combin;

            VHDL
```

```
while (TRUE) {
    switch (state) {
        state S0:
            x = a + b;
            state = S1;
            break;
        state S1:
            if (i1)
                y = a - c;
            if (i1)
                state = S3;
            else
                state = S2;
            break;
        ...
    }
}

            C
```

Figure 8-13:
Register-transfer
simulators in
VHDL and C.

When implementing hardware in a general-purpose programming language, you must add some simple mechanisms to keep the simulation going. The C program you write will be a special-purpose simulator designed to simulate one particular piece of hardware. In contrast, VHDL and Verilog use simulation engines which take in a separate description of the hardware to be simulated; as a result, these simulation engines can be used to simulate any sort of hardware model.

Figure 8-13 shows register-transfer simulator fragments in VHDL and C. In both languages, the simulation mechanism is based on a global variable (named state here) which holds the machine's current state; the states are defined by the constants S0,

S1, etc. In VHDL, the machine's state (as well as the state of each register) is held in a pair of signals: one for the present value and another for the next value. When the CLOCK signal is applied to the simulation, the next-state signal is copied into the state signal. The machine's next-state and output logic is contained in a separate process. A case statement tests the current state. Each when clause may set outputs either conditionally or unconditionally and sets the next state variable. C uses a similar mechanism: the switch statement acts as the next-state and output logic in the FSM—given a present state, it selects a case. However, the C program must provide its own clock by repeatedly executing the register-transfer code within a while loop: one execution of the while corresponds to one clock cycle. Registers are held in global variables in this scheme. Appendix C describes the construction of C register-transfer simulation models in more detail.

C is widely used in chip design for initial design capture. The fact that it does not capture the parallel nature of hardware description can be an advantage when creating an initial functional description of a chip. C becomes more important as levels of integration increase and chips implement more complex functions. The **SystemC** language has been created as an industry standard for system-level modeling. SystemC provides some mechanisms for the simulation of parallel processes, but it has a much simpler simulation engine than do VHDL or Verilog. The **SpecC** language is another industrial system-level modeling language. SpecC has more complex simulation semantics than does SystemC but is also aimed at the system-level design problem.

8.3 Register-Transfer Design

A register-transfer system is a pure sequential machine. It is specified as a set of memory elements and combinational logic functions between those elements, as shown in Figure 8-14. We don't know the structure of the logic inside the combinational functions—if we specify an adder, we don't know how the logic to compute the addition is implemented. As a result, we don't know how large or how fast the system will be. To make the sequential part more abstract, we can specify the logic over symbolic, rather than binary values—for example, specifying the output of a function to range over $\{\alpha,\beta,\chi,\delta\}$. The registers are generic flip-flops—we don't worry about clocking disciplines at this stage of design. Once the register-transfer has been completed, it can be mapped into a sequential system with a clocking discipline appropriate to the memory elements used.

The most common register-transfer description is the **block diagram**. A typical block diagram is shown in Figure 8-15. A block diagram is a purely structural

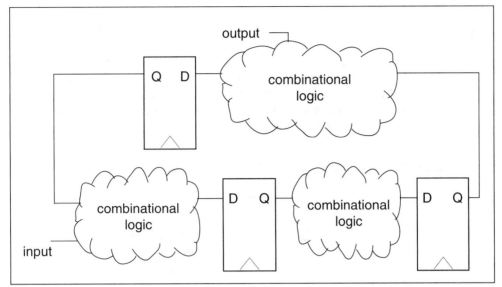

Figure 8-14: A register-transfer system.

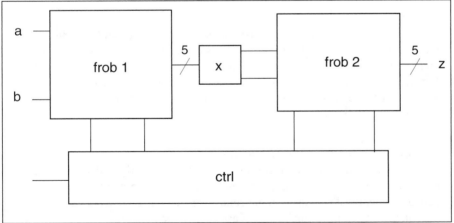

Figure 8-15: A typical block diagram.

description—it shows the connections between boxes. (The slash and 5 identify each of those wires as a bundle of five wires, such as the five bits in a data word.) If we know the boxes' functions, we can figure out the function of the complete system. However, many designers who sketch block diagrams are cavalier about defining their primitive elements and drawing all the wires. Wires may go into boxes with unclear functions, leaving the reader (for example, the person who must implement those blocks or the person who must figure out why the system doesn't work) at a loss. The block diagram also may not show all connections or may leave out some small bits of logic. While such omissions may make the major elements of the system easier to identify in the figure, it renders the diagram problematic as a specification of the design.

Register-transfer designs are often described in terms of familiar components: multiplexers, ALUs, etc. Standard components can help you organize your specification and they also give good implementation hints. But don't spend too much time doing logic design when sketching your block diagram. The purpose of register-transfer design is to correctly specify sequential behavior.

8.3.1 Data Path-Controller Architectures

Figure 8-16:
Data and control
are equivalent.

```
if i1 = '0' then
    o1 <= a;
else
    o1 <= b;
end if;

        control
```

o1 <= ((i1 = '0') and a) or ((i1 = '1') and b);

data

One very common style of register-transfer machine is the **data path-controller** architecture. We typically break architectures into data and control sections for convenience—the data section includes loadable registers and regular arithmetic and logical functions, while the control section includes random logic and state machines. Since few machines are either all data or all control, we often find it easiest to think about the system in this style.

The distinction between data and control is useful—it helps organize our thinking about the machine's execution. But that distinction is not rigid; data and control are equivalent. The two VHDL statements in Figure 8-16 correspond to the same combinational logic: the if statement in the control version corresponds to an or in the data version that determines which value is assigned to the o1 signal. We can use Boolean data operations to compute the control flow conditions, then add those conditions to any assignments to eliminate all traces of the control statement. The process can be reversed to turn data operations into control.

Operators such as adders are easily identifiable in the architectural description. As shown in Figure 8-17, some hardware is implicit. The if statement defines conditions under which a register is loaded and the source of its new value. Those conditions imply, along with control logic to determine when the register is loaded, a multiplexer to route the desired value to the register. We generally think of such multiplexers as data path elements in block diagrams, but there is no explicit mux operator in the architectural description.

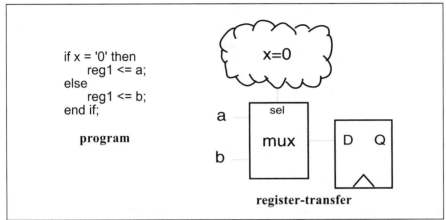

Figure 8-17:
Multiplexers are hidden in archi-tectural models.

Very few architectures are either all control (simple communications protocols, for example) or all data (digital filters). Most architectures require some control and some data. Separating the design into a controller and a data path helps us think about the system's operation. Separating control and data is also important in many cases to producing a good implementation—we saw in Chapter 6 that data operators have spe-cialized implementations and that control structures require very different optimiza-tion methods from those used for data. Figure 8-18 shows how we can build a complex system from a single controller for a single data path, or by dividing the nec-essary functions between communicating data path-controller systems.

8.3.2 ASM Chart Design

The **ASM chart** [Cla73] is a very useful abstraction for register-transfer design because it helps us to avoid over-specifying the partitioning of logic and to concen-trate on correctness. An ASM chart, like the one in Figure 8-19, looks like a flow-chart, but unlike a flowchart, it has precisely-defined hardware semantics. An ASM chart is a specification of a register-transfer system because it defines what happens on every cycle of operation. It is more of a functional specification than a block dia-gram—the flow of control from state to state is clearly shown. And, unlike a block diagram, it doesn't imply any partitioning of the functions to be executed.

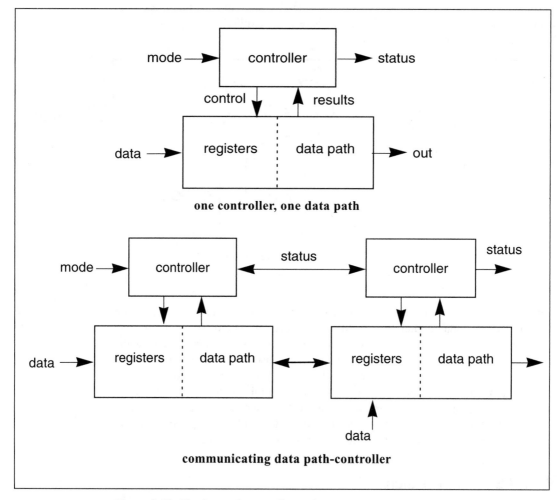

Figure 8-18: *The data path-controller architecture.*

An ASM chart specification is particularly well-suited to data path-controller architectures. The operations in the boxes, which are generally regular data operations, are
executed in the data path; the ASM chart's boxes and edges give the state transition
graph for the controller.

The most basic element of the ASM chart is the state; an example is shown in Figure
8-20. A state is represented in the ASM as a rectangle, with an optional name for the
state given in a nearby circle. The state is decorated with a number of operations

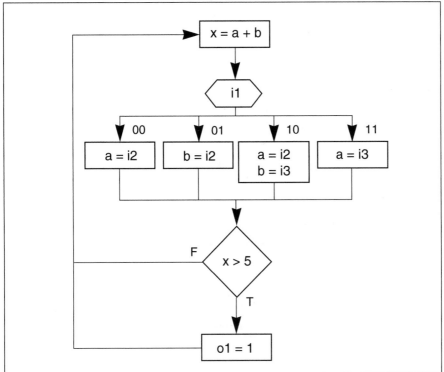

Figure 8-19: *An ASM chart.*

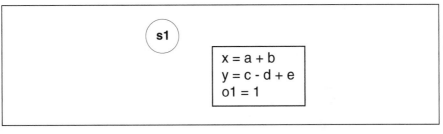

Figure 8-20: *An ASM state.*

shown inside the box. All the operations in a state are executed on the same cycle—a state in the ASM chart corresponds to a state in the register-transfer system. (Actually, an ASM state corresponds to a state in the system's controller and to many states in the complete system, since data operations like x = a + b can induce many states, depending on the values of a and b. The ASM chart is a powerful notation because it simply and cleanly summarizes the system's state.)

You can put as many operations as you want into a state, though adding more operations will require more hardware. The only proviso is that a single variable can be the target of an assignment only once in a state—this is known as the **single assignment**

Figure 8-21:
How to implement
operations in an
ASM state.

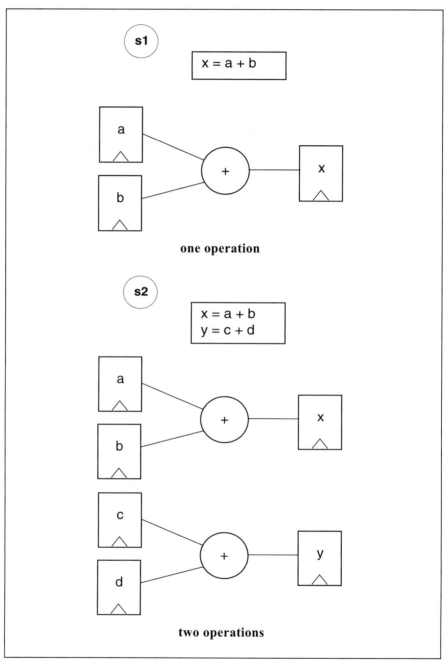

requirement. Consider the two states of Figure 8-21. State s1 specifies a single operation. A block diagram which implements this state would include the registers for the three variables and a single adder. State s2 requires two additions. To implement this state, we must include two adders in the block diagram, since both additions must be done in the same cycle.

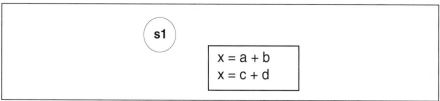

Figure 8-22:
Multiple assignment in an ASM state.

What effect does assigning twice to the same variable in a state have? In Figure 8-22, the variable x is assigned to twice, which requires loading the x register twice in a single clock cycle. We obviously can't load a register twice per cycle in a strict sequential system.

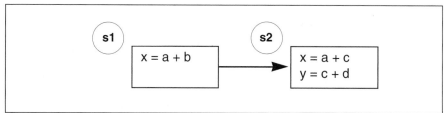

Figure 8-23:
Sequential states in an ASM chart.

It is, however, perfectly acceptable to assign a value to a variable in distinct states. In Figure 8-23, x is assigned to in both states s1 and s2. It is perfectly fine to load a register in two successive cycles. Presumably, some other part of the system or the outside world looks at x between the two assignments. But even if this ASM chart is a poorly thought-out specification, it is a valid register-transfer system.

The presence of multiple states makes it a little harder to design an efficient block diagram. Let's assume that our registers all have load signals to allow us to set their values only when desired. All the operations within a state require distinct hardware units, called **function units**, to implement the operations like +, since those operations must be done simultaneously. The simplest way to implement operations in sequential states is to assign a different function unit to each operation. That option is extremely wasteful—the ASM of Figure 8-23 would require three adders, even though at most two are used at one time. In practice, we will want to share function units across states, as shown in Figure 8-24. The minimum number of function units we need in the block diagram is the number required in any one state. To reuse a function unit, we put multiplexers on its inputs, as shown in the figure. The top adder can have its second input selected to be either b or c. The second adder doesn't need multiplexed inputs because it is only used once.

We can now start to understand how an ASM chart can be implemented as a data path plus controller. Figure 8-24 shows a system that implements the ASM chart fragment of Figure 8-23. The data path section includes the registers and the logic for the regular operations specified in the ASM states. The controller includes a controller which

Figure 8-24: How sequential ASM states are implemented as data path and controller.

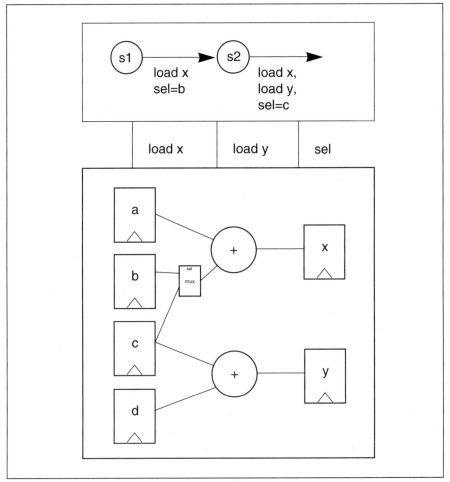

determines the sequence of actions to be performed. The states of the controller determine when we move from ASM state s1 to ASM state s2. In each state, the controller sets control signals which tell the data path what to do: in the first state, it sets the load signal for x and the multiplexer select signals to send the right operands to the adder; in the second state, it sets the load signals for x and y, along with a new set of mux select signals. Strictly speaking, the data path can go into many states, depending on the values of its registers. But the structure of states and transitions in the controller component mirrors the structure of the ASM chart.

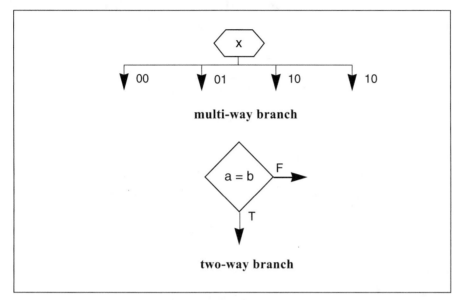

Figure 8-25:
Symbols for
branches in an
ASM chart.

An ASM chart with only unconditional transitions has limited charms. Branches are represented as diamonds of any shape—typically the four-sided diamond is used for two- or three-way branches, while the six-sided diamond is used for more numerous branches. The condition for the branch is given in the diamond, and each transition is labeled with the values which cause that condition to be true. The branch condition may be a direct test of a primary input, such as $i1 = 0$, where all the logic for the test is in the controller; it may also be computed in the controller. For example, to test $x = y$, we may subtract y from x in the data path, test the result to check for 0, and send a single signal from the data path to the controller giving the result of the test.

The condition is tested on the same cycle as the state preceding the branch. With this definition, the ASM chart corresponds to a traditional Moore machine. Figure 8-26 shows how the structure of a Moore machine corresponds to the structure of a state plus branch in an ASM chart. The Moore machine remembers its present state and accepts its inputs; from that information, its next-state logic computes the machine's next state. Similarly, given an ASM state, a given value at a branch selects an ASM transition which leads to the next ASM state.

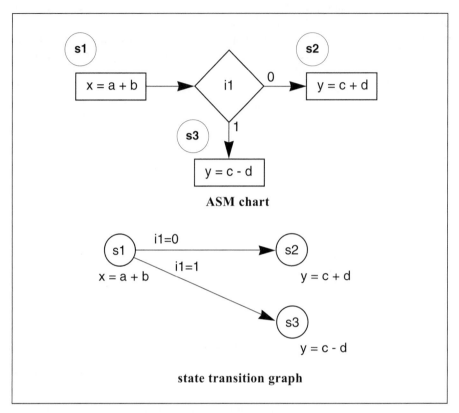

Figure 8-26:
Implementing an
ASM branch in a
Moore machine.

Specifying a Mealy machine in an ASM chart requires conditional outputs. As shown in Figure 8-27, conditional outputs in an ASM chart are given in rounded boxes. A conditional output is not a state and does not consume a clock cycle. If, in the figure, the branch leading to the conditional output is taken, the $y = c + d$ action occurs on the same cycle as the $x = a + b$ action. This corresponds to computing the output value of the Mealy machine based on the FSM's inputs, as well as its present state. Compare the Mealy controller FSM of Figure 8-27 to the Moore controller of Figure 8-26: the Moore controller, for example, executes $y = c + d$ in state s2, while the Mealy controller executes that action on one transition out of state s1. The restrictions on conditional ASM outputs are the same as those on Mealy machines—if a Mealy machine or ASM is connected to external logic which creates a combinational cycle between its conditional outputs and its inputs, the resulting logic is not a legal sequential system.

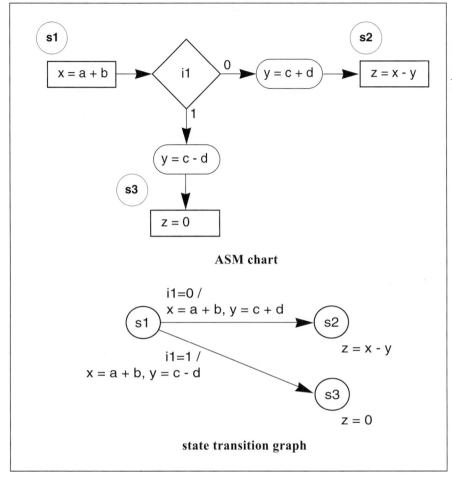

Figure 8-27: A conditional output in an ASM chart.

Since a Mealy machine is a general sequential machine, ASM chart notation lets us specify arbitrary register-transfer systems. The ASM chart is particularly powerful when the system performs a mixture of control and data operations. For a system which is mostly control, like the 01-string recognizer of Example 5-3, an ASM chart is no easier to use than a state transition graph. If we specify a data-rich system, like a CPU, with a state transition graph, we must simultaneously design the data path upon which those operations will be executed. The ASM chart lets us write down the operations without worrying about whether a particular operation goes into the data path or controller, or the exact structure of the data path. Once we have made sure the ASM chart is correct, we can refine the design to specify the data path structure and control signals. Think of pulling the operations out of the ASM states to produce the

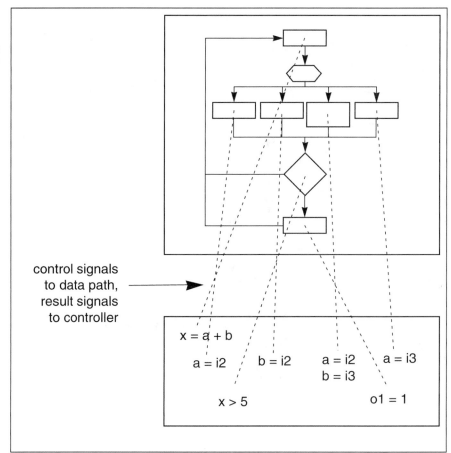

Figure 8-28:
*Extracting a data
path and control-
ler from an ASM
chart.*

data path, as shown in Figure 8-28. The data path operations drag behind them the control signals required to select the appropriate actions at each cycle; the signals that cross the data path-controller boundary, along with the system inputs and outputs, form the controller's inputs and outputs. The skeleton of the ASM chart left when the data path operations are extracted gives the structure of the controller's state transition graph.

8.4 High-Level Synthesis

A register-transfer isn't the most abstract, general description of your system. The register-transfer assigns each operation to a clock cycle, and those choices have a profound influence on the size, speed, and testability of your design. If you think directly in terms of register-transfers, without thinking first of a more abstract **behavior** of your system, you will miss important opportunities. Consider this simple sequence of operations:

```
x <= a + b;
y <= c + d;
if z > 0 then
    w <= e+ f;
end if;
```

How many clock cycles must it take to execute these operations? The assignments to x and y and the test of z are all unrelated, so they could be performed in the same clock cycle; though we must test z before we perform the conditional assignment to w, we could design logic to perform both the test and the assignment on the same cycle. However, performing all those operations simultaneously costs considerably more hardware than doing them in successive clock cycles.

High-level synthesis (also known as **behavioral synthesis**) constructs a register-transfer from a behavior in which the times of operations are not fully specified. The external world often imposes constraints on the times at which our chip must execute actions—the specification may, for example, require that an output be produced within two cycles of receiving a given input. But the behavior model includes only necessary constraints on the system's temporal behavior.

The primary jobs in translating a behavior specification into an architecture are **scheduling** and **binding** (also called allocation). The specification program describes a number of operations which must be performed, but not the exact clock cycle on which each is to be done. Scheduling assigns operations to clock cycles. Several different schedules may be feasible, so we choose the schedule which mini-mizes our costs: delay and area. The more hardware we allocate to the architecture, the more operations we can do in parallel (up to the maximum parallelism in the hard-ware), but the more area we burn. As a result, we want to allocate our computational resources to get maximum performance at minimal hardware cost. Of course, exact costs are hard to measure because architecture is a long way from the final layout: adding more hardware may make wires between components longer, adding delay which actually slows down the chip. However, in many cases we can make reasonable cost estimates from the register-transfer design and check their validity later, when we have a more complete implementation.

8.4.1 Functional Modeling Programs

A program that models a chip's desired function is given a variety of names: functional model, behavior model, architectural simulator, to name a few. A specification program mimics the behavior of the chip at its pins. The internals of the specification need have nothing to do with how the chip works, but the input/output behavior of the behavior model should be the same as that of the chip. Appendix C describes the mechanics of writing executable behavior models; for the moment, we just need to understand the relationship between program models and hardware.

```
o1 <= i1 or i2;
if i3 = '0' then
        o1 <= '1';
        o2 <= a + b;
else
        o1 <= '0';
end if;
```

Figure 8-29 shows a fragment of a simple VHDL functional model. This code describes the values to be computed and the decisions to be made based on inputs. What distinguishes it from a register-transfer description is that the cycles on which these operations are to occur are not specified. We could, for example, execute o1 <= '1' and o2 <= a + b on the same cycle or on different cycles.

Reading inputs and producing outputs for a functional model requires more thought than for a register-transfer model. Since the register-transfer's operations are fully scheduled, we always know when to ask for an input. The functional model's inputs and outputs aren't assigned particular clock cycles yet. Since a general-purpose programming language is executed sequentially, we must assign the input and output statements a particular order of execution in the simulator. Matching up the results of behavioral and register-transfer simulations can be frustrating, too. The most important information given by the functional model is the constraints on the order of execution: *e.g.*, y = x + c must be executed after x = a + b. A **data dependency** exists between the two statements because x is written by the first statement and used by the second; if we use x's value before it is written, we get the wrong answer. Data flow constraints are critical pieces of information for scheduling and binding.

8.4.2 Data

The most natural model for computation expressed entirely as data operations is the **data flow graph**. The data flow graph captures all data dependencies in a behavior which is a **basic block**: only assignments, with no control statements such as if. The following example introduces the data flow graph by building one from a language description.

Example 8-1: *Program code into data flow graph*

The first step in using a data flow graph to analyze our basic block is to convert it to single-assignment form:

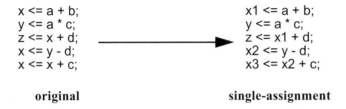

```
x <= a + b;           x1 <= a + b;
y <= a * c;           y <= a * c;
z <= x + d;           z <= x1 + d;
x <= y - d;           x2 <= y - d;
x <= x + c;           x3 <= x2 + c;
```

original   **single-assignment**

Now construct a graph with one node for each data operator and directed edges for the variables (each variable may have several sinks but only one source):

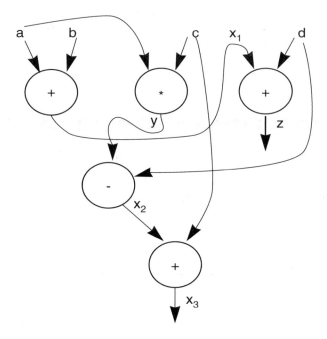

The data flow graph is a **directed acyclic graph** (DAG), in which all edges are directed and there is no cycle of edges that form a path from a node back to that node. A data flow graph has primary inputs and primary outputs like those in a logic network. (We may want to save the value of an intermediate variable for use outside the basic block while still using it to compute another variable in the block.) We can execute this data flow graph by placing values for the source variables on their corresponding DAG edges. A node *fires* when all its incoming edges have defined values; upon firing, a node computes the required value and places it on its outgoing edge. Data flows from the top of the DAG to its bottom during computation.

How do we build hardware to execute a data flow graph? The simplest—and far from best—method is shown in Figure 8-30. Each node in the data flow graph of the example has been implemented by a separate hardware unit which performs the required function; each variable carrier has been implemented by a wire. This design works, but, as we saw in Section 8.3.2, it wastes a lot of hardware. Our execution model for data flow graphs tells us that not all of the hardware units will be working at the same time—an operator fires only when all its inputs become available, then it goes idle. This direct implementation of the data flow graph can waste a lot of area—the deeper the data flow DAG, the higher the percentage of idle hardware at any moment.

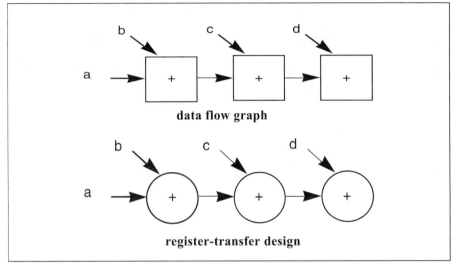

Figure 8-30: An *overgenerous implementation of a data flow graph.*

We can save hardware for the data operators at the cost of adding hardware for memory, sequencing, and multiplexing. The result is our canonical data path-plus-controller design. The data path includes registers, function units, and multiplexers which select the inputs for those registers and function units. The controller sends control signals to the data path on each cycle to select multiplexer inputs, set operations for multi-function units, and to tell registers when to load. We have already seen how to design the data path and controller for an ASM chart, which has fixed scheduling. The next example shows how to schedule and bind a data flow graph to construct a data path-controller machine.

Example 8-2: From data flow to data path-controller

We will use the data flow graph of Example 8-1. Assume that we have enough chip area to put one multiplier, one adder, and one subtractor in the data path. We have been vague so far about where primary inputs come from and where primary output values go. The simplest assumption for purposes of this example is that primary inputs and outputs are on pins and that their values are present at those pins whenever necessary. In practice, we often need to temporarily store input and output values in registers, but we can decide how to add that hardware after completing the basic data path-controller design.

We can design a schedule of operations for the operations specified in the data flow graph by drawing cut lines through the data flow—each line cuts a set of edges which, when removed from the data flow graph, completely separate the primary inputs and primary outputs. For the schedule to be executable on our data path, no more than one multiplication and one addition or subtraction can be performed per clock cycle. Here is one schedule that satisfies those criteria:

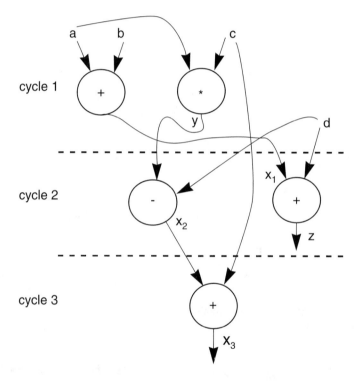

All the operations between two cut lines are performed on the same clock cycle. The next step is to bind variables to registers. Values must be stored in registers between clock cycles; we must add a register to store each value whose data flow edge crosses a cut. The simplest binding is one register per cut edge:

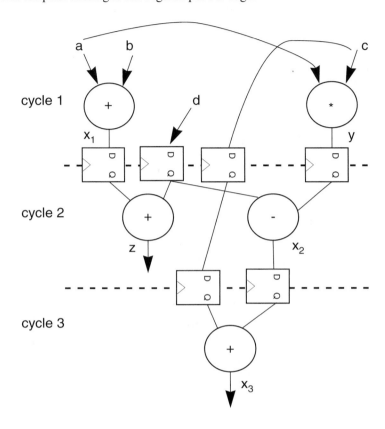

But, as with data path operators, one-to-one binding wastes area because not all values must be stored at the same time. In this graph, we can overwrite x_1's register with x_2's value since both are not needed at the same time. We can also share the three additions over one adder.

Now that we have scheduled operations, bound data operations to data function units, and allocated values to registers, we can deduce the multiplexers required and complete the data path design. The subtractor and multiplier each have their own unit, so their inputs won't require multiplexers. The adder requires multiplexers on each of its inputs, as does the register shared by x_1 and x_2. For each input to a shared function unit or register, we enumerate all the signals which feed the corresponding input on the operator; all of those signals go into a multiplexer for that input. For example, the left-hand inputs to the adder are a and the output of the x_1/x_2 register. Imagine laying all the addition operators in the registered data flow graph on top of each other, with the input lines for the addition are stretched to follow the operator. All the input lines which flow to the same point at the stacked-up additions require a multiplexer to make sure that exactly one value gets to that input at any given time.

The final data path looks like this:

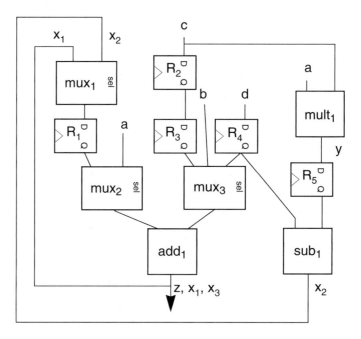

Note that when an input is used on two different cycles, we must add registers to save the value until the last cycle on which it is needed.

Now that we have the data path, we can build a controller which repeatedly executes the basic block. The state transition graph has a single cycle, with each transition executing one cycle's operation. The controller requires no inputs, since it makes no data-dependent branches. Its outputs provide the proper control values to the data path's multiplexers and function units at each step. The controller looks like this:

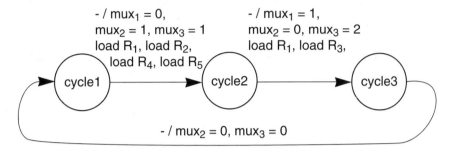

Once we wire together the data path and controller, the implementation is complete.

In the last example, we made a number of arbitrary choices about when operations would occur and how much hardware was available. The example was designed to show only how to construct a machine that implements a data flow graph, but in fact, the choices for scheduling—deciding when to execute an operation—and binding—deciding which hardware unit should store a value or execute an operation—are the critical steps in the design process. Now that we understand the relationship between a data flow graph and a data path-controller machine, we need to study what makes one data path-controller implementation better than another.

Obviously, scheduling and binding decisions depend on each other. The choice of a schedule limits our binding options; but we can determine which schedule requires the least hardware only after binding. We need to separate the two decisions as much as possible to make the design task manageable, but we must keep in mind that scheduling and binding depend on each other.

To a first approximation, scheduling determines time cost, while binding determines area cost. Of course, the picture is more complex than that: binding helps determine cycle time, while scheduling adds area for multiplexers, registers, etc. But we always evaluate the quality of a schedule by its ultimate hardware costs:

- **Area**. Area of the data path-controller machine depends on the amount of data operators saved by sharing vs. the hardware required for multiplexing, storage, and control.

- **Delay**. The time required to compute the basic block's functions depends on the cycle time and the number of cycles required. After the easy victories are won by obvious data hardware sharing, we can generally reduce area only by increasing delay—performing data operations sequentially on fewer function units.

- **Power**. The power consumption of the system can be greatly affected by scheduling and allocation, as we will see in Section 8.5.

There are many possible schedules which satisfy the constraints in a data flow graph. Figure 8-31 shows how to find two simple schedules. In this example we assume that we can perform as many additions as possible in parallel but no more than one addition in series per cycle—**chained** additions stretch the clock period. The **as-soon-as-possible** (ASAP) schedule is generated by a breadth-first search from the data flow sources to the sinks: assign the source nodes time 0; follow each edge out to the next rank of nodes, assigning each node's time as one greater than the previous rank's; if there is more than one path to a node, assign its time as the latest time along any path. The simplest way to generate the **as-late-as-possible** (ALAP) schedule is to work backward from the sinks, assigning negative times (so that the nodes just before the sinks have time -1, etc.), then after all nodes have been scheduled, adjust the times of all nodes to be positive by subtracting the most negative time for any node to the value of each node. The ASAP and ALAP schedules often do not give the minimum hardware cost, but they do show the extreme behaviors of the system.

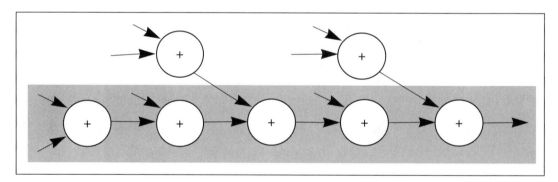

Figure 8-32: Critical path of a data flow graph.

The ASAP and ALAP schedules help us find the critical paths through the data flow graph. Figure 8-32 shows the critical path through our data flow graph—the long chain of additions determines the total time required for the computation, independent of the number of clock cycles used for the computation. As in logic timing, the critical path identifies the operations which determine the minimum amount of time required for the computation. In this case, time is measured in clock cycles.

Before we consider more sophisticated scheduling methods, we should reflect on what costs we will use to judge the quality of a schedule. We are fundamentally concerned with area and delay; can we estimate area and delay from the data path-controller machine implied by a schedule without fleshing out the design to layout?

Consider area costs first. A binding of data path operators to function units lets us estimate the costs of the data operations themselves. After assigning values to registers we can also estimate the area cost of data storage. We also compute the amount of logic required for multiplexers. Estimating the controller's area cost is a little harder because area can't be accurately estimated from a state transition graph. But we can roughly estimate the controller's cost from the state transitions, and if we need a more accurate estimate, we can synthesize the controller to logic or, for a final measure, to layout.

Now consider delay costs: both the number of clock cycles required to completely evaluate the data flow graph and the maximum clock rate. We have seen how to measure the number of clock cycles directly from the data flow graph. Estimating cycle time is harder because some of the data path components are not directly represented in the data flow graph.

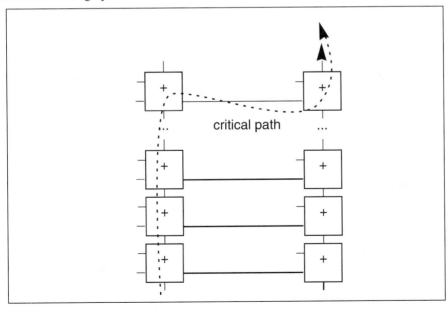

Figure 8-33:
Delay through
chained adders is
not additive.

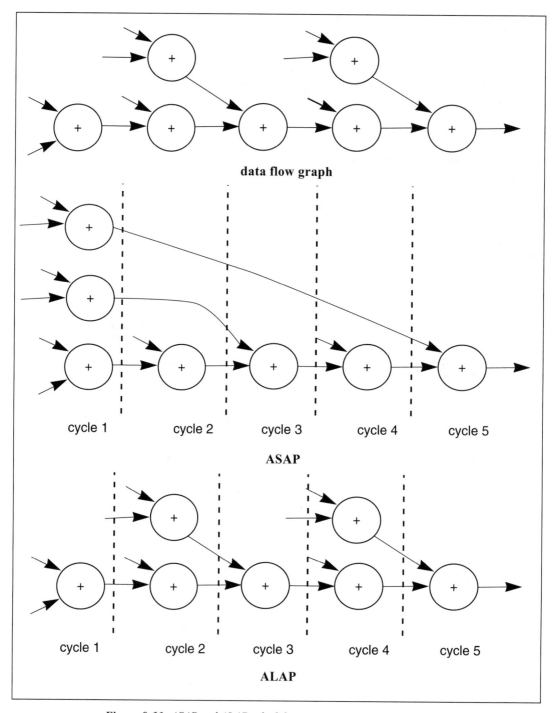

Figure 8-31: *ASAP and ALAP schedules.*

One subtle but important problem is illustrated by Figure 8-33: the delay through a chain of adders (or other arithmetic components) is not additive. The simplest delay estimate from the data flow graph is to assign a delay to each operator and sum all the delays along a path in each clock cycle. But, as the figure shows, the critical path through a chain of two adders does not flow through the complete carry chain of both adders—it goes through all of the first adder but only the most significant bit of the second adder. The simple additive model for delay in data flow graphs is wildly pessimistic for adders of reasonable size. For accurate estimates, we need to trace delays through the data path bit by bit.

If you are worried about delay, multiplexers added for resource sharing should concern you. The delay through a multiplexer can be significant, especially if the multiplexer has a large number of data inputs.

8.4.3 Control

One important reason to separate control from data is that arithmetic-rich and control-rich machines must be optimized using very different techniques to get good results—while optimization of arithmetic machines concentrates on the carry chain, the optimization of control requires identifying Boolean simplification opportunities within and between transitions. We typically specify the controller as a state transition graph, though we may use specialized machines, such as counters, to implement the control.

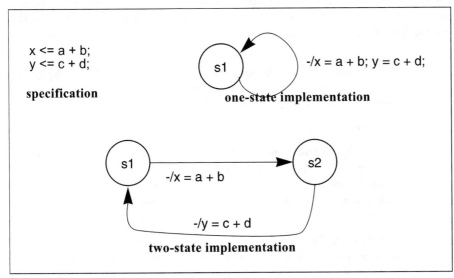

Figure 8-34: *How the controller changes with the data path schedule.*

In Chapter 5 we studied how to design a logic implementation of an FSM given a state transition graph. The high-level synthesis problem for control is one step more abstract—we must design the state transition graph which executes the desired algorithm. Consider the simple example of Figure 8-34. The two controllers are clearly not equivalent in the automata-theoretic sense: we can easily find one input sequence which gives different output sequences on the two machines, since the two machines don't even use the same number of cycles to compute the two additions. But even though the two controllers are not sequentially equivalent, they both satisfy the behavior specification.

How do we judge the quality of a controller which implements the control section of a program? That, of course, depends on our requirements. As usual, we are concerned with the area and delay of the FSM. The behavior specification may give us additional constraints on the number of cycles between actions in the program. We may have to satisfy strict sequencing requirements—when reading a random RAM, for example, we supply the address on one clock cycle and read the data at that location exactly one clock cycle later. We often want to minimize the number of cycles required to perform a sequence of operations—the number of cycles between reading a value and writing the computed result, for instance. To compute a result in the minimum number of cycles, we must perform as many operations as possible on each clock cycle. That requires both scheduling operations to take best advantage of the data path, as we saw in the last section; it also requires finding parallelism within the control operations themselves.

For now we will assume that the data path is given; in the next section we will look at how to choose the best trade-off between controller and data path requirements. The construction of a controller to execute a behavior specification proceeds as follows:

- Each statement in the behavior model is annotated with data path signals: arithmetic operations may require operation codes; multiplexers require selection signals; registers require load signals.

- Data dependencies are identified within each basic block.

- In addition, **control dependencies** are identified across basic blocks—a statement that is executed in only one branch of a control statement must be executed between the first and last states of that conditionally executed basic block. If the same statement appears in every branch, it is not dependent on the control signal and can be moved outside the control statement.

- External scheduling constraints, which reflect the requirements of the other machines to which this one will be connected, are added. External scheduling constraints are those which cannot be determined by looking at the behavior specification itself but which are required when the machine is connected to its intended working environment.

- Each program statement is scheduled—assigned an execution clock cycle which satisfies all the data and control dependencies.

- The controller's state transition graph can be constructed once the schedule is known.

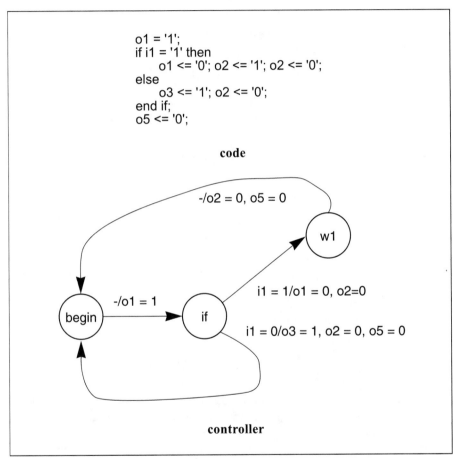

```
o1 = '1';
if i1 = '1' then
        o1 <= '0'; o2 <= '1'; o2 <= '0';
else
        o3 <= '1'; o2 <= '0';
end if;
o5 <= '0';
```

code

-/o2 = 0, o5 = 0

w1

-/o1 = 1

i1 = 1/o1 = 0, o2=0

i1 = 0/o3 = 1, o2 = 0, o5 = 0

begin

if

controller

Figure 8-35: Constructing a controller from a program.

Figure 8-35 shows how some opportunities for parallelism may be hidden by the way the program is written. The statements o1 <= '1' and o5 <= '0' are executed outside the if statement and, since they do not have any data dependencies, can be executed in any order. (If, however, one of the if branches assigned to o5, the o5 <= '0' assignment could not be performed until after the if was completed.) The assignment o2 <=

'0' occurs within *both* branches of the if statement and data dependencies do not tie it down relative to other statements in the branches. We can therefore pull out the assignment and execute a single o2 <= '0' before or after the if. If a statement must be executed within a given branch to maintain correct behavior, we say that statement is control-dependent on the branch.

Figure 8-36:
Rewriting a
behavior in terms
of controller oper-
ations.

```
                              x <= a - b;
                              if x < y then
                                    o1 <= '0';
                              end if;
```

behavior specification

```
source_1 <= a_source; source2 <= b_source; op <= subtract; load_x <= '1';
source_1 <= x_source; source_2<= y-source; op <= gt;
if gt_result then
      o1_mux <= zero_value;
end if;
```

controller operations

If we want to design a controller for a particular data path, two complications are introduced. First, we must massage the behavior specification to partition actions between the controller and data path. A statement in the behavior may contain both data and control operations; it can be rewritten in terms of controller inputs and outputs which imply the required operations. Figure 8-36 gives a simple example. The first assignment statement is replaced by all the signals required to perform the operation in the data path: selecting the sources for the ALU's operands, setting the operation code, and directing the result to the proper register. The condition check in the if statement is implemented by an ALU operation without a store. We must also add constraints to ensure that these sets of operations are all executed in the same cycle. (Unfortunately, such constraints are hard to write in VHDL and are usually captured outside the behavior model.) Those constraints are external because they are imposed by the data path—the data path cannot, for example, perform an ALU operation on one cycle and store the result in a temporary register for permanent storage on a later cycle. We also need constraints to ensure that the ALU operation for the assignment is performed on the same cycle as the test of the result, or the comparison result will be lost.

The second complication is ensuring that the controller properly uses the data path's resources. If we have one ALU at our disposal, the controller can't perform two ALU operations in one cycle. The resource constraints are reflected in the controller's pins—a one-ALU data path will have only one set of ALU control signals. We may, however, have to try different sequences of data path operations to find a legal implementation with both a good controller and the desired data path.

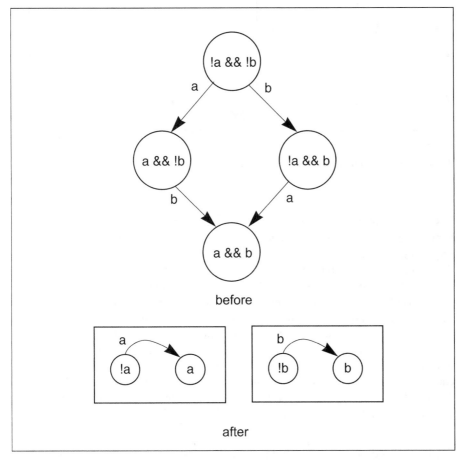

Figure 8-37:
Breaking a pair of tests into distributed control.

Rather than put all the control operations in one FSM, we may use **distributed control**, which splits the actions into communicating machines. Distributed control greatly simplifies the design of concurrent actions. Consider the example of Figure 8-37, where we want to wait for two signals, a and b, to go high. If we implement the test as one controller, we must create intermediate states for different event orderings: a but not b; b but not a. If we implement one machine for each test, each of those machines is very easy to describe. To wait for the desired condition, we simply wait for all the machines to be in their acceptance states, a simple AND condition. When more tests must be conducted in parallel, the number of intermediate states explodes.

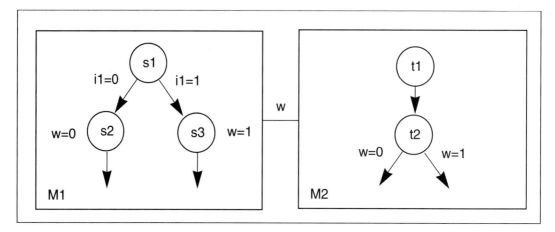

Figure 8-38: *Synchronized communication between two machines.*

Distributed control requires communication—the various machines must signal one another about what they have done or want to do. Communication introduces new scheduling problems. We could implement all communication by handshaking, but this asynchronous communication would both increase the area of the controllers and increase their state counts. We would like to use synchronous communication whenever possible—one machine throws a value and the receiving machine is ready to catch it. Synchronous communication requires simultaneous scheduling of the communication events in the two machines: operations must be assigned to states so that when one machine is in the state to send a value, the receiving machine is in the proper state to receive. Figure 8-38 shows two machines which communicate synchronously: when M1 sends a value through the wire w during state s2, M2 makes a decision based on the incoming value. If we add or delete states in M1 so that w is written on a different cycle, we must modify M2, and vise versa.

Finally, a word about controller implementation styles. You may have learned to implement a controller as either a **hardwired machine** or a **microcoded machine**. As shown in Figure 8-39, a hardwired controller is specified directly as a state transition graph, while a microcoded controller is designed as a microcode memory with a microprogram counter. (The microcoded controller also requires control logic to load the μPC for branches.) It is important to remember that these are implementation styles, not different schedules. The hardwired and microcoded controllers for a given design are equivalent in the automata-theoretic sense—we can't tell which is used to implement our system by watching only its I/O behavior. While one may be faster, smaller, or easier to modify than another for a given application, changing from one style to another doesn't change the scheduling of control operations in the controller.

You should first use control scheduling methods to design the controller's I/O behavior, then choose an implementation style for the machine.

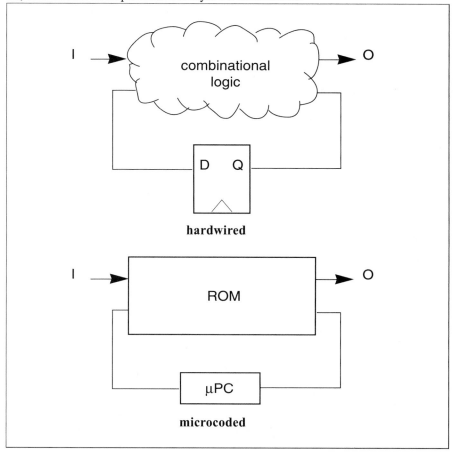

Figure 8-39:
Controller implementation styles.

8.4.4 Data and Control

So far, we have designed the data path and controller separately. Dividing architecture design into sub-problems makes some issues clearer, but it doesn't always give the best designs. We must consider interactions between the two to catch problems that can't be seen in either alone. Once we have completed an initial design of the data path and controller individually, we need to plug them together and optimize the complete design.

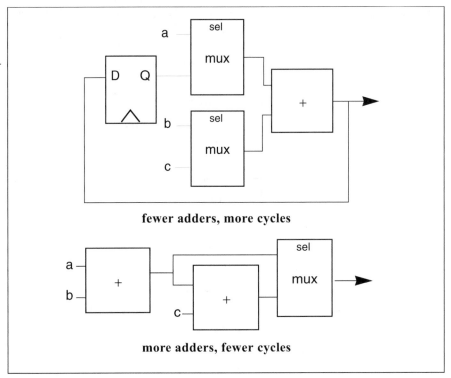

Figure 8-40:
*Adding hardware
to reduce the num-
ber of clock cycles
required for an
operation.*

fewer adders, more cycles

more adders, fewer cycles

The first, obvious step is to eliminate superfluous hardware from the data path. A schedule may have been found for a controller that doesn't require all the hardware supplied by the data path. A more sophisticated step is to add hardware to the data path to reduce the number of cycles required by the controller. In the example of Figure 8-40, the data path has been designed with one adder. The *true* branch of the if can be executed in one cycle if another adder is put into the data path. Of course, the second adder is unused when the false branch is executed. The second adder also increases the system's clock period; that delay penalty must be paid on every cycle, even when the second adder is not used. Whether the second adder should be used depends on the relative importance of speeding up the true branch and the cost in both area and delay of the second adder.

Another important optimization is adjusting the cycle time through the combined system. Even though the delay through each subsystem is acceptable, the critical path through the combination may make the cycle time too long. Overly long critical paths are usually caused by computations that use the result of a data path operation to make a control decision, or by control decisions that activate a long data path operation. In the example of Figure 8-41, the critical path goes through the carry chain of

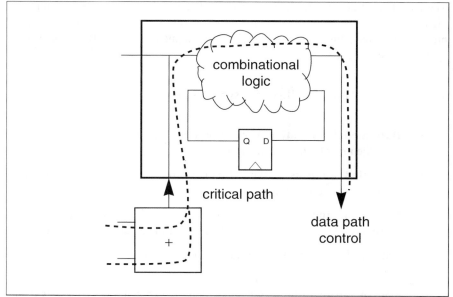

Figure 8-41:
Delay through a data path controller system.

the ALU and into the next-state logic of the controller. We can speed up the clock by distributing this computation over two cycles: one cycle to compute the data path value, at which point the result is stored in a memory element; and a second cycle to make the control decision and execute it in the data path. One way to view the effect of this pipelining is that it moves the control decision ahead one cycle, increasing the number of cycles required to compute the behavior. However, it may not always be necessary to add cycles—if the adder is free, rather than move the control decision forward, we can move the addition back one cycle, so that the result is ready when required by the controller.

8.4.5 Design Methodology

High-level synthesis allows designers to concentrate on the architectural design, rather than spend a great deal of time mapping the architecture to logic or layout. High-level synthesis can produce useful productivity gains when used to automate the transformation of the architecture to register-transfer form, but high-level synthesis can aid productivity in other ways, as described in the next example.

Example 8-3: *The IBM High-Level Synthesis System*

The IBM High-Level Synthesis System (HIS) [Ber95] was one of the first industrial high-level synthesis systems. It accepts design descriptions in VHDL or Verilog and produces register-transfer designs which can be input to the IBM BooleDozer logic synthesis system.

The main steps followed by HIS during synthesis include:

- **Data model generation.** Control flow and data flow graphs are generated from the input high-level description.

- **Data flow analysis.** Variable lifetimes are determined, explicit clocking constraints are analyzed, etc.

- **Scheduling and allocation.** The operations in the behavior are scheduled, registers and multiplexers are allocated, control signals are generated, etc.

- **Data path optimizations.** Data path optimizations try to efficiently share resources.

- **Control optimizations.** Behavioral don't-care conditions, state assignment, etc.

One major use of HIS is as a front-end to logic synthesis which allows designers to directly manipulate the architecture. Considerable time can be saved when changes to an architecture are made (as is common), since the designer need not translate the architectural changes to logic. However, high-level synthesis has also seen several other uses: as a fast synthesizer for designs input to logic emulation machines; as a fast synthesizer for verification systems; as a fast mapper for cycle-based simulation; and for early estimation and analysis. In all these cases, the ability to generate a design from a higher-level description saves time otherwise required by the designer to map the architecture into logic.

8.5 Architectures for Low Power

In this section, we will discuss two important architectural methods for reducing power consumption: **architecture-driven voltage scaling** and **power-down modes**. The first method increases parallelism, while the second method tweaks the sequential behavior to eliminate unnecessary activity.

8.5.1 Architecture-Driven Voltage Scaling

As was noted in Section 3.3.5, the power consumption of static CMOS gates varies with the square of the power supply voltage. The delay of a gate does not decrease as quickly as power consumption. Architecture-driven voltage scaling [Cha92] takes advantage of this fact by adding parallelism to the architecture to make up for the slower gates produced by voltage scaling. Even though the parallel logic adds power, the transformation still results in net power savings.

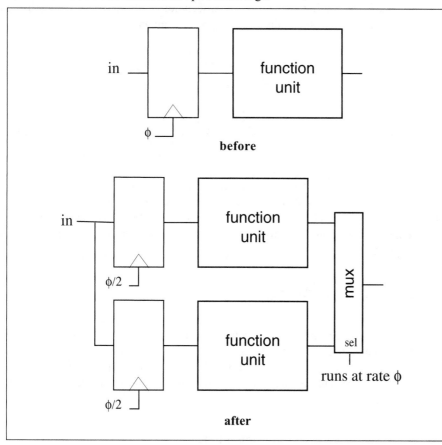

Figure 8-42: *Increasing parallelism to counteract scaled power supply voltage.*

This effect can be understood using the generic register-transfer design of Figure 8-42. A basic architecture would evaluate its inputs (clocked into registers in this case) every clock cycle using its function unit. If we slow down the operating frequency of the function unit by half, we can still generate outputs at the same rate by introducing a second function unit in parallel. Each unit gets alternate inputs and is responsible for generating alternate outputs. Note that the effective operation rate of the system is different in different components: the outputs are still generated at the original clock rate while the individual function units running at half that rate. Parallelism does incur overhead, namely the extra capacitance caused by the routing to/from the function units and the multiplexer. This overhead is, however, usually small compared to the savings accrued by voltage scaling.

Parallelism can also be introduced by pipelining. If the logic has relatively little feedback and so is amenable to pipelining, this technique will generally reduce in less overhead capacitance than parallel-multiplexed function units.

The power improvement over a reference power supply voltage V_{ref} can be written as [Cha92]:

$$P_n(n) = \left[1 + \frac{C_i(n)}{nC_{\text{ref}}} + \frac{C_x(n)}{C_{\text{ref}}} \right]\left(\frac{V}{V_{\text{ref}}} \right), \qquad \text{(EQ 8-1)}$$

where n is the number of parallel function units, V is the new power supply voltage, C_{ref} is the reference capacitance of the original function unit, C_i is the capacitance due to interprocessor communication logic, and C_x is the capacitance due to the input/output multiplexing system. Both C_i and C_x are functions of the number of parallel function units.

8.5.2 Power-Down Modes

We saw in Section 4.6 and Section 5.5 that glitch reduction reduces power consumption by eliminating unnecessary circuit activity. A power-down mode is a technique for eliminating activity in a large section of the circuit by turning off that section. Power-down modes are common in modern microprocessors and are used in many ASICs as well.

A power-down mode is more control-oriented than is architecture-driven scaling. Implementing a power-down mode requires implementing three major changes to the system architecture:

- conditioning the clock in the powered-down section by a power-down control signal;

- adding a state to the affected section's control which corresponds to the power-down mode;

- further modifying the control logic to ensure that the power-down and power-up operations do not corrupt the state of the powered-down section or the state of any other section of the machine.

The conditional clock for the power-down mode must be designed with all the caveats applied to any conditional clock—the conditioning must meet skew and edge slope requirements for the clocking system. Static or quasi-static memory elements must be used in the powered-down section for any state that must be preserved during power-down (it may be possible to regenerate some state after power-up in some situations). The power-down and power-up control operations must be devised with particular care. Not only must they put the powered-down section in the proper state, they must not generate any signals which cause the improper operation of other sections of the chip, for example by erroneously sending a *clear* signal to another unit. Power-down and power-up sequences must also be designed to keep transient current requirements to acceptable levels—in many cases, the system state must be modified over several cycles to avoid generating a large current spike.

8.6 Systems-on-Chips and Embedded CPUs

VLSI manufacturing can now easily put tens of millions of chips on a single die at reasonable cost. These manufacturing advances have enabled designers to create **systems-on-chips (SoCs)** that include all the components necessary for systems that only recently included multiple chips. A system-on-chip often contains a wide variety of elements: random logic, memory, CPUs, and analog circuitry. Systems-on-chips are challenging to design as much because of the diversity of elements that must be put on the chip as for the sheer size of the chips.

A system-on-chip is not useful unless it can be designed in a reasonable amount of time. If all the subsystems of an SoC had to be designed by hand, most SoCs would not be ready in time to make use of the manufacturing process for which they were designed. SoC design teams often make use of **intellectual property (IP)** blocks in order to improve their productivity. An IP block is a pre-designed component that can be used in a larger design. There are two major types of IP:

- **Hard IP** comes as a pre-designed layout. Because a full layout is available, the block's size, performance, and power consumption can be accurately measured.

- **Soft IP** comes as a synthesizable module in a hardware description language such as Verilog or VHDL. Soft IP can be more easily targeted to new technologies but it is harder to characterize and may not be as small or as fast as hard IP.

There are several challenges in IP-based design. The first is in designing IP blocks that are generally useful. It takes a considerable amount of work to fully test an IP block and document it so that it can be used by another designer; not all components on a chip are frequently-enough used to make it worthwhile to package them as IP. The second is choosing an interface standard for IP blocks. IP blocks must clearly be able to talk to each other and to custom logic in order to be useful; however, no one interface standard is best for all types of systems. The third challenge is ensuring that the IP block works properly in the SoC. The logical function, performance, power consumption, and testability of the IP block must all be verified in the context of the system.

One critical type of intellectual property for SoC design is the **embedded CPU**. An embedded processor can be programmed to perform certain functions on the chip, much as an embedded processor is used in a board design. Embedded CPUs have been used on chips for many years: early embedded processors were mostly 8-bit CPUs used for basic sequencing; today, powerful 32-bit CPUs can be embedded on a system-on-chip. The fact that not just the CPU but also its cache, main memory, and I/O devices can be integrated on the same chip make embedded processors especially attractive?

Embedded CPUs are increasingly popular on SoCs for several reasons. First, many sophisticated applications are best implemented in software. Multimedia applications like MP3 audio and MPEG video are examples of functions that are difficult to implement without some amount of embedded software. Second, many complex systems must run embedded software in order to implement their applications. For example, digital audio systems must run digital rights management software that is available only in binary form. Many systems-on-chips also use Linux, Windows CE, the Palm OS, or some other OS to provide file management and networking. Third, embedded CPUs help decrease design time. Because the embedded processor is a relatively well-understood component, the design of the software can be somewhat decoupled from the hardware design.

The next example describes a system-on-chip for multimedia applications that makes use of a sophisticated embedded CPU.

Example 8-4: The TriMedia TM-1300 Programmable Media Processor

The TriMedia TM-1300 is a system-on-chip for multimedia applications built around a **VLIW** embedded processor. A VLIW (very-long instruction word) processor allows multiple functions to be specified in each instruction. VLIW processors are widely used in multimedia systems because they allow compilers and programmers to take advantage of the data parallelism built into many multimedia algorithms. The TM-1300's VLIW CPU has a 128-bit register file that holds 32-bit operands. It has 27 functional units, with room in each instruction to schedule operations on five of those function units per instruction. The VLIW processor can be programmed in C or C++ using a programming environment hosted on a PC.

The chip has 5.6 million transistors in a six-metal layer 0.25 μm process. It operates at 2.5 V at the core and 3.3V at the pins. It runs at 143, 166, 180, and 200 MHz at minimum voltage. The chip is 58 mm^2 and its package is a 292-pin ball grid array.

Here is the TM-1300 system block diagram:

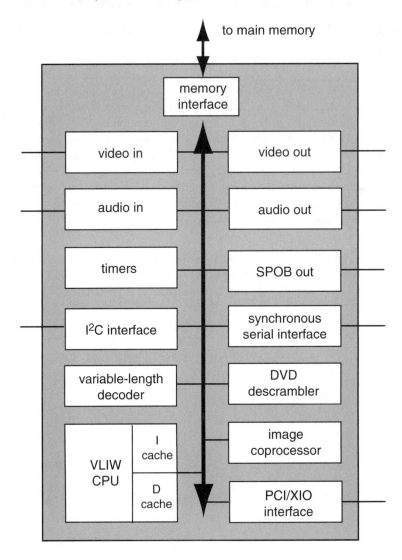

And here is a photomicrograph of the chip:

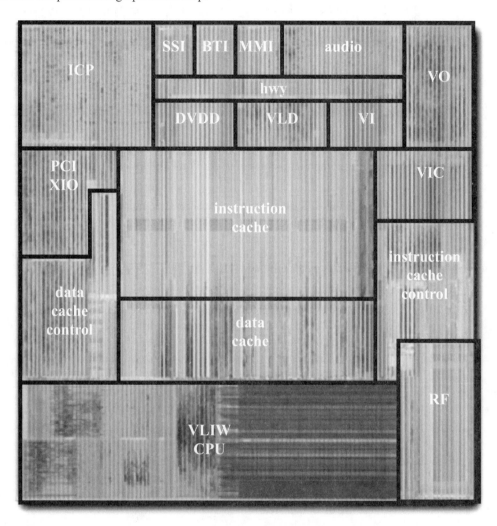

courtesy Philips

The system-on-chip includes a variety of I/O devices and accelerators as well as the VLIW CPU. The I/O devices are chosen to satisfy a variety of needs for multimedia systems. Audio and video input and output are clearly important; the I^2C interface talks to an industry standard bus that is often used for chip-to-chip communication of low-rate control information. Accelerators are chosen to augment the VLIW CPU's capabilities and speed up critical operations, such as variable-length (Huffman) coding and descrambling of encrypted DVD bit streams.

There are several important issues in system-on-chip design:

- **I/O devices**. Systems-on-chips inherently require input and output devices. Some I/O devices may be available as IP blocks. Others may not be available from outside sources, either because the block does not command a high enough price to be worth selling or because it is so strategically important that the designers will not release it.

- **Mixed-signal design**. A system-on-chip that interacts with the world probably needs mixed-signal elements that combine analog and digital processing. However, many digital VLSI manufacturing processes are not especially good at analog components. It may be possible to put some types of analog interfaces on a largely digital chip. In other cases, it may be necessary to put the mixed-signal processing on a separate chip manufactured with a different process.

- **Memory systems**. Many systems-on-chips are geared toward memory-intensive applications. The on-chip memory system may include caches for the embedded CPUs as well as on-chip main memory. The on-chip memory may be flash memory, SRAM, or embedded DRAM. The chip may also need to connect to off-chip memory. The memory system must be carefully designed to meet performance and power requirements as well as to make best use of silicon area.

- **CPU selection**. A wide variety of CPUs are available as both soft and hard IP for use in systems-on-chips. Some processors are **configurable**—the designer can choose a number of CPU design parameters, such as added instructions and data path bit width, then generate a custom CPU optimized for the application at hand. The choice of CPUs is determined by a variety of factors ranging from program performance to software compatibility through development environment.

- **Hardware/software co-design**. Many systems-on-chips require sophisticated architectures: multiple CPUs, custom memory systems, and I/O devices. These architectures must be designed to meet hard real-time dead-lines and strict power requirements. Hardware/software co-design is a sophisticated problem that requires careful measurement and design judge-ment [Wol00,DeM01]. We will briefly describe co-design in Section 10.11.

8.7 Architecture Testing

Making sure an architecture is testable is a balancing act, just as is making sure that it runs fast enough. The simplest way to make a system run fast usually requires too much hardware, so we look for judicious ways to reuse hardware without compromis-ing performance. Similarly, brute force application of extra testing hardware usually makes the system both too big and too slow. Luckily, we can usually make the system more easily testable with relatively simple fixes.

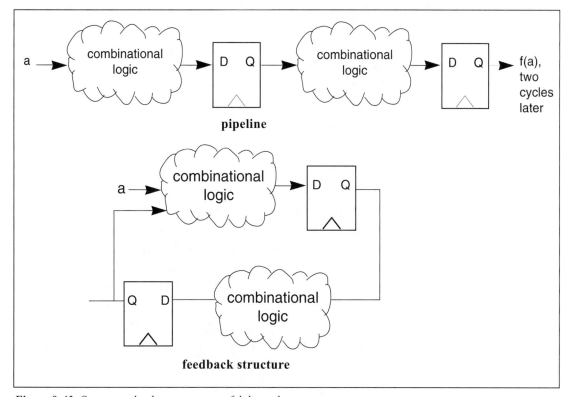

Figure 8-43: *Some scan latches are more useful than others.*

We studied LSSD design in Section 5.7. Scan latches add both area and delay. We can reduce the cost of scan design by using **partial scan**—making only some of the memory elements in the system scannable. Figure 8-43 shows why scan latches are more useful in some locations than others. The value of a latch that is in the middle of the pipeline is guaranteed to be available at the primary outputs after n clock cycles. The value at the pins will be determined by the combinational logic between the latch and the pin, but we can reverse-engineer the latch's value (perhaps with some ambiguity caused by the combinational functions performed). The situation in a general sequential machine, like the FSM shown, is more complex. Some latch values may be immediately accessible, while others may not show their effects at the pins for many cycles. A scan latch for a value that can be directly viewed is much less useful than a scan latch for the value that recirculates before becoming visible.

Registers become harder to test as their distance from primary inputs or outputs increases [Fri76]. We can identify high-payoff locations for scan latches by building a **register graph** [Che89], as shown in Figure 8-44. Nodes in the graph represent memory elements; an edge is drawn between two nodes if there is any combinational path between the two memory elements. The shortest distance to a given node from any node which is a primary input is called the **sequential depth** of that node; the graph's sequential depth is the largest sequential depth of any of its nodes. Cycles in the graph show feedback paths for state—memory elements in a cycle compute their value at least partially from other internal information, rather than from the primary inputs. Self-loops—edges that connect a node to itself— identify latches whose inputs are computed from their own outputs. Memory elements which participate in cycles, such as the FF2-FF3 cycle, tend to be harder to test (though self-loops are relatively easy to test). Furthermore, memory elements which are far away from the primary inputs are also hard to test. If we allow FF1 to be directly loaded or read (either by normal operation or by scanning), then FF3 can be loaded in one cycle, but loading FF2 requires two cycles. We can add partial scan registers to reduce the distance from a primary input/output to a memory element.

We can bind variables to registers to improve testability [Lee92]. Two binding rules help improve testability. First, make sure that as many registers as possible are assigned at least one variable which is a primary input or output of the behavior. Making even one variable assigned to a register a primary input or output ensures that the register will be directly connected to the pins. Second, minimize the sequential depth of your register graph. Draw the register graph for each binding and choose the one with the smallest sequential depth for any register. In many cases, a binding can be found which has about the same hardware cost but a much smaller sequential depth.

An alternative to applying test vectors from an external tester is **built-in self-test** (BIST). BIST is especially attractive for large chips which require long test

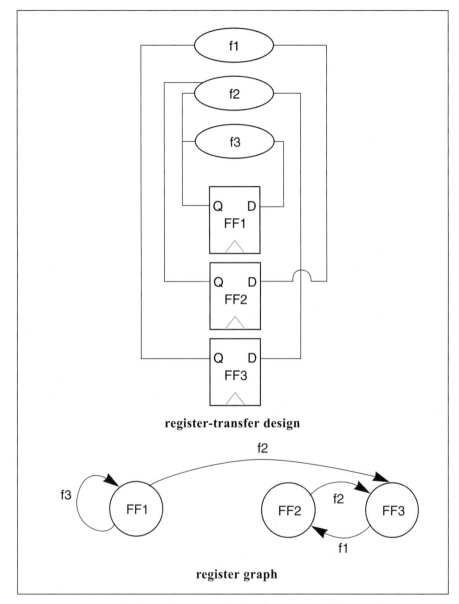

Figure 8-44: The register graph.

register-transfer design

register graph

sequences: because the internal testing circuitry runs at on-chip speeds, it can apply test vectors much more quickly than can an external tester. However, because we don't want to devote an extraordinarily large amount of chip area to the test circuitry, BIST doesn't apply custom test sequences created by an ATPG program. Instead, most BIST strategies use pseudo-random sequences as the test sequence.

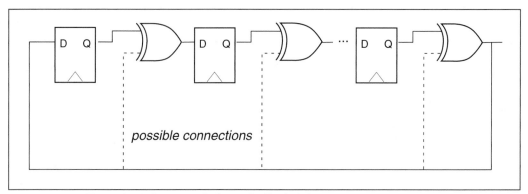

Figure 8-45: *Structure of a linear feedback shift register.*

A **linear feedback shift register** (LFSR) can be used to generate a pseudo-random sequence. One possible structure of an LFSR is shown in Figure 8-45. The memory elements hold the current pseudo-random value; XORs between stages compute the next value. Not all of the dotted connections are actually made—by making different feedback connections, we can generate different pseudo-random sequences. (If a feedback connection is not made, the corresponding XOR is not necessary.) An LFSR can also be used to store and compress a sequence of binary words, a technique commonly known as **signature analysis**. (Signature analysis was originally developed by Hewlett-Packard for printed circuit board testing.) If we want to record a sequence of values, we can add those values as additional inputs to the XORs, causing them to be added into the pseudo-random sequence. This scheme loses information—there may be several sequences of inputs which produce the same value in the LFSR. However, a relatively small LFSR can give a very low probability of aliasing, making the LFSR a very good compression scheme.

Figure 8-46: *A logic block configured for built-in self-test.*

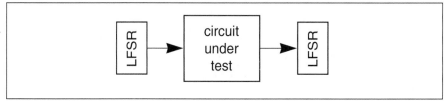

The testing configuration of a built-in self-test system is shown in Figure 8-46. One LFSR is used to generate inputs for the logic to be tested while another LFSR is used as a signature register. A multiplexer is used to switch between normal and test modes by switching the circuit under test's inputs between the primary inputs and the LFSR.

A fault simulator is used to simulate the circuit under test's response to the sequence generated by the input LFSR. The signature register's value can either be compared against a single signature which indicates correct operation or the register can be made available at the chip's pins.

8.8 References

We will discuss scheduling and binding algorithms in more detail in Chapter 10. Landman et al. [Lan96] describe a CAD system for architectural power optimization. Lee et al. [Lee92] discuss how to automatically allocate registers to improve system testability; Papachristou and Chiu [Pap90] discuss related techniques for built-in self-test. The built-in self-test techniques used in the Intel 80386 have been described by Gelsinger [Gel86, Gel87]. General principles of built-in self-test are described by McCluskey [McC86] and Abramovici et al. [Abr90]. Lyon et al. describe design-for-testability of the Motorola 68HC16Z1 embedded controller [Lyo91]; Bishop et al. describe testability considerations in the design of the Motorola MC68340 peripheral [Bis90].

8.9 Problems

8-1. Find an assignment of variables to multiplexers for the example of Figure 8-23 which gives the minimum number of multiplexer inputs.

8-2. Give a pair of ASM charts with conditional outputs which, when connected together, form an illegal sequential system.

8-3. How would you translate a register-transfer structure into a legal two-phase latched sequential machine?

8-4. Design an ASM chart for the traffic light controller of Example 5-4.

8-5. Design an ASM chart to repeatedly receive characters in 7-bit, odd parity format. You can assume that the bits are received synchronously, with the most significant data bit first and the parity bit last. The start of a character is signaled by two consecutive zero bits (a 1 bit is always attached after the last bit to assure a transition at the start of the next character). At the end of a character, the machine should write the seven-bit character in bit-parallel form to the *data* output and set the output *error* to 1 if a parity error was detected.

8-6. Design the ASM chart for a programmable serial receiver, extending the design of Question 8-5. The receiver should be able to take seven-bit or eight-bit characters, with zero, one, or no parity. The character length is given in the *length* register—assume the register is magically set to seven or eight. The parity setting is given in the *parity* register: 00 for zero parity, 01 for zero parity, and 10 for one parity.

8-7. Design a data path and an FSM controller for the ASM chart of Figure 8-19. Assume that *i1* is a primary inputs, *o1* is an external output, and *a*, *b*, and *x* are registers. Show:

> a) a block diagram showing the signals between the data path and the controller and the connections of each to the primary inputs and outputs;
>
> b) the structure of the data path;
>
> c) the state transition graph of the controller, including tests of primary inputs and data path outputs and assignments to primary outputs and data path inputs.

8-8. Draw data flow graphs for the following basic blocks:

> a) c <= a + b; d <= a + x; e <= c + d + x;
>
> b) w <= a - b; x <= a + b; y <= w - x; z <= a + y;
>
> c) w <= a - b; x <= c + w; y <= x - d; z <= y + e;

8-9. For the basic block below:

```
t1 <= a + b;
t2 <= c * t1;
t3 <= d + t2;
t4 <= e * t3;
```

> a) Draw the data flow graph.
>
> b) What is the longest path(s) from any input to any output?
>
> c) Use the laws of arithmetic to redesign the data flow graph to reduce the length of the longest path.

8-10. This code fragment is repeatedly executed:

```
if i1 = '1' then
    c <= a - b;
    d <= a + b;
else
    c <= a + e;
end if;
```

i1 is a one-bit primary input to the system, e is an n-bit primary input, and a, b, c, and d are all stored in n-bit registers. Assume that e is magically available when required by the hardware.

a) Design a data path with one ALU which executes this code.

b) Design a controller which executes the code on your data path.

8-11. How does the delay of a multiplexer grow with the number of data inputs? How does its area grow?

8-12. A full adder has three-input multiplexers on each of its data inputs. What are the relative sizes (in transistors) of the adder and multiplexers?

8-13. Design a new data path and controller for the code of Question 8-10. Use an ASAP schedule with no other constraints on the hardware.

8-14. Schedule the data flow graph of Figure 8-31 assuming that two additions can be chained in one clock cycle:

a) show the ASAP schedule;

b) show the ALAP schedule.

8-15. Schedule the data flow graph of Figure 8-31 assuming that no more than one addition can be performed per clock cycle.

8-16. The carry out of an adder is used as an input to a controller. How might you encode the controller's states to minimize the delay required to choose the next state based on the value of the carry out?

8-17. Draw a register graph for the scheduled and allocated design of Example 8-2. What is the largest sequential depth of any register in this design?

9

Chip Design

Highlights:

Chip design methodologies.

Two example chip designs.

9.1 Introduction

This chapter works through two design problems: a simple kitchen timer chip and a microprocessor data path. These designs are simple relative to the size of chips that can be fabricated today, but they will help us see how the many techniques we have discussed fit together—the sequence of decisions required to design an integrated circuit. Before describing these designs, we'll start with a survey of design methodologies.

9.2 Design Methodologies

The exact sequence of steps you follow to design a chip will vary with your circumstances: what type of chip you are designing; size and performance constraints; the design time allowed; the CAD tools available to you; and many other factors.

Figure 9-1: A generic integrated circuit design flow.

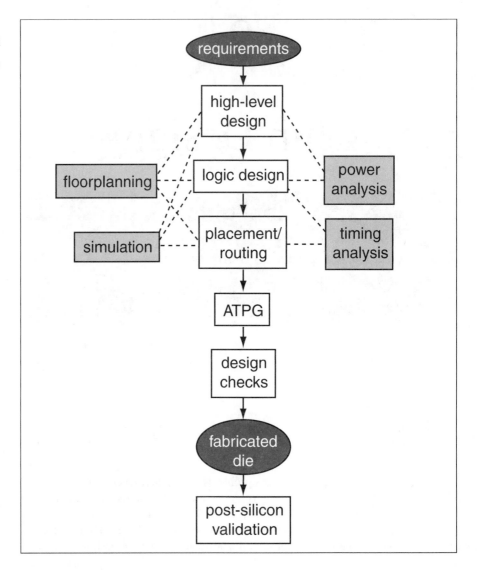

A **design methodology** is frequently called a **design flow** since the flow of data through the steps in the methodology may be represented in a block diagram. Figure 9-1 shows a generic design flow for VLSI systems. While all design methodologies will vary from this in practice, this flow shows some basic steps that must be considered in almost any design. The initial **requirements** for a system are often specified in English and may be vague; while many designs are follow-ons to previous designs, any new features must be described in some way that may lead to misunderstandings.

Problems in translating the requirements into architectures can and should be caught early to avoid the embarrassment of implementing the wrong chip. High-level design may be performed manually or using high-level synthesis tools, but somehow an initial set of functions must be translated into a register-transfer design. Similarly, logic and physical design may be performed by CAD tools, manually, or in some combination. Automatic test pattern generation (ATPG) generates test vectors for manufacturing test. ATPG is, of course, no substitute for the creation of functional test vectors which will be used by simulators to validate the design at all levels of abstraction. Several sorts of design checks, including design-rule checking, electrical checking, and timing analysis are all important at the end of the design process to be sure that no fundamental errors have been inadvertently introduced. Once die are returned from manufacturing, they must be evaluated to be sure that the design not only runs at the proper speed, but runs at a range of power supply voltages and other checks that ensure adequate yields. The importance of post-silicon electrical testing will be discussed in Example 9-2. Figure 9-1 shows several analysis steps connected to synthesis-oriented design steps by dotted lines. Simulation, floorplanning, timing analysis, and power analysis (among other analysis steps) are all important and must be performed at several different levels of abstraction. Early design stages rely on estimates which may be supplied by tools; those estimates must be verified as the design is refined and, if necessary, used to drive the redesign to meet requirements that have been missed.

Let's consider this process in more detail:

- **Architecture.** If a chip is a rework of an existing design—a design shrink, a few added features, etc.—then the architectural design is simple. But when designing something new, a great deal of work is required to transform the requirements into a detailed microarchitecture ready for logic design. Architectural design requires construction of a microarchitectural simulator that is sufficiently detailed to describe the number of clock cycles required for various operations yet fast enough to run a large number of test vectors. A test suite must also be constructed that adequately covers the design space; if the design is a rework of a previous architecture, then the vectors for that system are a starting point which can be augmented with new tests. Architectural design requires extensive debugging for both functionality and performance; errors that are allowed to slip through this phase are much more expensive to fix later in the design process.

- **Logic design and verification.** Logic design may be performed manually or using logic synthesis tools. In either case, the design will probably go through several refinement steps before completion. Initial design verification steps will concentrate on logical correctness and basic timing properties. Once the basic structure of the logic has taken shape, scan registers can be inserted and power consumption can be analyzed. A more detailed set of timing checks can also be performed, including delay, clock skew, and setup/hold times. In extreme cases, perhaps because of a limited number of choices in the gate and register libraries, it may be necessary to make more drastic changes to the logic to correct problems found late in the logic design process.

- **Physical design.** Physical design starts with floorplanning to determine the overall structure of the layout. If the logic was designed in large blocks, it may be necessary to partition those large blocks into smaller pieces at this point. Placement and routing will generate layouts of blocks, or layouts can be designed by hand. Once the layout is complete, the wiring parasitics must be extracted and back-annotated to the logic design. The back-annotated design can then be simulated to verify that layout did not violate any timing constraints. Hopefully, problems can be fixed with minor modifications to the layout but changes to the logic design may be required.

- **Back-end checks.** ATPG must be performed late to ensure that minor design changes did not inadvertently cause testability problems. Similarly, design-rule and electrical checks of the complete layout are an important sanity check to ensure that shorts or opens were not introduced late in design.

As we saw in Chapter 8, large chips are often designed with intellectual property (IP) blocks. IP-based design offers advantages over reducing design time: the blocks have been functionally verified; and they provide more accurate estimates of area, speed, and power early in the design process. S Providers of libraries generally refer to two major categories of pre-defined components: **hard macros** are complete layouts, while **soft macros** are specified as logic or perhaps behavior-level blocks. Hard macros, because they are complete designs, provide accurate information about area, delay, and power as soon as they are selected. On the other hand, hard macros are harder to port to new processes and harder to modify to meet variations in customer requirements. Soft macros, by comparison, are not as well-characterized but can be more easily adapted to changing needs. Macros have a broad range, ranging from memory arrays through UARTs, to central processing units.

Integrated circuits are notoriously hard to debug after fabrication. The few hundred pins on a large chip cover only a tiny fraction of the state contained in the multiple millions of electrical nodes in a large VLSI IC. While it may be possible to deduce the internal behavior of the chip, some errors may be difficult to detect and may also mask other flaws. **Voltage contrast** [Ben95] is a technique for observing the chip's internal behavior using a scanning electron microscope (SEM). The electron beam of the SEM is reflected differently off electrical nodes at high and low voltages; a raster scan of the chip by the SEM results in a picture of the voltages across the chip. Voltage contrast also requires expensive equipment but practitioners have found it very valuable in tracking down certain types of bugs.

One constant through all circumstances is the importance of good design documentation. You should write down your intent, your process, and the result of that process at each step. Documentation is important for both you and the others with whom you work:

- Written descriptions and pictures help you remember what you have done and understand complex relationships in the design. A paper trail also makes the design understandable by others. If you are hit by a truck while designing a complex chip, leaving only a few scribbled notes on the backs of envelopes and napkins, the grief of your loved ones will be matched by the anguish of your employer who must figure out how to salvage your incomplete design.

- **Design reviews** are very valuable tools for finding bugs early. A design review is a meeting in which the designer presents the design to another group of designers who comment on it. In preparation for review, the designer prepares documentation which describes the component or system being designed: purpose of the unit, high-level design descriptions, detailed designs, procedures used to test the design, etc. During the design review, the audience, led by a review leader, listen to the designer describe this information and comment on it (politely, of course). Many bugs will simply be found by the designer during the course of preparing for the meeting; many others will be identified by the audience. Design reviews also help the various members of a team synchronize—at more than one design review, two members of the same design team have realized that they had very different understandings of the interface between their components.

Even after the chip is done, documentation helps fix problems, answer questions about its behavior, and serves as a starting point for the next-generation design. You will find that a little time spent on documentation as you go more than pays for itself in the long run.

One thing to keep in mind is that methodology helps ensure that things aren't overlooked. A large chip is complex with many opportunities for error. Unfortunately, even some small errors can completely disable a chip, causing expensive and frustrating delays. Methodologies are put in place to minimize the chance of error. Each company generally develops its own design methodology based on its experience, including its earlier mistakes. Different methodologies can work equally well so long as they are followed carefully and with an understanding of their intent.

The next two examples give much more specific examples of design flows for two fairly different categories of chip design: ASIC and CPU design.

Example 9-1: Design methodology for IBM ASICs

ASICs are in general designed in a partnership of the ASIC's customer and its manufacturer—the manufacturer handles most design tasks closely tied to manufacturing, while the customer takes care of elements of the design unique to the customer's needs.

The first steps in the IBM ASIC design flow [Eng96] require cooperation between the customer and the manufacturer's design house:

- **Design entry.** Designs are entered in a hardware description language such as VHDL or Verilog. Schematic entry is also supported.

- **Logic synthesis.** IBM logic synthesis tools are used to map the design into a gate-level design in the IBM cell library. Logic synthesis also ensures that the design is appropriate for LSSD.

- **Simulation.** The design can be supported either at the functional or gate level. Gate-level netlists can be back-annotated with timing information for delay simulation.

- **Floorplanning.** Floorplanning can be used to estimate wiring capacitance, area, and wiring congestion. Floorplanning also allows the user to create bit-slice designs.

- **Test structure verification.** This step ensures that the design satisfies a set of IBM-defined rules that ensure the design—including RAM, etc.—is in compliance with the requirements for LSSD.

- **Static timing analysis.** This step analyzes the worst-case clock speed for the implementation.

- **Formal verification.** The design is checked for equivalence with a Boolean specification using efficient algorithms for solving the Boolean equivalence problem.

- **CMOS checks.** This step checks fan-out, I/O, boundary scan, and other low-level circuit checks.

- **Design hand-off.** When the design is ready for physical design, a netlist, timing assertion data, pad placement, and floorplanning information are given to the IBM design center.

At this point, physical design is handled by the manufacturing center:

- **Front-end processing.** Clock trees and test logic are generated at this step and static timing analysis is used to check performance.

- **Pre-layout sign-off.** This step allows the customer to ensure that no errors have been introduced by front-end processing.

- **Layout.** Detailed layout is performed by automatic tools guided by professional designers. Layout can be performed either on a flat or hierarchical design.

- **Post-layout sign-off.** Verifying the logic and timing at this step ensures that no errors were introduced during layout.

- **Tape-out to manufacturing.** Automatic test pattern generation is used to generate test vectors and the mask data is generated and sent to the manufacturing line.

Die are delivered to the customer after fabrication and manufacturing testing.

The next example illustrates the design of a CPU. CPU design projects take advantage of previous CPU designs, but also break new ground to meet more aggressive requirements.

Example 9-2: Design methodology for the HP 7100LC

The HP 7100LC CPU contains approximately 905,000 transistors. The design methodology for the 7100LC [Bas95] was designed to support the design decisions on the microprocessor.

The control logic for the 7100LC was designed using commercial tools for logic synthesis (from Synopsys) and for placement-and-routing (from Cadence). The previous-generation CPU, the 7100, had used a PLA-based methodology. The control logic equations from the PLA-based design could be reused but timing budgets had to be more carefully allocated and enforced in the synthesis-based methodology. While PLAs have easy-to-estimate delays which are roughly equal for all outputs, synthesized logic can show widely varying changes both from output to output and design iteration to design iteration. The designers judged the overall results of this technique to be good. The 7100LC added integer superscalar execution and memory and I/O control, yet it occupies about half the area of the 7100 PLA-based control.

The CPU went through several levels of verification before fabrication of the first samples. Behavioral modeling was used extensively to verify the design. The behavioral models were created in Verilog. The Verilog model was no faster than one created in a proprietary HP simulation environment, but allowed the design team to use industry-standard tools for timing verification, synthesis, etc. The CPU, memory controller, and I/O controller were verified separately for reasons of efficiency. Functional models were created in C or other high-level languages to stimulate these blocks during stimulation. Watchdog code was also inserted to flag errors during simulation. In addition to behavioral simulation, switch-level models of the implementation were also created to verify the transistor-level circuit extracted from the implementation.

Considerable effort goes into the post-silicon validation of any part as complex as a CPU. The 7100LC design team found electrical verification—including timing, shorts, etc.—to be very important. Only the most obvious electrical problems cause total system failure. Many electrical problems are reflected in lowered yields—chips do not run at certain frequencies, supply voltages, etc. The design team also put a great deal of effort into making sure that the chips could be quickly tested with high coverage during manufacture to ensure that systems could be reliably constructed from the chip.

The next example describes the design methodology used in a system-on-chip.

Example 9-3: *Design of the Viper digital video chip*

The Viper is a system-on-chip [Dut01] designed by Philips for digital television and set-top boxes. This system-on-chip includes 35 million transistors and was fabricated in a 0.18 μm process. The Viper includes two CPUs: a TriMedia VLIW processor similar to that used in the TM-1300 of Example 8-4; and a MIPS CPU. The chip also contains a number of I/O devices and a synchronous DRAM memory controller.

The Viper was designed to meet the Philips DVP standard. This standard defines architectures and interfaces that make it easier to design scalable digital video systems-on-chips. Early in the design, a number of standardized blocks such as interrupts and debug interfaces were designed so that they could be used at several places in the chip.

A number of register-transfer design rules were followed during RT design:

- All high-bandwidth peripherals used the Philips DVP standard for DMA.

- All low-bandwidth peripherals and MMIO used the Philips PI-bus standard specified by DVP.

- Miscellaneous interface signals (reset, interrupts, etc.) were implemented according to the DVP standard.

- Signals such as resets and interrupts were implemented using standard modules.

The design was verified at both the register-transfer and gate levels. A regression set of tests was executed every week. This test took about 72 hours to execute using 60 CPUs. Emulators were also used to verify the design.

The chip was designed to be fully scan testable. Both structural stuck-at tests and functional tests were used. Some large memories and caches were created with built-in self-test logic. The tests were developed module by module, with each designer responsible for achieving 99% or better fault coverage on his or her component. Busses and interconnect were tested with interconnect tests.

The entire Viper could be synthesized with logic synthesis in about eight hours running on multiple CPUs. After logic synthesis, the scan chains were inserted. The netlist was then partitioned and laid out. The physical hierarchy used for layout was not the same as the logical hierarchy used for register-transfer design. The design was partitioned into chiplets of 200K cells or fewer, with a total of nine chiplets in Viper. Three hard IP blocks were used: the TriMedia CPU, the MIPS CPU, and a custom analog block. Signals between chiplets were connected via abutment.

Timing closure was achieved in two stages. First, each chiplet was analyzed for timing and budgets for input and output timing were set up. Next, the entire chip was analyzed and optimized. Clock connections between chiplets were carefully designed and verified to be sure that the clocks were phase-aligned and met their required timing. A great deal of work went into design-for-manufacturability, including design rule checking, removing antennas, and doubling vias were possible.

9.3 Kitchen Timer Chip

A kitchen timer is a simple example of an ASIC: the function is specialized, and high performance is not essential. (Of course, we might want to implement this function with a microcontroller. Assume for the sake of argument that manufacturing cost or other considerations push us to an ASIC implementation.) Let's walk through the

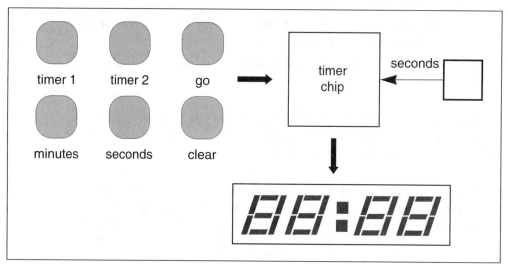

Figure 9-2: *The kitchen timer chip in situ.*

complete design of this chip, from defining its function to assembling the final layout and validating its correctness. We will use a design process which minimizes design time, inevitably at the expense of area and delay. That means we will use both standard cell layout synthesis and logic synthesis as much as possible. Readers who wish to design the world's fastest or smallest kitchen timer are invited to improve upon this design.

9.3.1 Timer Specification and Architecture

The first step in design is specifying what the chip should do; that requires understanding how the chip fits into the larger system of which it is inevitably a part. Figure 9-2 shows the kitchen timer in its system context. The kitchen timer supplies two independently-running countdown timers. Each timer can be set by incrementing minutes or seconds; the timer starts running on command. Once a timer has reached zero, it operates a buzzer, which turns off when any button is pressed. The two timers share a single display. The front panel is designed to minimize the number of buttons required—while buttons aren't costly, they do take up space which makes the timer large and unwieldy.

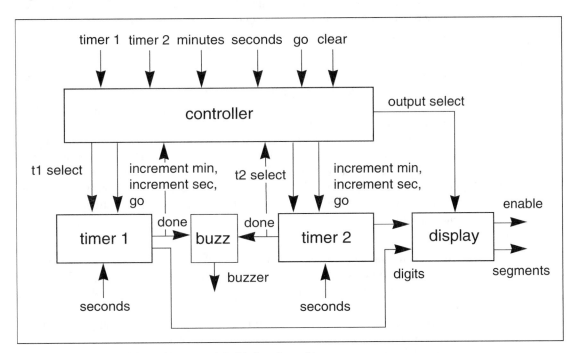

Figure 9-3: *A sketch of the architecture of the kitchen timer chip.*

In this case, it is hard to think about what the chip does without imagining at least a simplified architecture. Figure 9-3 gives a first cut of the timer chip's internal architecture. (For simplicity, we will assume that the buttons have been debounced off-chip.) Two identical timer units perform the timing functions; each is attached to a seconds clock, which comes from off-chip. A separate controller machine translates button commands into control signals for the timers and the display. It will probably be easier to use a separate machine to determine when to ring the buzzer, rather than use the button controller (though we can wait to make that decision until after we have formalized the chip's function). The display has four seven-segment digits; rather than use 28 pins to display all those segments simultaneously, we will multiplex the display—drive one digit at a time at a rate sufficient rapidly to keep all the digits visible.

With pictures of the chip's environment and of the relationship between the timers and the controllers in mind, we can start to specify in detail what the chip must do. The most concrete and useful way to specify the chip's function is to build a functional simulator which emulates the input-output behavior of the chip. The timer chip's functional simulator will, for example, accept button values (on/off) and the seconds clock as inputs; it generates the display signals as outputs. The program will accept one set of inputs, compute the appropriate outputs using its internal state, then wait for the next set of inputs.

The functional simulator isn't a register-transfer simulation because it doesn't break each action into cycle-by-cycle operations. The way the program works may bear little resemblance to the way the chip works—for example, the program may use integers for the timers, then convert those integers to seven-segment values. But all that matters at this point is that you capture the intended function for the chip and you run the program to be sure that what you specified is what you really meant. You will find that casting the chip in the concrete form of a program will be an enlightening experience: operations you thought you understood will be revealed to be vague; hidden assumptions will be uncovered; and hints of implementation difficulties will be encountered.

You should thoroughly exercise your functional simulator to be sure that you have really specified the behavior you desire. The functional model, when finished, is the golden standard of what the chip is supposed to do. Your goal should be to apply the functional simulator's inputs to the final chip (with appropriate timing, of course) and see the outputs predicted by the functional model. English descriptions of complex digital systems are often vague and sometimes self-contradictory, but you can always run the functional simulator to see what it does. Appendix C gives a functional model of the kitchen timer written in the C programming language and some sample results of executing that model. You can, of course, use a hardware description language like VHDL or any programming language to construct your model. If you wish to simulate an abstract architecture, you may use a hardware simulation system which can efficiently model the architecture's subsystems and their communication.

9.3.2 Architecture Design

Once we understand what the chip is supposed to do, we can start designing the architecture. We must, of course, work top-down from the specification to architecture to logic, but we also want to develop an initial floorplan to identify potential area and delay problems.

Our first decision is how to represent timer values—that decision has profound implications on the amounts of both logic and wiring. Our choices are unsigned, binary-coded decimal (BCD), and seven-segment. Internal seven-segment coding is a bad idea—while it doesn't require decoding for the display, it uses a lot of latches to store the values and logic to implement increment and decrement operations for seven-segment-coded numbers. Unsigned integers require the fewest latches and least logic for adders, but they require large decoders to generate seven-segment digit values. BCD representation is a good compromise, balancing the requirements of arithmetic and display decoding and being easy to understand as well.

The decision to use BCD should immediately signal a change in the architecture's decomposition into components from the proposal of Figure 9-3. We want to minimize the number of long wires running from the timers to the display controller. Though we eventually reduce the number of wires to seven, we must first go through two stages of selection: first to select between the two timers, then to select the currently displayed digit. We want to organize the layout to make those decisions as physically close to the timers as possible. Stretching all the timer signals across the chip costs both area and delay. That suggests grouping the two timers in a single layout component, so that the timer digits can be paired off and multiplexed with very short wires. As shown in Figure 9-4, that gives us a hierarchy of physical components different from our logical component hierarchy—we want to think of the system as a pair of timers, but we implement the chip using a single component which includes both timers and the multiplexers.

The need to minimize wiring also suggests a change to the definition of the timer component. Rather than send all of each timer's digits to the display machine, which then selects which one to display, we can send only one digit from the currently selected counter. Extra logic inside the *timers* component can keep track of which timer is currently selected for display and select the proper digit of that timer.

Before we make a preliminary floorplan, we need to understand better what these components must do. We can outline the function and pinout of each. Every component receives ϕ_1, ϕ_2, and a reset signal, which clears all latches to 0.

Figure 9-4: The timer chip's component hierarchy.

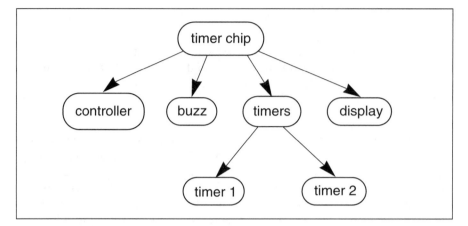

- **Timer** A *timer* input selects the timer to which a command signal is to be applied. Each timer holds the current timer value in BCD, can increment minutes or seconds on command, and decrement by one second when the *seconds* clock is raised. Each timer can be cleared on command. Each timer starts running when its *go* signal is raised and generates a *done* signal when the timer has counted down to zero. One display from the currently selected timer is selected for display; one digit, selected by *digit_select,* of that timer is sent to the *digit* output. All command inputs remain high for one cycle.

  - *Inputs*: incr_seconds[2], incr_minutes[2], seconds, go[2], digit_select[2].

  - *Outputs*: done, digit[4].

- **Timers ctrl** The timer controller multiplexes and demultiplexes control signals based on which timer is currently selected. It also remembers which timer is currently selected.

  - *Inputs*: timer, incr_seconds[2], incr_minutes[2], seconds, go[2], digit_select[2].

  - *Outputs*: done, digit[4].

- **Display** The display controller machine cycles through the four digits to be displayed. It sends the digit selection to the timers. The machine accepts the BCD digit, decodes it to seven-segment form, and generates a digit enable signal for the display.

  - *Inputs*: digit[4], output_select.

  - *Outputs*: digit_select[2], enable[4], segments[7].

- **Buzz** This machine accepts a signal to turn on the buzzer. It then enables the buzz signal until it receives a *stop*.

  - *Inputs*: done, clear.

  - *Outputs*: buzzer.

- **Controller** The system controller interprets commands from the buttons. It accepts signals from the six buttons. Once a *timer 1* or *timer 2* button is pressed, subsequent command buttons are applied to that timer, until another timer is selected. The controller generates a *timer* selection signal for the *timers* component on every cycle, so that *timers* applies commands to the signal specified in the current value of the *timer* signal. The *clear_in* signal both clears the selected timer and sends a signal to the *buzz* component to turn off the buzzer. The *go_in* signal starts the selected timer. *Minutes_in* and *seconds_in* request that the minutes and seconds, respectively, of the selected be incremented.

  - *Inputs*: timer_1, timer_2, minutes_in, seconds_in, clear_in, go_in (referred to as *all buttons* in Figure 9-5).

  - *Outputs*: timer, incr_minutes, incr_seconds, go, clear, output_select.

We can now draw block diagrams that reflect our refined design. We should draw a block diagram not just for the chip, but for each major subsystem which is made up of components significantly larger than gates. Figure 9-5 shows the block diagram of the timers component and of the complete chip. Thorough documentation also requires a block diagram for a *timer*, which we leave to the reader. The *timers ctrl* block in the *timers* component routes control signals like incrementing to the proper timer and multiplexes the timers' digits.

Now we can estimate the sizes of the components and count the wires between them. For simplicity, we will estimate component sizes in terms of two-input gates + latches, which is good enough to produce a rough layout sketch.

- **One timer:** 14 latches (ϕ_1 and ϕ_2) and about 30 gates.

- **Timers:** 28 latches and 70 gates, counting the additional logic required to route signals to and from the individual timers.

- **Display:** Four latches and about 15 gates.

- **Buzz:** Two latches and two gates.

- **Controller:** Eight latches and about 20 gates.

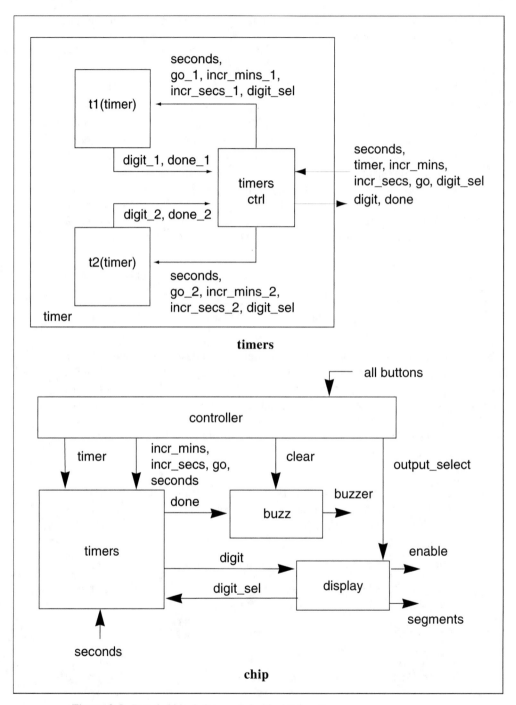

Figure 9-5: *Detailed block diagrams for the kitchen timer.*

How do we make these kinds of estimates? By looking at, or guessing at the logic functions implemented. If one component is an adder, for example, we can rewrite the sum and carry equations in terms of two-input gates and then simply count the gates. Our timers are slightly harder to estimate accurately because they include increment, decrement, and non-binary carry logic. For a state machine, we can estimate the number of states and guess at the number of transitions per state by counting the number of inputs to the FSM. The number of states gives us an estimate of the number of latches required, while the number of transitions in the machine gives us a rough idea of the amount of logic required. Remember, we need only an estimate. While the estimate should be accurate enough to avoid suggesting wrong conclusions, it doesn't have to be precise to the last gate.

Using this information, we can sketch a floorplan for the chip, shown in Figure 9-6. This initial floorplan gives us an idea of the relative sizes of blocks and of the number of wires which must go between blocks. The number of signals between blocks is shown, but this floorplan sketch doesn't show more details of the routing. We generally want to make the chip as square as possible because a square packs the most chips onto a wafer. The *timers* block dominates this floorplan, so we can pack the other components around it to make a squarish layout.

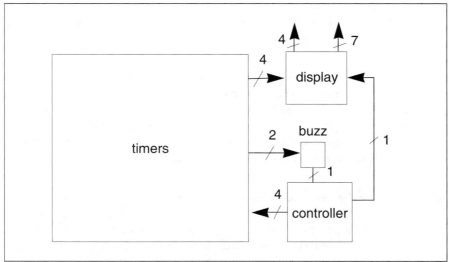

Figure 9-6: An initial sketch of timer chip floorplan.

This floorplan doesn't detail wiring. It may be helpful to create a more detailed floorplan which shows the global routing of signals. Such a wiring plan is especially important when large numbers of signals must be passed between components, such as data and address busses in a microprocessor. The global routing helps identify long wires which may cause performance problems. It also lets you estimate the sizes of routing channels, which can use a significant fraction of the chip area in a large lay-out. The kitchen timer has few enough wires that more detailed planning of the layout is overkill, but the importance of carefully estimating both component sizes and rout-ing area grows with the size and complexity of the layout.

Figure 9-11 shows an ASM chart for the controller module. The states t1 and t2 remember the timer to which the other button commands refer. Those states repeat-edly scan the buttons, testing for a press. Extra states are used for the minute, sec-ond, clear, and go buttons to wait for the button to become released before another command is taken.

9.3.3 Logic and Layout Design

Now that we have broken down the system into manageable components, we can start implementing them. In this case, each of the components has been implemented using standard cell placement-and-routing. Components like timers were specified as parts of a register-transfer structure. A logic optimizer designed the combinational logic from the register-transfer equations. Finally, a standard cell synthesis program created the layout. The FSMs were specified as state transition tables with manual state assignment; the same logic optimization and standard cell tools used for the structural register-transfer components were used to complete the design of the FSMs.

The timer pair is the most complex element of the system. To be sure that we build them correctly with minimum backtracking, it is best to start out simple. For example, you can first design a single digit as an up-counter, then a four-digit up-counter, add the down-counter logic, and finally complete the other features required of a single timer. Once a complete, single timer has been built and verified, the two-timer module can be constructed with confidence that it will work.

A **register-transfer generation language** is a big help in specifying logic blocks like the timers and the display controller. A register-transfer generator is like a layout generator in that the generator is a program which, when executed, creates a list of the registers, logic gates, and wires in the description. Generator languages are especially good for regular and semi-regular structures, such as the timer, because we can use loops to create the bit-wise copies of the logic and conditionals to create the necessary bit-to-bit variations.

Figure 9-7 shows part of the generator, written in the PDL++ language [Lip90A], for one digit of a timer. The loop is executed once for each bit in the digit—both the com-binational logic and the latches. The intermediate variables xxx and yyy hold the

```
for (i=0; i<bits_per_digit; i++) {
    mdg[i] = drlatch(xfer[i],phi1,reset);
    /* increment */
    xxx[i] = incr(mdg[i],cry[i]);
    if (i < bits_per_digit-1) {
        carry[i+1] = incrcarry(mdg[i],carry[i]);
    }
    /* decrement */
    yyy[i] = decr(mdg[i],brw[i]);
    if (i < bits_per_digit-1) {
        borrow[i+1] = decrborrow(mdg[i],borrow[i]);
    }
    plus[i] = tf_test(digitcarry,ZERO,xxx[i]);
    minus[i] = tf_test(digitborrow,rolldown[i],yyy[i]);
    ndig[i] = tf_test(clear,ZERO,tf_test(add_sub,plus[i],minus[i]));
    xfer[i] = drlatch(ndig[i],phi2,reset);
}
```

Figure 9-7: Part of a C register-transfer generator for a timer.

results of increment and decrement. The if statements which surround the generation of carry and borrow ensure that the carry/borrow out of the last bit is not generated; this is done strictly to avoid creating a dangling output which is used nowhere else. The tf_test() function adds multiplexing logic, using the first argument as the selector, to determine which value to feed into the ϕ_2 latch declared by the drlatch() function. When this program is executed, the declarations of logical functions, latches, etc. cause execution of hidden functions which write the appropriate logic structures: logic gate functions, latches, and the connections between them. For a regular structure like the counter, writing a loop which generates the logic is much shorter and easier to get right than writing all the bits by hand. The complete logical design of a

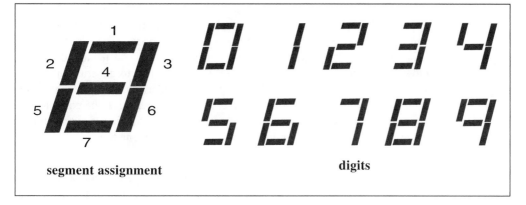

Figure 9-8: The seven-segment display.

timer is left as an exercise.

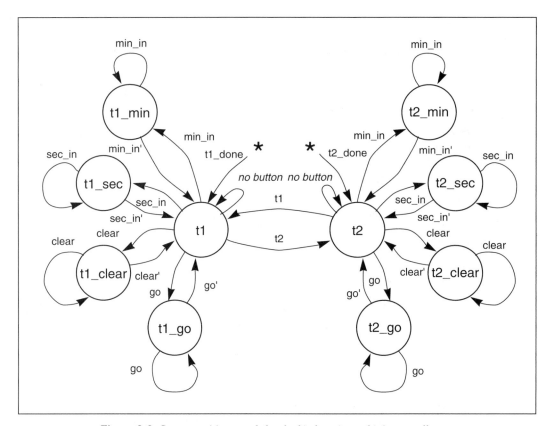

Figure 9-9: *State transition graph for the kitchen timer chip's controller.*

The *display* unit is an excellent illustration of the power of register-transfer specification plus logic optimization. The component has two functions: it must cyclically generate digit select signals to tell *timers* which digit to send; and it must decode the BCD digit into seven-segment form. Figure 9-8 shows how to generate digits on a seven-segment display. A BCD-to-seven-segment decoder has no obvious structure. We want the logic to be as small as possible, but we also want it to be correct. Since we have logic optimization, we can write the decoding function in the simplest possible way and rely on the optimizer to find an efficient implementation.

The controller's state transition graph—showing no output values on transitions for clarity—is shown in Figure 9-9. Its structure closely resembles that of the ASM chart, naturally. The machine has a total of ten states. If we use four bits to encode these states, we would have six states in the implementation with unspecified transitions. The latches we will use have a reset input; resettable latches add implicit transitions from every implementation state to the 0000 state. Judicious choice of state codes assigns 0000 to a natural reset state in the state transition graph. A global reset state is an important first step in testability. We may find after logic implementation and test generation that some changes to the state transition graph may be necessary to make the machine sufficiently testable.

Figure 9-10 shows part of the state transition table (in Berkeley KISS format) for the controller. The introductory lines declare the numbers of states, inputs, and outputs, along with the names of the primary inputs and outputs. The rest of the file gives one FSM transition per line, in the form *inputs*, *present state*, *next state*, *outputs*.

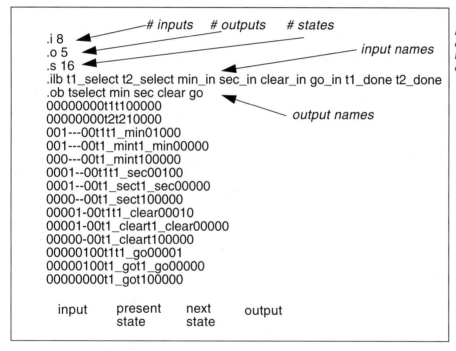

Figure 9-10: *Part of a state transition table for the controller.*

Once we have prepared the logic for all four components, we can synthesize their layouts. We used *misII* for logic optimization and *wolfe* (no relation to the author) for standard cell placement and routing; the standard cells were from the Mississippi State University (MSU) library. One property of the standard cell synthesis system we used is that it does not allow precise placement of the input and output pins along the edges of the cell. That means two things: that we must adapt our global routing to the pin placement given to us and that the placement may change if we modify the logic and re-synthesize. The fact that pins may change position from run to run give us

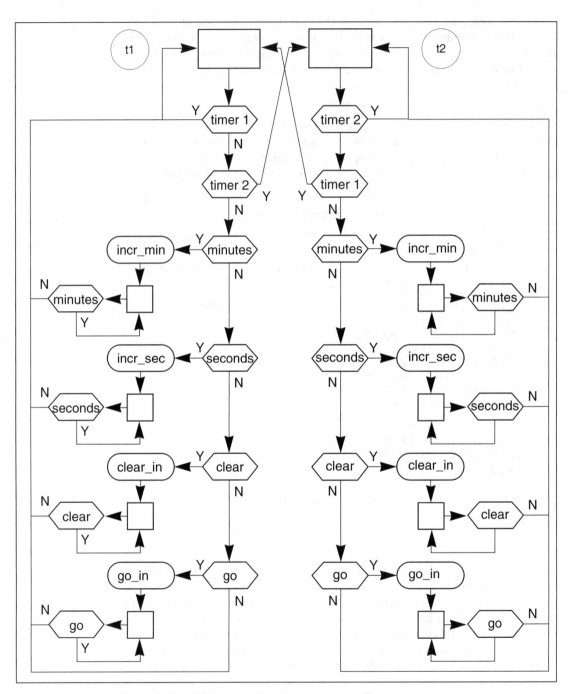

Figure 9-11: *ASM chart for the kitchen timer controller.*

added incentive to make sure the logic is correct before we proceed to final chip layout.

Whether you have synthesized a component or designed the layout by hand, it is important to verify each component before you insert it into the complete chip layout. You should perform functional simulation to make sure the component performs the right function. You should also extract parasitics and perform timing analysis or simulation to check delays. After calculating delays for each component, you should compare the values you observe to the delay values you assumed while planning your chip. If the delay through a component is very different from what you expected, the chip's critical delay path may not be where you think it is.

Once you are sure the components are correct, you can begin detailed floorplanning. You should check the actual sizes of the components to the sizes you estimated; in this case, our estimates were sufficiently accurate to not invalidate our floorplan sketch. You need to design routings for power, clocks, and signals.

Figure 9-12 shows the power and clock wiring plans for the chip. To design the power routing, you must know how much power each block requires. Once you have counted the worst-case current requirements of each block, you can easily determine the metal wire width required for the block's power lead. *Wolfe* generates layouts with V_{DD} on the left and V_{SS} on the right, so a floorplan without rotated blocks fulfills our requirements for planar routability.

Since this layout is fairly small, clock skew is not a major problem, but driving the clock load is more difficult. Each latch cell presents a load on the clock input of 99 fF of gate capacitance and 17 fF of wire capacitance, giving a total of 116 fF. With 56 latches in the system, the latches present a total capacitive load of about 6500 fF. The metal wires which distribute the clock through the cells add about 1700 fF of capacitance, making the total capacitance on the clock driver about $C_{clk} = 8200\ fF$. Let's specify rise and fall times of 1 ns on each phase. To find the size of the pullup and pulldown for the clock driver, we set $2.2R_{pd}C_{clk} = 1\ ns$ and $2.2R_{pu}C_{clk} = 1\ ns$. This gives $R_{pd} = R_{pd} = 55\ \Omega$. We compute the width of a minimum-length driver with this resistance by computing the ratio of the required resistance and the minimum-size transistor resistance. We find that the pulldown must be 436 λ wide and the pullup must be 1164 λ wide. This makes the driving inverter about as wide as the entire chip core, though very narrow; we need one such driver for each phase. The design of the driver chain leading up to this final stage depends on the circuits we use to get the clock signal on-chip. We can reduce the size of the clock driver by lengthening our rise and fall time specifications, which is an option in this low-performance ASIC but may not be feasible in all cases. The large clock buffer does not change our floorplan

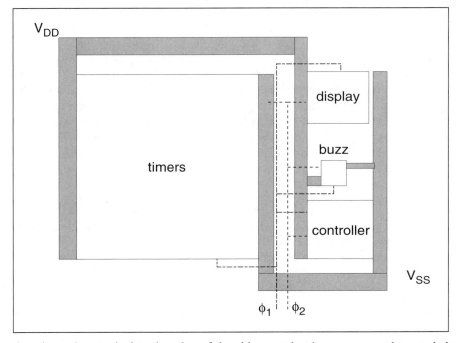

Figure 9-12:
*Power and clock
wiring plans for
the kitchen timer
chip.*

since it can be attached to the edge of the chip core, but it may present layout challenges in more complex designs.

You should also design a global routing for the signals before sitting down at a terminal to complete the layout. The first step is to define channels. Our standard cell router uses metal 1 for horizontal tracks and metal 2 for vertical tracks, which is a good choice for the chip's channels as well. Figure 9-13 shows one set of channel definitions; the arrows show the direction of horizontal tracks. Once we know where the channels are, we can design a global routing which assigns each signal to a sequence of channels. Once signals are assigned to channels, we can assign signals to tracks in the channels relatively independently, only making sure that the switchboxes connecting the channels do not become too messy due to bad track assignments.

The chip's global wiring, without cells or the clock driver, is shown in Figure 9-14. This plot includes the power lines, clock lines, and all signals. All signals connected to input or output pins have been routed to points at the edge of the layout. Most of the routing is done in the channel between *timers* and the other blocks, though some of the signals must be wrapped around from other sides of the cells. In a few places we violated the metal 1 horizontal-metal 2 vertical rule; since this routing problem is relatively simple, there are many places where we can route wires to avoid switching layers. Even regions with relatively few wires can be deceptively tricky, however; you

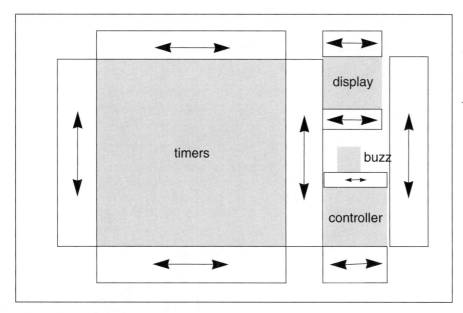

Figure 9-13:
Channel defini-
tions for the
kitchen timer
floorplan.

should sketch out your wiring on a piece of paper first. A global routing layout tool can be very useful, but if you don't have access to one, you must draw the rectangles for the wires yourself. Chip-level layout design is difficult because it is hard to get a good view of a long wire: on the one hand, you must be able to see a large section of the layout to know where your wire is going; on the other hand, you must see the end-points close-up to accurately place the wire. Most layout editors allow several windows to be opened onto a single layout, with changes made in any window immediately reflected in all the windows. This feature is extremely handy for designing long wires, since you can have one close-up window for each end of the wire and a third window showing the entire layout.

A layout of the timer chip core—the logic minus the pad frame, clock driver, and routing between them—is shown in Figure 9-15. When we add the pad frame, we connect the pad ring's V_{DD} and V_{SS} signals to the core's power supplies, connect the clocks, and route the input and output signals between the terminals at the layout's edge to the pads.

9.3.4 Design Validation

After you have assembled a complete layout, you should make sure it works. If your layout editor does not support incremental design rule checking, you must use a sepa-

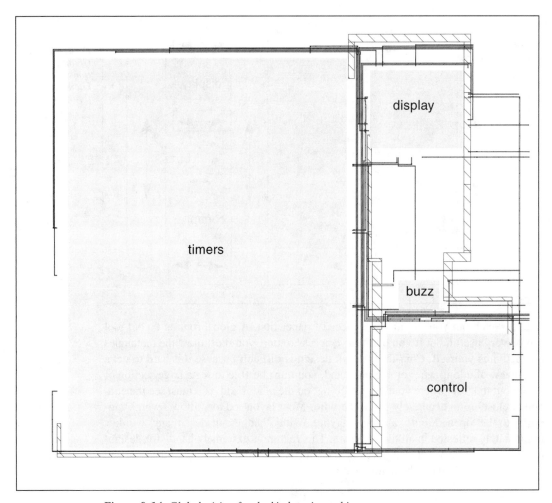

Figure 9-14: *Global wiring for the kitchen timer chip.*

rate program to check for design rule errors. Your next step is to extract a switch-level circuit from the chip layout and run a number of tests:

- A switch-level simulation of the extracted circuits will catch a number of problems: previously undiscovered design errors, wiring errors, charge sharing bugs, etc. You should compare the results of this simulation to your earlier simulations to the extent possible. It may not be feasible to construct an exact correspondence between your higher-level simulations and the extracted circuit, but you should convince yourself that the final circuit conforms to your expectations.

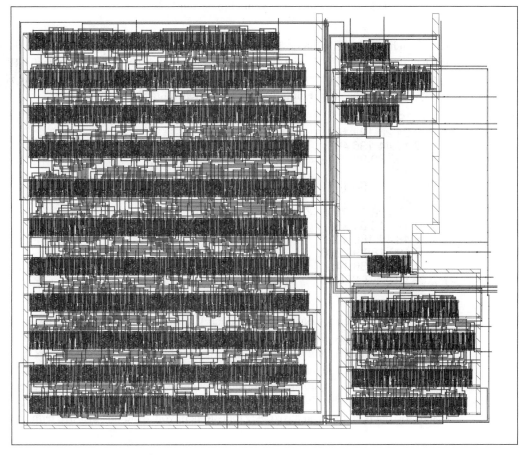

Figure 9-15: *Layout of the kitchen timer chip.*

- You should analyze the performance of the complete chip using a timing analyzer or a timing simulator. This final check ensures that overly long logic delays or excessively large parasitics have not stretched the chip's minimum clock period beyond acceptable bounds.

- You should use an automatic test pattern generation program to generate a set of test vectors or a fault simulator to develop your own tests. If you cannot obtain nearly 100% fault coverage, you may need to make some memory elements scannable or make other logic changes to increase the chip's testability.

- Make a check plot of the complete layout and look at it yourself. The check plot serves as a sanity check to catch errors not caught by your tools; it also makes a nice memento of your chip design.

Once you have verified your design, you can generate a layout file for shipment to the fabrication facility. The layout description is usually in terms of rectangles. **CIF** (Caltech Intermediate Form) [Mea80] is a commonly accepted layout format. Here is a very small part of the CIF description of the kitchen timer:

```
L CSN;
 B 24 28 12 410;
 B 16 44 136 402;
 B 192 80 96 68;
 B 128 28 64 14;
 B 48 28 168 14;
L CSP;
 B 104 28 76 410;
 B 128 16 64 388;
 B 48 44 168 402;
 B 192 188 96 286;
 B 16 28 136 14;
 94 GND! 0 12 CMF;
```

A **B** command describes one rectangle. Each **L** command declares a new layer for subsequent rectangles. The **94** command defines and places a text string which will appear in a check plot but not in the masks used for fabrication. Chip designers used to have to design their layouts one rectangle at a time, sometimes by typing in textual descriptions like this, or perhaps by drawing rectangles on huge sheets of graph paper. Electric erasers were highly-prized layout tools in the early days of integrated circuit design. Be thankful that you have CAD tools to help you avoid this massive tedium.

9.4 Microprocessor Data Path

As a contrast to the highly application-specific design of the kitchen timer, we will consider designing a microprocessor data path. A microprocessor also requires control, bus interface, etc., but many of the VLSI challenges lie in the datapath. To design a data path, we must: choose a register-transfer structure that efficiently executes the instructions; choose a clocking scheme; design circuits; and implement those circuits as layout. This exercise will show how closely related those decisions can be, even for such a simple machine.

9.4.1 Data Path Organization

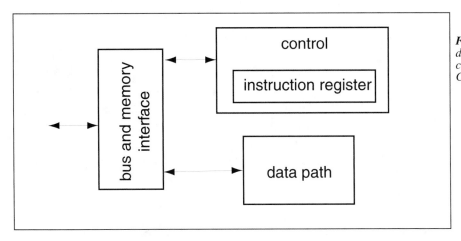

Figure 9-16: The data path in the context of the CPU.

The data path is a critical element of any CPU. We do not have the space to thorougly review microprocessor architecture [Pat98], so we will concentrate on a simple data path for a basic microprocessor. Figure 9-16 shows how the data path fits into the CPU. The memory and bus interface reads and writes the data path's register file from memory. The control logic sets the control signals that tell the data path what to do, based on the instruction register's value and other elements.

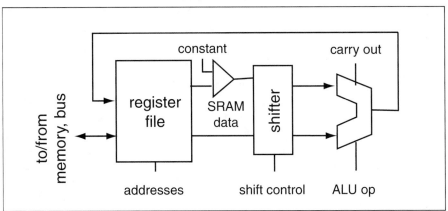

Figure 9-17: Structure of a basic microprocessor data path.

Figure 9-17 shows the structure of a generic microprocessor data path. This data path has three major elements: a register file, a shifter, and an ALU. We saw in Chapter 6 how to design shifters and ALUs. We will use an SRAM for our register file. One of the shifter inputs has a multiplexer so that a constant can be substituted for a register argument. Each block has control inputs and/or outputs that connect to the control unit as well as data inputs and outputs that stay within the data path.

One important question is how many ports the register file needs. On the one hand, adding ports to the register file increases the number of data path operations we can perform per cycle since it allows us to read and write more operands from the register file. On the other hand, adding ports to the register file slows down the SRAM's operation. We must carefully evaluate the design to ensure that we do not increase the number of operations per cycle at the expense of the cycle time.

To better understand the implications of the instruction set on the register file design, consider a register-to-register ALU operation, such as an addition. The number of register file ports limits the number of cycles it will take to perform this operation:

- **one port:** three cycles (read operand 1, read operand 2, write result);
- **two ports:** two cycles (read operand1 and operand2, write result);
- **three ports:** three cycles (read operand 1 and operand2 and write result).

The delay of an SRAM increases roughly linearly with the number of ports. There is a noticeable cost to increasing the available parallelism in the data path. Clearly, the circuit characteristics of SRAM have a profound impact on the CPU's microarchitecture and its instruction-level performance. We will use a two-port SRAM in our data path; to determine where the latches should be placed to store the temporary values, we must consider timing, which we will do in the next section.

The number of words in the register file is typically limited by the data path clock cycle. Using an overly large register file causes the access time of the SRAM to stretch the data path's clock cycle beyond an acceptable limit. SRAM register files are small compared to SRAMs in commodity memories, which allows two simplifications. First, register files often use single-ended memory cells, which transmit only one polarity of the stored value. Second, if the register file is very small, the sense amps may be dispensed with. In this case, logic is still needed to precharge the bit line, but the bit line may be used directly as the bus connection.

9.4.2 Clocking and Bus Design

We next need to consider the timing of operations in the data path. We will use the clocking discipline to analyze when operations happen. The key element in the data

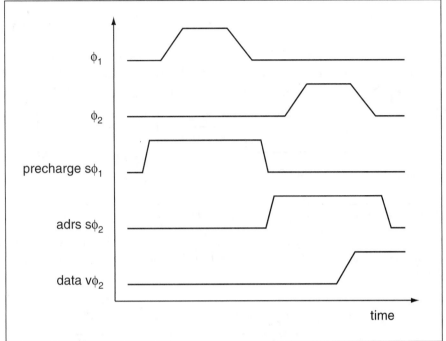

Figure 9-18:
SRAM timing.

path timing scheme is the SRAM because it is precharged. If we precharge the register file on ϕ_1 and read or write it on ϕ_2, how must the rest of the system be clocked? Figure 9-18 shows the timing of the SRAM using the clocking discipline. The address inputs should be $s\phi_2$. The data output is not a strict clocking discipline signal: it is $v\phi_2$ but not $s\phi_1$.

Furthermore, we cannot take the SRAM output, send it through the shifter and ALU, and write it back to the SRAM at the end of the phase. The output of the ALU, which would be the input to the SRAM write port, would be $v\phi_2$. Since it would arrive even later than the signal that drives it, namely the $v\phi_2$ read value from the SRAM, ϕ_2 would have to be stretched even further to complete the write.

To solve these problems, we need to add latches in the proper places in the data path. We need to latch both inputs to the shifter/ALU, not just the one saved from the previous SRAM access, in order to ensure that they both have the same clocking type. We also need to add a latch to the data write input of the SRAM so that we can give the data input adequate time to drive the SRAM bit lines. Figure 9-19 shows a data path with latches at the appropriate places and clocking types to show that this is a valid clocking scheme. There are in fact several possible retimings of this data path: for instance, we could move the ϕ_1 latch before the ALU inputs. The exact placement of

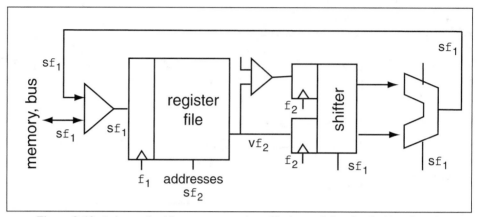

Figure 9-19: A datapath with a two-port register file that satisfies the clocking discipline.

the latches is determined by the relative delays of the register file, shifter, and ALU; these values in turn depend on the word width and the number of words in the register file.

9.4.3 Logic and Layout Design

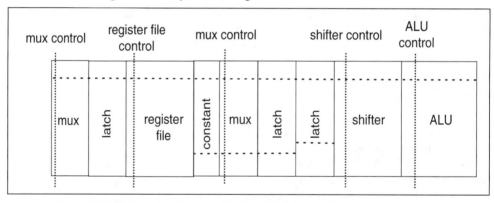

Figure 9-20: A floorplan for a one-bit slice of the data path.

We can now design a floorplan and a wiring plan for the data path. It is generally easier to sketch the floorplan first, ignoring details of wiring, then consider which layers are best for the data path layout and modify the floorplan if necessary. Figure 9-20 shows a floorplan for a one-bit slice. The major connections through the data path are shown: the ALU output back to the register file latch; and the outputs of the register file to the shifter latch inputs. Register selects, read/write signals, and ALU opcodes flow vertically through the data path.

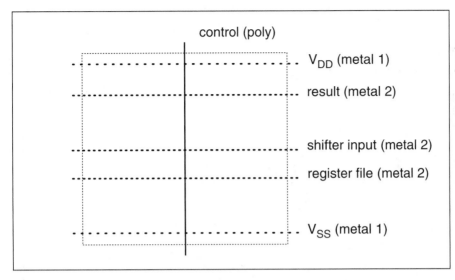

control (poly)

V_DD (metal 1)

result (metal 2)

shifter input (metal 2)

register file (metal 2)

V_SS (metal 1)

Figure 9-23:
Wiring plan for a
data path bit slice.

Figure 9-23 shows a wiring plan for the data path. Metal 1 is the only reasonable choice for power routing. Since data busses have only a few connections, we can reasonably route them in metal 2. Since metal 2 can be connected only to metal 1, it is not a good choice for signals with many connections. Since metal runs horizontally, we use poly to send control signals up and down the data path.

Another possible color plan uses metal 1 for all horizontal wires and metal 2 for the vertical control signals. While that choice gives the control lines less resistance, it makes connections to and from the control lines more difficult. Most of the control lines' connections will be to transistor gates; if a metal 2 wire were used to distribute the signal, a connection into a cell would require a metal 2-metal 1 via and a metal 1-poly via. As Figure 9-24 shows, signals in a bit may be driven at many points, while control signals are driven from a single point. Putting many wide transistors in the bit slice to drive the horizontal signals will make the data path much too large. Even if the transistors in the control line buffers are large, there is only one driver for each wire, while there are n copies of the horizontal driver transistors in an n-bit data path. If you must use poly to route signals, running control signals in poly is the preferred choice.

Figure 9-24:
Delays and
driving capa-
bilities in data
paths.

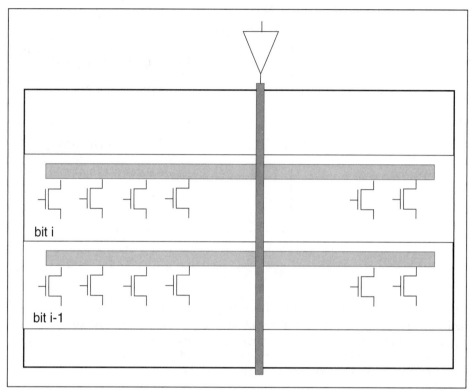

9.5 References

West Bend makes a well-designed line of kitchen timers, popular with both cooks and pilots. The timer chip described here loosely follows the specification of that family of timers, though it replicates none of them. Chip designs for which the designers have taken the time to describe the design and verification process: the DEC Alpha processor [Dob92]; an HP Precision Architecture CPU [For87]; the WE 32106 math accelerator [Mau88]; and the VAX 6000 [Cal90]; the VAX 9000 [Hoo90]; the PowerPC 601 [Bro93]; and the Intel Pentium™ [Sai93]. Lager [Shu91] and Cathedral [Lan91] are chip-level synthesis systems optimized for the design of signal-processing ASICs. The book edited by Chandrakasan et al. discusses a number of topics in high-performance microprocessor design.

9.6 Problems

9-1. Modify the register-transfer design of the kitchen timer so that when the *minute* or *second* buttons are held down, the appropriate digits are incremented at a reasonable pace.

9-2. Write a register-transfer description of the ones-seconds digit of a timer.

9-3. Write a register-transfer description of one four-digit timer.

9-4. Design the seven-segment decoding function:

   a) Write a truth table whose rows are the digits 0-9 and whose columns are the on signals for the seven segments.

   b) Write seven Boolean functions, one for each segment, giving the segment's on signal in terms of the BCD digit's bits.

9-5. Design a transistor-level circuit for a one-bit slice of a data path register file. Show one bit of constant ROM, one bit of SRAM, and the precharge and read/write logic.

9-6. Is a single-ended SRAM cell faster or slower than a double-ended cell? Justify your answer.

9-7. Draw a block diagram of a data path that uses a three-port register file. Show that your design satisfies the clocking discipline.

10

CAD Systems and Algorithms

Highlights:

CAD system organization and databases.

Simulation.

Placement and routing.

Performance analysis.

Logic synthesis and state assignment.

Automatic test pattern generation (ATPG).

High-level synthesis and hardware/software co-design.

10.1 Introduction

Building a chip requires using a variety of CAD tools, both to analyze the design and to synthesize parts of the design. A modern CAD system is built from several million lines of code, so it isn't reasonable to assume that a designer will understand every nuance of each tool. However, it is important that you understand the concepts underlying the major CAD tools you use. You need to understand what a tool can and cannot do to take the greatest advantage of it.

Understanding the capabilities of a CAD tool requires learning its underlying models and its algorithms. Any tool must operate on an abstraction; logic optimizers, for example, have almost no knowledge of layout, except perhaps for parasitic capacitance estimates. The properties of the tool's model are intimately linked to the algorithm used—we select a model that can be efficiently solved. Understanding the algorithms themselves is equally important. CAD algorithms are carefully chosen to be efficient because chip design problems can grow to be extremely large, requiring hours or days of CPU time to solve. The algorithm may have properties that affect the solution of your problem.

This chapter concentrates on algorithms which are less likely to be applied by hand—while some algorithms are intuitive and easy to apply, other algorithms are really suitable only for program implementation. This isn't an exhaustive analysis of CAD algorithms; rather, we will discuss the techniques in enough detail to help you reason about your CAD tools' behaviors.

We will cover a variety of synthesis and analysis tools: switch-level simulation; layout synthesis; layout analysis; timing analysis; logic synthesis; sequential machine optimizations; test generation; and architectural optimizations. But first, a brief word about design representation in CAD systems.

10.2 CAD Systems

Since we use CAD tools at many stages of the design process, those tools must be able to communicate with each other. As a result, we must inevitably think of our tools as part of a CAD system, even if that system is an ad-hoc collection of programs from disparate sources. There are two ways to organize a collection of CAD tools. In a **database** system, information is stored in a central set of files and access to that information is provided through a set of standard routines. A database-driven CAD system is sometimes referred to as a **hub-and-spoke s**ystem because all tools talk to the central database and see a common representation for the design. In a **translator** system, tools do not share a common design representation. Instead, the output of one program is translated by a special-purpose program into a format acceptable to the next program. A complete set of translators between every set of programs requires *n(n-1)* translators, which require not only a great deal of patience to build but also allow for many opportunities for errors and dropped information. On the other hand, the more unified the design representation in a database, the more restrictions are imposed on the CAD tools which use that information.

Most CAD design data may be thought of as structural description—graphs, geometric objects, etc. (In contrast, most programming tools appear to the user to be manipulating text files.) Layout representations require geometric information, typically rectangles, but other primitive shapes such as triangles may be used due to the requirements of layout algorithms. Circuit information may be stored as either net lists or component lists. Net lists are ubiquitous because they span the design hierarchy from transistors to register-transfer systems. Functional behavior is often stored as either a simple data flow graph or as a more general **control-data flow graph** [Sno78], which includes both data flow and sequencing information.

Design data generated at one stage should be kept for later checking. One important function of integrated CAD systems is **back-annotation**—the amendment of a description, usually a circuit or gate net list, with new information. For example, simulation and delay analysis performed before layout design must make some sort of assumption about parasitic resistance and capacitance. Once layout is complete, the parasitics can be measured and added to the original net list. Back annotation may require the addition of new components, since the original design may not have given any resistance or capacitance values for many nets.

10.3 Switch-Level Simulation

Logic gate models for simulation are important, but they may give misleading simulation results for systems built from MOS transistors. Even if we reduce the transistor to an ideal switch, a unidirectional logic gate model cannot model charge sharing, which is a result of the bidirectionality of MOS transistors. While we could in principle catch charge sharing bugs by full-blown circuit simulation, it is useful to have an intermediate model which is efficient enough to let us simulate a large chip at a fairly low level. Switch-level simulators are at the lowest level of digital simulation.

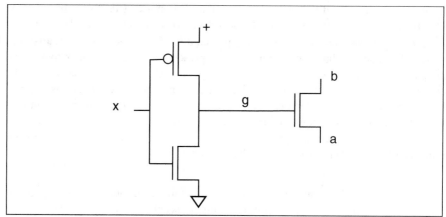

Figure 10-25: An example circuit for switch-level simulation.

The opening and closing of switches changes the circuit interconnections during operation. Switch simulation must consider sets of wires which may be connected by channels. In Figure 10-25, node g may be connected to either power supply, but the transistor gate which terminates that wire isolates g from nodes a and b. Electrical node values must be solved iteratively—for example, first solving for the value of g, then solving for the values of a and b, depending on whether g forces the transistor to be opened or closed.

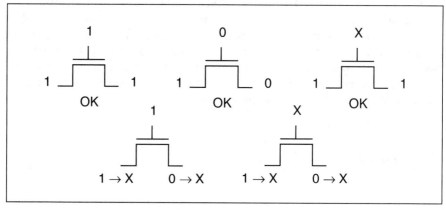

Figure 10-26: Rules for evaluating switch-level simulations (after Bryant [Bry87A] © 1987 IEEE).

With logic gate models, we assume that wires are ideal; in switch simulation we must assume that all wires have capacitance, since we are using switch simulation precisely to catch charge sharing bugs. Figure 10-26 illustrates how capacitance changes node evaluation by showing the change to the source and drain node values when the transistor gate changes value [Bry87A]. When an n-type transistor is open, its source and drain nodes can be at different values; if the switch is closed and its source and drain were previously at the same value, everything is okay. If the source and drain were at different values, what happens depends upon the ratio of the capacitances at the source and drain nodes. If the capacitances have about equal values, both nodes go to X—the capacitors at V_{SS} and V_{DD} share charge to arrive at a voltage of about $(V_{DD} - V_{SS})/2$. If one capacitance is considerably larger than the other, the value on that node wins—charge is still shared but the small capacitor can't absorb enough charge to significantly degrade the logic value.

COSMOS [Bry86,Bry87C] computes the switch network's behavior over time by writing a system of equations derived from the interconnection of switches, then solving those equations. Each electrical node has a size, representing its capacitance; each transistor has a strength, representing its conductance. A channel graph represents how transistors connect electrical nodes: the nodes in the graph represent the electrical nodes, and edges are introduced between nodes connected by a transistor channel. A channel-connected subnetwork of the channel graph is a component of that graph, along with the transistors for which these nodes are sources or drains. The switch circuit's steady state response depends on how values in the network set transistor gates, creating paths from one node to another. Any set of closed switches that form a complete path to a power supply node help determine the value of a node. The steady-state response at node n is determined by examining the paths to n: if all completed paths drive n to the same value, the node takes on that value, otherwise it is assigned X. Strengths are taken into account by generating a series of equations, starting from the highest strength and working downward; the effects of a stronger path can override the results of a weaker one. The network of Boolean equations can be solved by Gaussian elimination [Bry87B].

IRSIM [Sal89] is a switch-level simulator which uses RC models to produce more accurate estimates of timing and charge sharing. IRSIM has been used for both timing and power analysis.

10.4 Layout Synthesis

Layout is among the most tedious of design tasks, so it is natural to want to automate layout design as much as possible. Unfortunately, the problems that make layout hard for people also make it hard for programs. The most successful methods for automating layout have been special-purpose to some degree—they make simplifying assumptions about layout design so that algorithms can be used to perform the design task. The ideal, of course, is a completely general algorithm that produces small, high-speed chip layouts in a minuscule amount of CPU time; what most people settle for is an algorithm that does a good job on a problem that is most like the one they have to solve.

Automatic layout methods vary in the quality of the layout produced (both in chip area and delay), in the degree of automation, and the total manufacturing cost of the design. Two key abstractions make these tasks much more tractable: building the layouts out of components and wires, rather than directly from rectangles, and decomposing layout design into separate placement and routing phases.

It is important for the student of placement and routing to remember that the practice of these subjects is much more advanced than their theory. Many workable heuristics have been found for design tasks which do not have a good formal problem statement, let alone an efficient algorithm. The situation is less than ideal—many layout problems are still not solved, and many solutions are barely workable. But practical solutions exist for a wide variety of layout problems encountered in daily life.

As discussed before, most automatic layout systems make two simplifying assumptions: that the layout can be divided into separate component and wiring areas; and that layout can be split into placement and routing phases. Both assumptions, while seemingly obvious, are critical to developing effective layout algorithms. Allowing wires only in component-free areas allows layout algorithms to build much simpler and efficient models of wires to manipulate in wiring algorithms. Separating placement and routing makes it much simpler to define metrics on each phase for which optimization methods can be defined.

Routing is usually divided into global and detailed phases. We can divide the chip into a number of routing areas that can be routed (more or less) independently. Global routing chooses the areas through which the wires must be run to make the required connections. Detailed routing finishes the design of the wire segments in each routing area. The results of placement and routing can be improved by making some exceptions to these rules. Cells can define feedthrough areas through which wires can be safely run. They can also provide several electrically connected terminals so that the routing algorithm can choose the most convenient one. But these extensions do not violate the principle that the routing algorithm need only worry about interference between wires, not between wires and components.

A number of possible placement and routing algorithms can be used in various combinations. Each algorithm has its strengths and weaknesses. The algorithms are chosen to meet the requirements of the physical design style—the organization of components and wiring areas on the chip.

10.4.1 Placement

The separation of placement and routing immediately raises an important problem: how do we judge the quality of a placement? Because the wires are not routed at the end of placement, we cannot judge the placement quality by the two obvious measures: area of the layout and delay. We cannot afford to execute a complete routing for every placement to judge its quality; we need some metric which estimates the quality of a routing that can be made from a given placement. Different placement algorithms use different metrics, but a few simple metrics suggest important properties of placement algorithms.

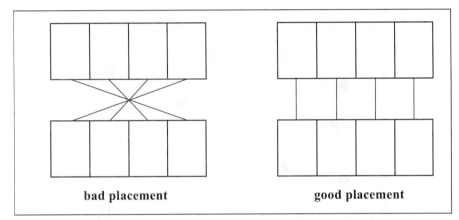

Figure 10-27:
Wire length as a
placement metric.

bad placement   good placement

One way to judge the area of a layout before routing is to measure the total length of the wires on the chip. While total wire length is not absolutely correlated to chip area, we would expect that a placement which gave much longer wires would eventually result in a larger layout. Of course, we can't measure total wire length without routing the wires, but we can estimate the length of a wire by measuring the distance between the pins it connects. As shown in Figure 10-27, we can estimate wire length by measuring the **Manhattan** or **Euclidean distance** between the pins. Euclidean distance is given by a standard ruler; Manhattan distance is the sum of the x and y distances between the points. Distance is only an estimate—it assumes that the wire between two pins will not leave the pins' bounding box. But it is a reasonable estimate of wire length, which in turn estimates area, and pin-to-pin distance can be computed very quickly from a placement.

We need different placement algorithms for different layout styles. Placement for brick-and-mortar layout is relatively hard. Placement is much simpler for standard cell layout because the components are all roughly the same size and (as we will see below) the interactions between placement and routing are limited.

Standard cell layouts use components of roughly equal size, which makes it easy to use a number of constructive and iterative algorithms. The simplest placement method is **pairwise interchange**, which can be used with a variety of placement metrics. The algorithm starts with a placement—either generated at random or by some other algorithm—and then repeatedly swaps components to improve it. At each iteration the algorithm selects a pair of components for consideration; if swapping their positions improves the chip's cost (total wire length or some other measure), then the components are swapped. The algorithm may stop after some number of iterations or when the improvement at a step is sufficiently low. Pairwise interchange is rarely used as the sole placement method, but may be used as an improvement step along with some other algorithm.

***Figure 10-28:**
Placement by partitioning.*

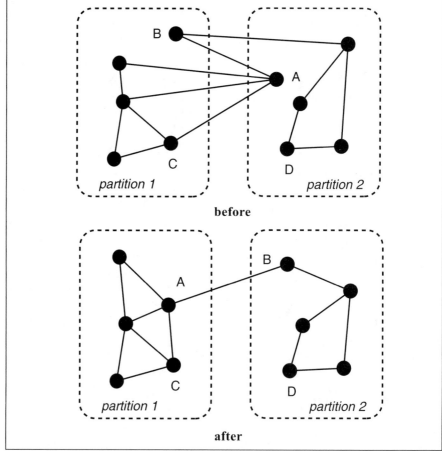

Placement by partitioning is more aggressive and gives good results without chewing up excessive CPU time [Bre77,Dun85]. Partitioning is a heuristic strategy to minimize total wire length and wiring congestion by placing strongly connected components close together. The partitioning process for a **min-cut, bisecting** criterion is illustrated in Figure 10-28. The circuit is represented by a graph where nodes are the components and edges are connections between components. The goal is to separate the graph's nodes into two partitions that contain equal (or nearly equal) number of nodes and which have the minimum number of edges (wires) that cross the partition boundary to connect nodes (components) in separate partitions. With A and B in their positions in the **before** configuration, there are five wires crossing the partition boundary. Swapping the two nodes reduces the net cut count to one, a significant reduction. However, moves cannot be considered in isolation—it may require moving several nodes before it is clear whether a new configuration is better.

The Kernighan-Lin algorithm [Ker70] is one well-known graph partitioning algorithm. It iteratively tries to find and move out-of-place elements in the partitions. Given an element e in partition 1 that might be moved into partition 2, we can compute some measures that indicate whether e should be moved:

- I_e is the number of nets connected to e that have all their connections in partition 1.

- E_e is the number of nets connected to e that have all their other connections in partition 2.

If e is moved from 1 to 2, all the nets counted in E_e will no longer cross the cut, while the nets counted in I_e will cross. If $D_e = E_e - I_e$, the gain obtained by exchanging e and f is given by $g = D_e + D_f - c_{ef}$, where c_{ef} is a correction factor to avoid double-counting shared nets. The Kernighan-Lin algorithm exchanges sets of nodes between partitions to reduce the number of crossing nets; by exchanging sets of nodes, rather than pairs, it can improve the partitioning even when moving any single pair of nodes will not give improvement.

Since dividing all the chip's components into two partitions doesn't give very fine direction for placement, the partitioning is usually repeated by creating subpartitions of the original partitions. Partitioning can be recursively applied until each partition contains one node, but it is usually stopped before then and another algorithm is used to place components in the small partitions—a standard cell layout may be partitioned to create one partition per row, then a separate placement method may be used to place gates within the rows.

Simulated annealing is a much more computationally expensive approach that can give very good results. Programs such as TimberWolf [Sec85] place standard cells by simulated annealing. It takes its name from an analogy to the annealing of metals—the metal is heated to very high temperatures, in which atoms move at very high energies, and the metal is then cooled, causing atoms to tend to fall into their proper places. As in pairwise interchange, an initial placement is improved through a succession of moves. However, components to be moved are selected at random and even moves that do not decrease the cost of the placement are randomly selected at some probability. Selecting unprofitable moves is a strategy to search more of the solution space. An annealing schedule is used that initially accepts a large fraction of bad moves, analogous to high temperature, then reduces the probability of acceptance to gradually cool the design to a good placement.

10.4.2 Global Routing

A global router's job is to assign each net a route through a set of areas. Global routing should try to balance the use of the available routing channels to avoid wasting chip area. If the sizes of routing channels are fixed, it must keep close track of the channels' capacities to be sure that none are oversubscribed.

Optimal routing of a single net is much easier than optimally routing a sequence of nets. Unfortunately, good formulations for simultaneous routing are few and far between. Most layout systems work by routing nets one wire at a time, choosing the order of nets to route by some strategy that hopefully gives good results.

Figure 10-29:
Maze routing.

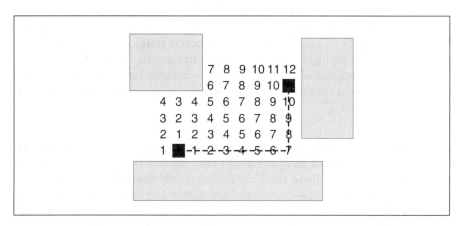

Maze routing (often called the Lee/Moore algorithm) [Lee61] will find the shortest path for a single wire between a set of points, if any path exists. The algorithm, illustrated in Figure 10-29, is actually quite simple. The route is assumed to occur in a unit-size grid. The algorithm works in two phases. First, points on the grid are iteratively labeled with their distance from a source point. In the first iteration, all points at distance 1 from the source are labeled with 1; at the second iteration all unlabeled points adjacent to the 1-points are labeled with 2, and so on. The labeling stops when the sink of the net to be routed is found. Second, a route is traced back from the sink to the source: the route is drawn so that never goes to a point with a higher-order label. There are generally many possible paths back to the source; only one is shown in the figure. It is simple to prove that the traced route is as short as possible and that, if no path can be found back to the source, then none exists.

The maze search algorithm automatically handles routing around obstacles. Figure 10-29 shows how labeling stops at the edge of an obstacle (a component or the chip edge). The distances computed automatically reflect the cost of avoiding the obstacle and the traced route automatically avoids all obstacles. The algorithm's simplicity is a strong argument in its favor. The algorithm does, however, have a high cost in both memory and CPU time. The grid map costs considerable memory: if distances are stored in chip units, then an $n \times m$ chip requires a map of $n \times m$ integers of length lg $max(n,m)$ bits. It is possible to reduce each grid point to two bits [Ake67]—the algorithm does not need to know the exact distance from the source, only the direction to travel to move closer to the source—but even with this improvement, the array is large. The time required to route a net is $O(n + m)$: computing the distances and tracing the path are equally hard.

Line search routing, which was described in Section 7.2.2, is a heuristic method that works well in many problems. It is not guaranteed to find the shortest path, or any path at all, even if such a path exists, but it requires much less memory and CPU time than maze routing.

Both maze and line search routing are sensitive to net order. Since nets are routed one at a time, routing one net first may block another net routed later, even though it is possible to route both by using different paths. Routing programs often get around the net-ordering problem by using one of these algorithms to produce an initial routing (using some initial wire order) and then improving the solution. Improvement usually involves ripping up and rerouting one or more nets. There are several improvement strategies in use: ripping up nets that limit the size of a channel; ripping up all the nets in a chosen region; or ripping up nets at random. Improvement methods are most important in gate array routing, where the size of the channels is fixed. The last resort for improving a global route is manual intervention.

10.4.3 Detailed Routing

A channel router assumes a rectangular routing region of varying height and pins along two edges. (End pins are generally handled as special cases.) We can now discuss more sophisticated routing algorithms than the left-edge algorithm. Figure 10-30 gives a simple **dogleg** channel routing algorithm. Each net is broken into segments that join adjacent pins on the net. To avoid adding too many doglegs, the *range* parameter forces several contiguous segments of a net to be routed together. The dogleg algorithm uses the left-edge criterion to fill up the tracks. The left-edge algorithm does not give an optimal routing when there can be several horizontal segments on a track. The result of dogleg routing depends on the order of nets considered and the number of doglegs allowed, among other factors.

A number of algorithms and heuristics have been introduced for improving the routing, by finding better track assignments for the nets, routing to minimize the number of vias on a net, and so on. Most recent algorithms bear a weaker resemblance to the left-edge algorithm; we give two examples below.

The greedy channel router of Rivest and Fiduccia [Riv82] routes the channel from left to right, assigning all nets that cross the current column to tracks. It uses a number of greedy heuristics to choose assignments. At each column, the router uses the rules illustrated in Figure 10-31 to assign nets to tracks:

1. Make connections to pins at the top and bottom of the column.

2. Add jogs to try to reduce nets that occupy more than one track to a single track.

3. Add jogs to the remaining nets that occupy several tracks to try to reduce the distance between their tracks. This step simplifies later collapsing of the nets.

4. Add jogs to move a net up if its next connection will be made to a top pin, or down if its next connection will be to a bottom pin.

5. If any pin in the current column could not be connected, add a track to the channel.

YACR2 [San84] tries to minimize the number of vias on nets as well as the number of tracks required. As shown in Figure 10-32, the algorithm first assigns nets to tracks; it temporarily satisfies vertical constraints by adding blank space between pins. After routing this simplified channel, it eliminates the added blank space by adding jogs to the wires in the vertical constraint. It uses a simple maze router to reroute these wires;

```
dogleg_route(set_of_nets nets, set_of_pinpos bottom_pins,
             set_of_pinpos top_pins, int range) {
    /* Nets is the set of nets to route; bottom_pins and top_pins are
    the locations of the pins along the channel's top and bottom;
    range controls when adjacent dogleg segments are forced
    onto the same track. */

    /* Each element of the ith element of the nets_in_track array holds a set of
     nets that occupy the ith track. */
    set_of_nets nets_in_track[MAX_TRACKS];
    /* Holds first available track. */
    int track;

    /* sort nets by left edge, then break net into segments, with one segment
    between
    each pair of adjacent pins on the net */
    sorted_nets = sort(nets,leftmost_pin);
    fragment(sorted_nets);
    /* process nets from left to right */
    track = 1;
    while (sorted_nets != Empty) {
        current_net = first(sorted_nets);
        current_segments = first_n_segments(current_net,range);
        /* route all the current_segments at once */
        if (!conflict(current_segments,net_in_track[track]) &&
                    lowest_constrained(track)) {
            /* OK to add segments of this net on this track */
            nets_in_track[track] = nets_in_track[track] + current_segments;
            /* remember that you already routed these segments */
            current_net = current_net - current_segments;
            /* do housekeeping chores */
            if (empty(current_net))
                pop(sorted_nets);
            /* should we try the next track? */
            if (full_track(nets_in_track[track],sorted_nets)
                track = track + 1;
        }
    }
}
```

Figure 10-30: A simple dogleg routing algorithm.

it can make both horizontal and vertical polysilicon segments to try to minimize the number of vias. Its via-minimization method is heuristic, but gives good results in practice.

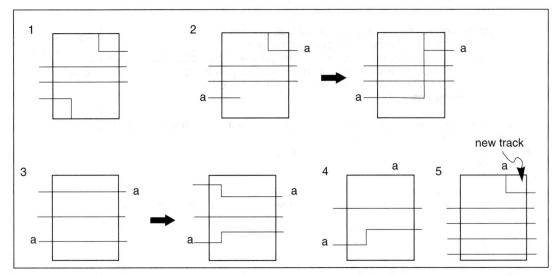

Figure 10-31: *Rules used by Rivest and Fidducia's greedy router (after Rivest and Fiduccia [Riv82] © 1982 IEEE).*

Figure 10-32:
How YACR2 cre-
ates short horizon-
tal segments
without vias.

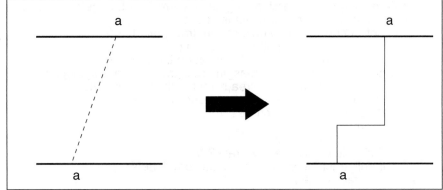

10.5 Layout Analysis

In addition to layout synthesis, we need layout analysis tools: we must check hand-designed cells for design rule violations as well as to verify our layout synthesis tools; we also need to extract circuits and parasitics from layouts for simulation and functional verification.

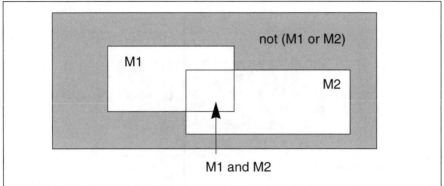

Figure 10-33:
Boolean combina-
tions of masks.

We saw in Chapter 2 how to analyze layouts using Boolean mask operations plus grow and shrink operations. Figure 10-33 illustrates Boolean combinations of masks.

The most efficient algorithms for mask operations use a **scan line** mechanism [Szy88]. Each polygon is represented by its bounding line segments, with each segment labeled to mark the polygon's interior. Each horizontal segment is labeled with a direction—when facing in the direction of the sweep, the direction gives the side of the segment on which the polygon's interior lies. In Figure 10-34, M2's interior lies between the left-hand edge on the top and the right-hand edge on the bottom.

We process the layout by sweeping the scan line across it. At each point along the scan line, we count the number of left-hand and right-hand edges we have seen, from which we can determine whether we are inside or outside a polygon. For example, point *a* is inside the M1 polygon because the scan line has crossed one left-hand edge, while *b* is outside the polygon because we have crossed one left-hand and one right-hand edge. We can generate Boolean mask combinations along the scan line; for example, we add a left-hand edge on the M1 *AND* M2 mask as soon as we have crossed one left-hand edge (and no right-hand edges) of polygons on both masks. We can generate grow-shrink masks by keeping track of a window around the scan line.

Using mask operations to extract a transistor circuit is very straightforward: we identify transistors from their active regions; we identify vias that connect layers; we can then use a scan line algorithm to assign a unique number to each electrical node by first assigning a different number to each polygon on each layer, then equating node numbers when we find a via that connects two polygons. Capacitance extraction is relatively straightforward, by measuring the area and perimeter of the polygons. Resistance extraction is much more difficult. Various algorithms are used to approximate resistance from polygon shape [Hor83B].

Figure 10-34:
Scan-line analysis of masks.

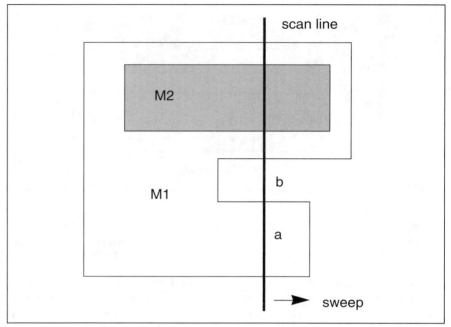

10.6 Timing Analysis and Optimization

In this section we move up one level in the abstraction hierarchy and consider two circuit problems: how to analyze a network to find the longest delay; and how to synthesize the transistor sizes in a network to satisfy performance requirements while minimizing area. We briefly discussed some principles behind timing analysis in Section 4.4. Now we will consider the models and algorithms in more detail.

Timing analysis differs from simulation in that it is *value-independent*. A simulator can accurately model delays, but only for one configuration of values on wires. The delay through a circuit can vary widely with the input and state values. In Figure 10-35, we can observe the delay path starting from the A input only when we change the A input from 0 to 1—that transition causes a transition on every gate output along the path. Even changing A from 1 to 0 will not exercise the critical delay path. If you fail to simulate the input combination that gives the longest delay, you may develop an optimistic estimate of your chip's performance.

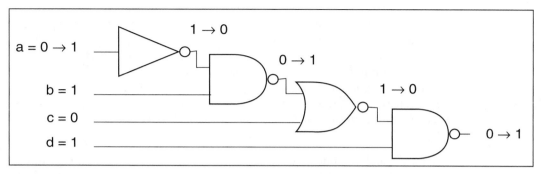

Figure 10-35: *Delay depends on signal values.*

Though the two steps are often combined, timing analysis can be divided into building a graph labeled with delays, then searching the graph to find the longest path from
an input to an output. The delay graph must model $0 \rightarrow 1$ and $1 \rightarrow 0$ delays separately,
because the value of a signal may terminate a path. (Other input combinations can
block propagation of the input transition.) Both transistors and wires add resistance
and capacitance to the circuit, but transistors are distinguished because their gate values determine whether signals pass between the source and drain. Either simple RC
or slope models may be used for transistors.

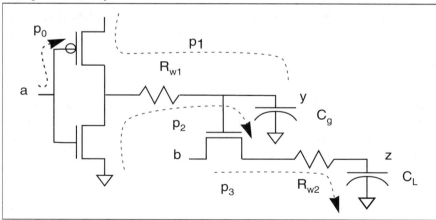

Figure 10-36:
*Paths for timing
analysis.*

Figure 10-36 shows the paths which must be analyzed to compute the delays through
a simple switch circuit. Both a and b are primary inputs, which we will assume rise
instantly. (The more realistic model of a resistor for each primary input and capacitor
for each primary output is easy to add to the analysis.) First, we compute a **primitive
path delay** for every primitive path from a primary input or power supply connec-

tion to a transistor gate or primary output. This example has four primitive paths: p0, from a to the inverter gates (since the a input rises or falls instantaneously, this path adds no delay); p1, from VDD to y; p2, from VSS to y; and p4, from b to z The rise time and fall time at node y, which controls the pass transistor, are easy to compute: if a changes at time 0, then the time at which y rises is $\delta_{ax,r}$ and the time it falls is $\delta_{ax,f}$. The delays from b to z are $\delta_{bz,r}$ and $\delta_{bz,f}$. The **timing analysis graph** has a node for each primary input, primary output, and for each power supply node, and an edge for each primitive path labeled with the path's rise and/or fall delay.

We must search the timing analysis graph to find the longest path from a primary input to a primary output. Before considering the search algorithm, let us consider the computations that must be done during the search. The path from b to z is controlled by the pass transistor; when node y falls, the path is cut off. (We will assume that complete circuit doesn't suffer from charge sharing problems.) When y rises, it completes the connection from b to z. We can compute rise and fall times for z, remembering that z rises when y rises and b is 1: the rise time is $\delta_{ax,r} + \delta_{bz,r}$, since z cannot start rising until after y rises and switches on the pass transistor; similarly, the fall time is $\delta_{ax,f} + \delta_{bz,f}$.

The actual search can be conducted by either depth-first or breadth-first search of the timing analysis graph. The graph is supposed to be free of cycles, since we have broken all cycles in the sequential system at the memory elements, but depth-first search is able to detect inadvertent cycles in the combinational logic. At each node, the delay is computed from the primitive path delays and the arrival times of signals that control the path. If there is more than one path to a node, then the rise and fall times are independently taken to be the latest arriving rising and falling signals, respectively.

We introduced false paths in Section 4.4.2. Boolean false paths are identified in a combinational logic network rather than a switch network like Figure 10-36. Determining whether a path is false is NP-complete, but algorithms can quickly find false paths in practical networks [Dev91].

Another source of pessimistic delay estimates in transistor circuits is misdiagnosis of the direction of signal flow. (Signal flow may be opposite current flow, but is an appealing concept because signals always flow from sources, such as the power supply.) Such paths are also called false paths, though they are not the same type of path as that found in a Boolean logic network. Figure 10-37 shows what happens in a barrel shifter when we assume that signals can flow in either direction through a transistor—a search for all possible paths from inputs to outputs finds many paths that never occur and are much longer than the longest true path. In reality, the drivers at the inputs force signals to flow in only one direction through each transistor in the circuit;

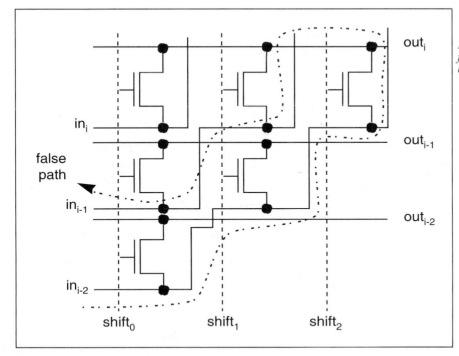

Figure 10-37: *A false path in a barrel shifter.*

when we examine only paths formed by the properly directed edges across transistors, we find the true paths. Signal direction can be determined deductively in virtually all cases, as illustrated in Figure 10-38: the signal directions through t_1, t_3, and t_4 are obvious. The direction of t_1 forces the direction of t_2; finally, the direction through t_5 can be deduced.

Timing analysis can be performed hierarchically if an abstract model of a cell replaces the cell's circuit, as shown in Figure 10-39 [Wol89]. At each primary input, the environment specifies a driving resistance. The model itself includes a resistance and capacitance inside each primary input terminal. Similarly, the environment specifies a load resistance and capacitance at a primary output, which is driven by a resistor within the cell. If there is a path from a primary input to a primary output within the implementation, it is replaced by a single graph edge labeled with the rise and fall delays; this edge represents the total delay from the primary input's load to the primary output's driver.

Transistor sizing algorithms can individually tune the widths of all the transistors in a circuit to meet a delay requirement. Tilos [Fis85] uses advanced mathematical programming techniques to solve for transistor widths. Such circuits make very efficient

Figure 10-38:
Determining sig-
nal direction
through transis-
tors.

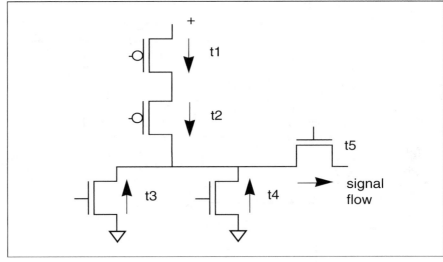

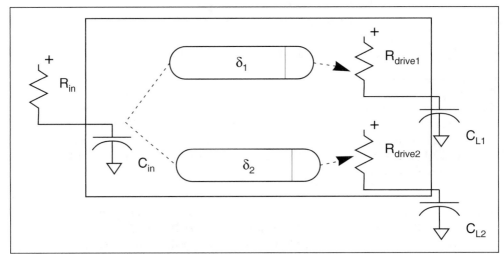

Figure 10-39: An abstract timing model.

use of layout area because transistors are made no larger than they have to be to meet the delay requirement. As a result, such circuits are also very power-efficient. However, transistor sizing can generate a circuit in which no two transistors are of the same size, which can be a monumental headache to lay out. Transistor sizing algorithms are most powerful when used with programs that generate custom layouts from transistor netlists. An alternative is to approximate the exact solution by choosing among standard cells [Hil89] of several different transistor sizes.

10.7 Logic Synthesis

Logic synthesis (or logic optimization) creates a logic gate network which computes the functions specified by a set of Boolean functions, one per primary output. While we could get a gate implementation for a function by directly constructing a two-level network from the sum-of-products specification, using one logic gate per product plus one for the sum, that implementation almost certainly won't meet our area or delay requirements. Logic synthesis is challenging because, while there are many gate networks which implement the desired function, only a few of them are small enough and fast enough. Logic synthesis helps us explore more of the design space to find the best gate implementation, particularly for relatively unstructured random logic functions.

Synthesizing a logic gate network requires balancing a number of very different concerns: simplification from don't-cares; common factors; fanin and fanout; and many others. To help manage these complex decisions, logic synthesis is divided into two phases: first, **technology-independent** logic optimizations operate on a model of the logic that does not directly represent logic gates; later, **technology-dependent** logic optimizations improve the network's gate-level implementation. The transformation from the technology-independent to the technology-dependent gate network is called **library binding**.

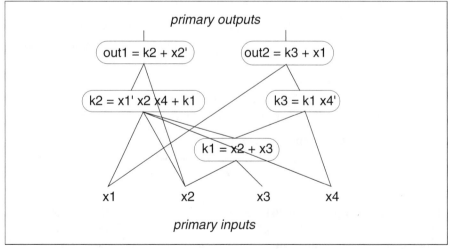

Figure 10-40: A Boolean network.

Figure 10-40 shows a **Boolean network** [Bra90], which is the standard technology-independent model for a logic network. Nodes in the network are primary inputs, primary outputs, or functions; the function nodes can be thought of as sum-of-product expressions. Edges show on which variables a node depends (each node is referred to by a variable name). The Boolean network has a structure reminiscent of a logic network, but the function nodes are more general than logic gates: they can have an unbounded number of input variables and express an arbitrary Boolean function. The real set of logic gates we have to work with may be limited in several ways: we may not have gates for every function of *n* variables; we may have several different gates for a function, each with different transistor sizes; and we definitely want to limit the fanin and fanout of a gate. We can create the overall structure of the logic network using the Boolean network, ignoring some details of the gate design, then fine-tune the implementation using technology-dependent optimizations on the gate network.

A few definitions are helpful: a function's **support** is the set of variables used by a function; the **transitive fanin** of a node is all the primary inputs and intermediate variables used by the function; the transitive fanin is also known as a **cone** of logic; the **transitive fanout** of a node is all the primary outputs and intermediate variables which use the function.

10.7.1 Technology-Independent Logic Optimization

Technology-independent optimizations can be grouped into categories based on how they change the Boolean network:

- **Simplification** rewrites (hopefully simplifying) the sum-of-products representation of a node in the network; most of the improvement usually comes from taking advantage of don't-cares.

- **Network restructuring** creates new function nodes that can be used as common factors and collapses sections of the network into a single node in preparation for finding new common factors.

- **Delay restructuring** changes the factorization of a subnetwork to reduce the number of function nodes through which a delay-critical signal must pass.

How do we judge whether an operation has improved the network? We can estimate area by counting literals in the Boolean network. (Remember from Section 3.2 that a literal is the true or complement form of a variable.) Counting the literals in all the functions of the Boolean network gives a reasonable estimate of the final layout area [Lig90]—each literal corresponds to a transistor if the functions in the network are implemented directly. Delay can be roughly estimated by counting the number of functions from primary input to primary output.

Simplification rewrites a single function in the network to reduce the number of literals in that network. In Figure 10-41, we have shown the function as a set of points: each variable is represented by a dimension with the coordinates 0 and 1; the function $x_1'x_2'x_3' + x_1x_2'x_3' + x_1'x_2x_3' + x_1x_2x_3$ is therefore represented by a three-dimensional space. A fully specified function is defined by two sets: the **on-set**, or the set of points (black in the figure) for which the function's value is 1; and the **off-set**, the points for which the function is 0. In the figure, the point $x_1 = 1, x_2 = 1, x_3 = 1$ is in the on-set and $x_1 = 1, x_2 = 1, x_3 = 0$ is in the off-set.

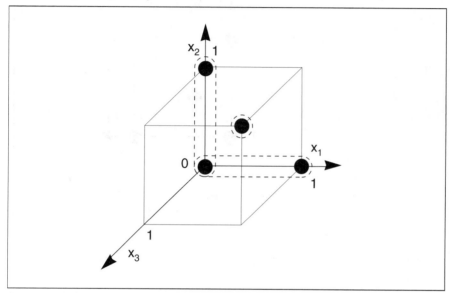

Figure 10-41: A three-input function shown in three-space and a cover of the function.

There are many ways to write this function as a sum of products. Each representation is called a **cover** because it must cover all the points in the function's on-set. Each member of the cover is a subspace of the function space, which is written algebraically as a product (or **cube**). For example, $x_1 = 1, x_3 = 0$ or $x_1 x_3'$ is a one-dimensional subspace of the function, and $x_2 = 0$ or x_2' is a two-dimensional subspace. The cubes in the cover must include all the points in the on-set and none of the points in the off-set. The number of literals is counted in the cube representation—the larger the subspace a cube covers, the fewer literals in the cube.

A function has many covers, some of which have more literals than others. The simplest and largest cover of a function is just the sum of all the points in the on-set, for example $x_1'x_2'x_3' + x_1x_2'x_3' + x_1'x_2x_3' + x_1x_2x_3$. A cover with both fewer cubes and fewer literals is $x_2'x_3' + x_1'x_3' + x_1x_2x_3$; the point $x_1'x_2'x_3'$ is covered by two cubes.

Simplification is important largely because most function specifications include don't-care values. Such a function is called **partially specified** function because, for some input values, the function value is not specified. Of course, the logic implementation must produce a 0 or 1 value for all inputs, but the specification states that we don't care which value the output assumes. Don't-cares let us put a point in the on-set or off-set, depending on what produces the smallest cover.

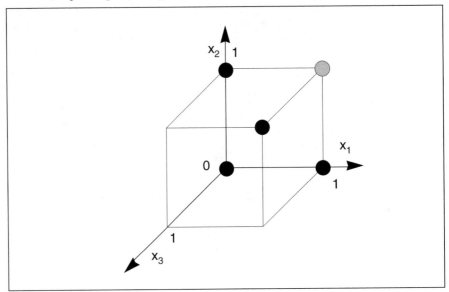

Figure 10-42: A partially specified Boolean function.

In Figure 10-42, the don't-care set consists of a single point $x_1 x_2 'x_3'$. If we assume that point is in the off-set, the cover is the same as before; if we put that point in the on-set, we can use the smaller cover $x_3' + x_1 x_2 x_3$ and save three literals.

Espresso [Bra84] is the best-known two-level logic optimizer. The algorithm starts with a cover given by the user (simply the cover used to describe the function) and iteratively reduces it in size using an **expand-irredundant-reduce** loop. An example is shown in Figure 10-43. The first step is to make each cube as large as possible without covering a point in the off-set. This step increases the number of literals in the cover, but sets the stage for finding a new and possibly better solution. The next step is to throw out redundant cubes—points may be covered by many cubes after expansion, so the algorithm throws out smaller cubes whose points are covered by larger cubes. Finally, the cubes in the cover are reduced in size, reducing the number of literals in the cover. In general, the new cover will be different than the starting cover because the *expand* and *minimize* steps find a new way to cover the points in the on-set. Hopefully, this new cover will also be smaller. The algorithm successively generates new covers until the cover can no longer be improved. While this solution is not

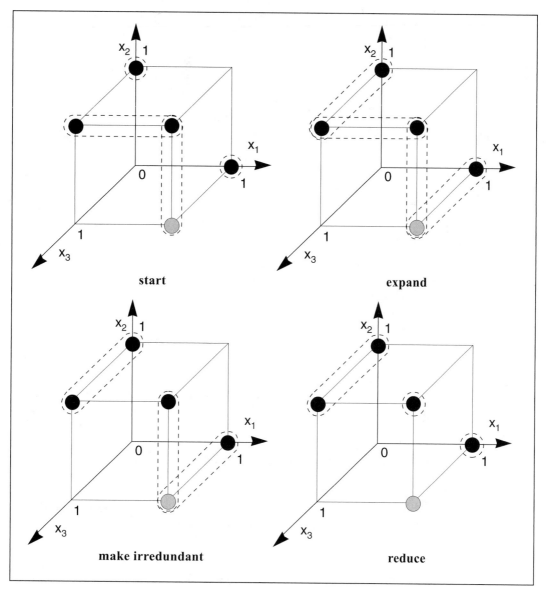

Figure 10-43: *An expand-irredundant-reduce cycle.*

guaranteed to be the global optimum, experience shows that the solution is near optimal for typical functions.

Figure 10-44:
Varieties of don't-cares in Boolean networks.

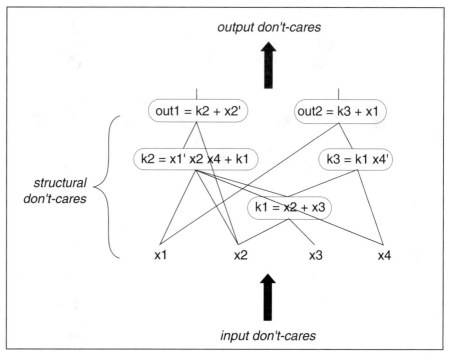

When simplifying a function in a Boolean network, we must consider more than the output don't-cares specified by the user. Figure 10-44 shows where different varieties of don't-cares occur in Boolean networks. **Structural don't-cares** [Bra90] are created by the internal structure of the Boolean network. A single function has no structural don't-cares because it has no internal structure, but a logic optimization system for Boolean networks must be able to extract and use structural don't-cares. There are two types of structural don't-cares: **satisfiability don't-cares** occur when an intermediate variable value is inconsistent with its function inputs; **observability don't-cares** occur when an intermediate variable's value doesn't affect the network's primary outputs.

An example of satisfiability don't-care optimization is shown in Figure 10-45. The g function defines the relationship between the primary inputs a and b and the intermediate variable y. We know that the value of y is always equal to the value of the g function, so for example, we know that the case $a = b = 0$, $y=1$ can never occur in the network. The case $y'g + yg'$ can be used as a don't-care condition for simplifying the f function.

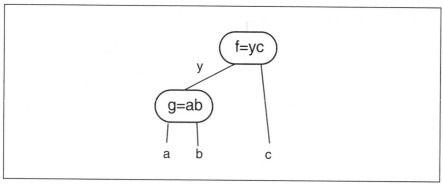

Figure 10-45:
*Satisfiability
don't-cares.*

Observability don't-cares are more complex to compute than are satisfiability don't-cares. As a simple example [Hac96], consider an OR gate with two inputs that are intermediate variables in the Boolean network. If the gate's *a* input is 1, then the gate's output is 1 independent of the value of its *b* input; the symmetric situation holds for the *b* input. These conditions define observability don't-cares for this simple case.

Simplification can have a minor influence on the structure of the Boolean network— simplifying a function may eliminate a variable from a function, deleting an edge from the network. However, we need optimizations that can radically restructure the network. We may want to decompose a network for two reasons. The parity function is an excellent example of the savings provided by recursively decomposing a function into subfunctions. Each function is used only once in the network, yet the multi-level network has many fewer literals than the flat, sum-of-products form, which has an exponential number of literals. In less regular functions, common factors are the predominant form of intermediate node. Particularly in networks with several primary outputs, a subfunction may appear in several different functions.

Factorization algorithms can find both types of intermediate nodes. Unfortunately, our factorization algorithms are not perfect, so it may be possible to **collapse** several nodes, then apply node extraction to the larger node to find a good structure for the network. Figure 10-46 shows a network that has been partially collapsed, leaving f_2 intact. We can work from the collapsed function to find a new structure for F that could not be found starting from the given multilevel structure.

Factorization algorithms are based on *division*—they first choose candidate functions for the intermediate nodes, then divide the candidate functions into the functions already in the network and measure whether the candidate function makes the network smaller. Division is the mechanism we use to determine how a function *f* splits

Figure 10-46:
*Partial collapsing
of a Boolean net-
work.*

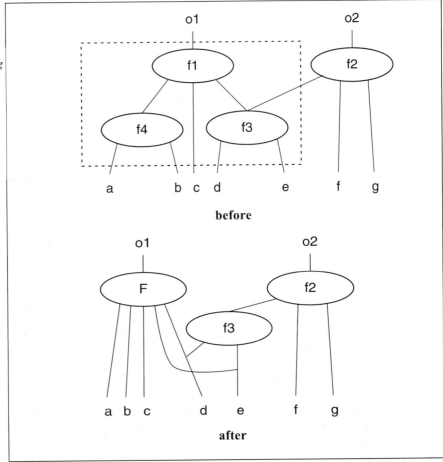

into two functions: the candidate function c and a cofactor function g. If we find that $g = f/c$, we can use c as an intermediate function to compute f.

We can develop surprisingly powerful algorithms for factorization using relatively weak **algebraic division** [Bra90]. Algebraic division relies only on the algebraic identities: commutativity ($a + b = b + a$); associativity ($(a+b) + c = a + (b+c)$); and distributivity ($a(b+c) = ab + ac$). In contrast, more general Boolean division uses algebraic identities plus the Boolean identities: $a \cdot 0 = 0$; $a \cdot 1 = a$; $a + 0 = a$; $a + 1 = 1$. Algebraic division is important because, even though it is less powerful than **Boolean division**, algebraic division requires much less CPU time. That means we can test many more candidate functions. Only a subset of the possible candidate common factors are algebraic divisors, but algebraic divisors include many useful intermediate functions for typical networks. We generally use algebraic factorization to find the general structure of the Boolean network, then use Boolean factorization to fine-tune the design.

Whether potential new factors are generated using algebraic or Boolean division, factoring the network takes three steps:

1. generate all potential common factors and compute how many literals each saves when substituted into the network;

2. choose which factors to substitute into the network;

3. restructure the network by adding the new factors and rewriting the other functions to use those factors.

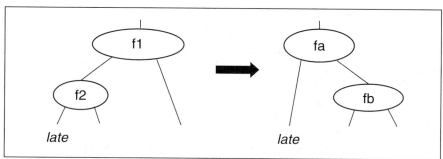

Figure 10-47:
Factorization for delay.

Boolean networks can also be optimized for delay. While delay estimates made on the network are rough, technology-dependent delay optimization can avoid creating logic with very long delay paths. As shown in Figure 10-47, a node can be re-factored to compensate for late-arriving signals. In the example, the latest-arriving signal must go through both *f1* and *f2* [Sin88]. We can collapse those two nodes into a single function, then extract new factors. But rather than choose a factor for its literal savings, we choose factors that are functions of early-arriving signals.

10.7.2 Technology-Dependent Logic Optimizations

Library binding (also known as technology mapping) transforms the Boolean network into a gate network using the gates in our library. Library binding is easy if we have only one or two types of gates, but modern libraries include 200 or more gates. We can optimize for both area and delay during technology mapping; some area and delay optimizations can be made only on the gate network.

Library binding can be viewed as a pattern matching process [Keu87]. We first translate the Boolean network into a canonical network built from NAND gates; we have saved the library gates in the canonical NAND form, so we can match library gates to pieces of our network to determine what gate best represents that section of the network. The library stores the canonical NAND network and area and delay costs for each gate; the costs are used to drive the mapping.

To bind our canonical network into a network of gates from our library, we need to cover it with the patterns of library gates:

- every NAND in the network under design must be covered exactly once;

- we want to select the minimum cost cover, where cost is some function of the area and delay parameters.

If we allowed our network and the library patterns to be arbitrary graphs, our covering problem would require subgraph isomorphism, which is NP-complete [Gar79]. However, if we break the network under design into trees, as shown in Figure 10-48, and require that all library gates be representable by trees, then we can use a much simpler algorithm. The tree requirement creates some problems around the points where we break our network and it means we cannot put XOR gates and other gates with reconvergent pattern networks into the library. However, these problems can easily be solved by a post-processing step which fixes up the network at the break points.

Figure 10-48:
Breaking a canonical NAND network into trees.

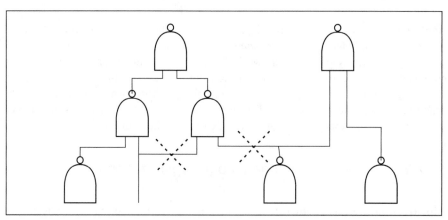

We can bind each tree separately using dynamic programming [Hil80]. The algorithm proceeds from the primary inputs (the leaves of the NAND tree). At each step, the lowest-cost match for a node provides an upper bound on the best cost. Consider the example of Figure 10-49. We proceed from the primary inputs to the primary outputs;

at each NAND gate output, we try to match that gate and some of the gates below it to a pattern in the library, accepting the lowest-cost match. At the first three levels, there is only one match possible, each match covering one NAND gate. At the fourth level there are two possible matches: a match by an inverter to the single NAND gate at the current point, or a match to that gate and most of the other gates below it. The cost of the AOI gate is given in the library; we must compare that cost to the total cost of all the previous best matches under the AOI match. The AOI gate is of lower cost, so we rip up the previous matches under the AOI and replace them with the single AOI pattern. In a deeper network, matches may be discarded several times, but when we get to the primary outputs, we are guaranteed to have the lowest-cost match.

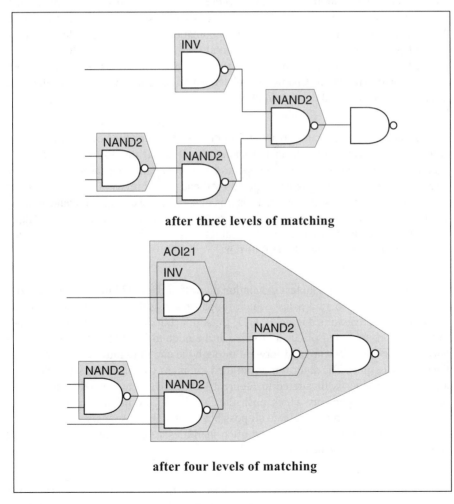

Figure 10-49:
Library binding by template match-ing.

Pattern matching does not take advantage of Boolean identities. Boolean library binding techniques [Mai91] select a section of the input Boolean network and try to cover it with library functions; don't-care conditions can allow additional matches to be found which would be missed by pattern matching.

10.8 Test Generation

An automatic test pattern generation (ATPG) program accepts a description of a machine and derives a sequence of test vectors which test for faults in that machine. Test generation programs exist for both combinational and sequential logic. At the heart of ATPG programs are test generation algorithms which derive a test for a particular fault. These algorithms operate on the same basic principles we used to generate combinational tests in Chapter 3 and sequential tests in Chapter 5, but they use efficient methods to find the test quickly.

The essence of test generation is identifying cases for which the good and faulty circuits give different results. Roth introduced the **D notation** [Rot66] to succinctly describe the relationship between a good and a faulty circuit. The cases when the two circuits give the same results are denoted by 0 and 1 for brevity. The case when the good circuit is 1 and the faulty circuit is 0 is denoted by D and the converse case by $\overline{D}$. Our goal during test generation is to propagate a D-value to a primary output where it can be observed. The set of D-values closest to the primary outputs during generation of a test is called the **D frontier**.

The PODEM combinational test generation algorithm [Goe81] implicitly enumerates the primary inputs to find a test which propagates a fault to the primary outputs. PODEM uses a five-valued logic: 0, 1, D, $\overline{D}$, and X. Initially, PODEM sets all signals in the network to X and places a D-value at the node to be tested. It then alternates between working backward and forward through the circuit to move the D-frontier to the primary outputs. The algorithm is outlined in Figure 10-51 and the search to propagate the D-frontier is illustrated in Figure 10-50. PODEM recursively searches for primary input assignments which push the D-value to the primary outputs. In the worst case, PODEM must evaluate all possible input values, as must any test generation algorithm. But its search method often identifies a test relatively quickly and it can be efficiently implemented.

Time-frame expansion is the intellectual basis for many sequential test generation algorithms. However, a variety of techniques may be used and heuristic algorithms are especially important, given the difficulty of the problem. HITEC [Nie91] applies a

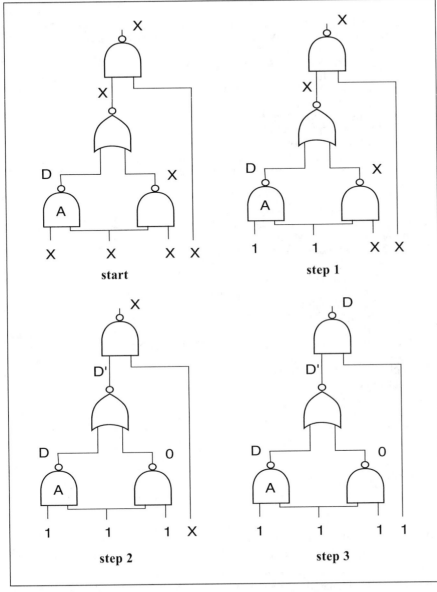

Figure 10-50:
Fault propaga-
tion during combi-
national testing.

variety of methods to generate sequential tests without requiring the system to have a global reset. STEED [Gho91] generates tests from a state transition graph: after identifying a test in the machine's combinational logic, it tries to justify and propagate the fault on the fault-free machine's state transition graph, then checks the faulty machine to be sure the test is valid. The fault-free propagation is very fast because it can be done directly on the state transition graph.

Figure 10-51:
The PODEM
algorithm.

```
boolean PODEM(logic_network network, gate_value_pair objective) {
    /* Try to justify the given value at the target gate in this network.
    Returns TRUE if it found a test. */
    if (propagated_to_po(network,target,value))
        return TRUE;
    if (no_test_choices_left(network))
        return FALSE;
    /* not done—keep looking */
    /* find to propagate something from the D-frontier */
    if (no_work_done_yet) {
        /* initially, the D-frontier contains only the fault itself */
        objective.gate = target; objective.value = value;
    }
    else
        objective = select_from_frontier(network);
    /* try to justify this objective from the PIs */
    imply(network,objective);
    if (PODEM(network,objective))
        return TRUE;
    /* try the opposite value */
    objective.value = complement(objective.value);
    imply(network,objective);
    if (PODEM(network,objective))
        return TRUE;
    /* can't propagate fault from here—return network to its old state */
    objective.value = X;
    imply(network,objective);
    return FALSE;
}
```

10.9 Sequential Machine Optimizations

The most important algorithm for sequential systems is state assignment, the general name for encoding algorithms, and so called because the most important signal to encode is the machine state. Finding a good state assignment for most FSMs is even more difficult for human designers, than logic optimization on control logic, so good state assignment algorithms are critical.

Kiss [DeM85] minimizes the symbolic state machine to find groups of states which should have closely related codes, then solves the constraints which define those groups to generate an encoding. Multiple-valued logic optimization can be used to generate groups of states which should be coded close together.

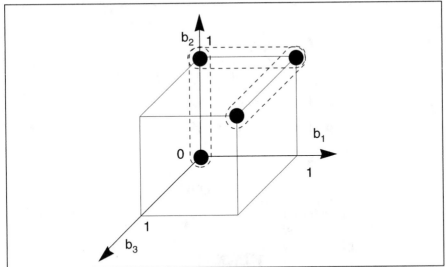

Figure 10-52:
Satisfying coding constraints by adding code bits.

Once those constraints are generated, a search algorithm assigns codes that satisfies the constraints. A state may appear in several groups, which means it must be coded to be near several other codes. The search starts with the minimum number of bits required to code the states, then adds state bits as required to ensure that all constraints can be met. For example, in Figure 10-52, the three axes represent the three bits of the state code. The three groups $\{s1, s2\}$, $\{s2, s3\}$, and $\{s3, s4\}$ should be coded such that the states in each group are distance-1 apart (since the groups are each of size two), but that requires the state code be three bits long.

Kiss optimizes for minimum number of cubes, which is an effective cost measure for two-level logic implementations such as PLAs but not as effective for multilevel logic. A variety of algorithms have been developed to generate state assignments optimized for multilevel implementation. For example, MUSE [Du91], Jedi [Lin91], and Mustang [Dev88] all estimate literal savings generated by encoding pairs of states close together, using varying strategies to estimate the effects of interactions between coding constraints in the multilevel implementation.

10.10 Scheduling and Binding

Figure 10-53:
Force-directed
scheduling.

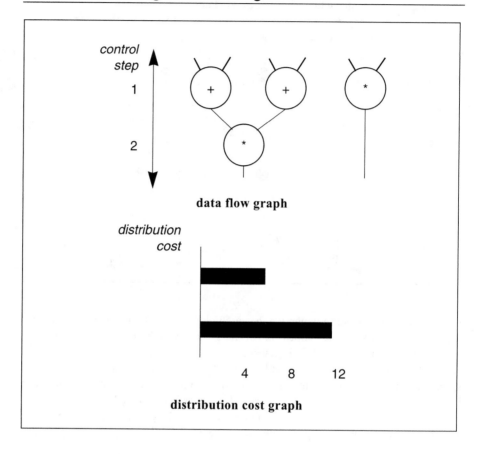

Figure 10-53:
Force-directed
scheduling.

While it is sometimes useful to simply find a feasible schedule for a set of constraints, scheduling algorithms usually try to bind operations to function units while scheduling to create both a low-cost register-transfer implementation and a fast schedule. Most scheduling algorithms optimize data path resources. An example is force-directed scheduling [Pau89], whose operation is illustrated in Figure 10-53. Each data operator is to be assigned a **control step**, or a cycle on which it is to be executed. Force-directed scheduling, like many data path scheduling algorithms, assumes that the data operators have been assigned cost values proportional to their area cost; in this example, we assume that an adder's cost is 1 and a multiplier's cost is 8. The algorithm starts by computing the first and last control steps on which an operation may be executed, which can be done by computing the as-soon-as-possible and as-late-as-possible schedules. Next, a distribution cost is calculated for each control step: the cost is the sum of the cost of each operator times the probability that the operator will be assigned that control step. In the example, the left-hand operations have no freedom, so the algorithm places the full cost of each operator in its control step; on the other hand, the right-hand multiplier may be assigned to control step 1 or 2, so its cost in each control step is half the multiplier cost. Force-directed scheduling then computes a force for each operator in each control step—the force is equal to the difference between the distribution value in that control step and the average of the distributions in all the control steps to which the operation can be assigned. This cost is proportional to the hardware cost added by placing that operator in that control step. The lowest force value gives the lowest hardware scheduling of that operator.

In the previous example, we assumed that each data operator took exactly one cycle. That is obviously a simplifying assumption—in reality, a multiplier and an adder have vastly different delays. Two important extensions to scheduling remove this restriction. **Multicycling** scheduling allows an operation to be executed over more than one control step (assuming, of course, the function unit implementing the operation has internal registers to hold the intermediate value between cycles). **Chaining** allows two operations to be performed in the same cycle [Pan87]. Another useful extension explored by several authors [Wak89A,Cam91] is to simultaneously schedule the different sets of operations taken by branches so that the scheduling and binding across the different branches are as similar as possible.

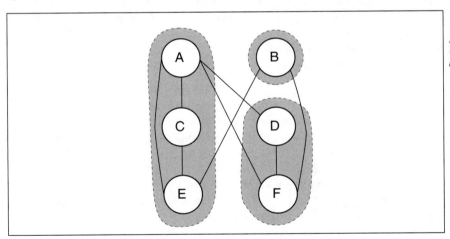

Figure 10-54:
Using cliques to
bind resources.

Binding maps a set of elements onto a set of resources; if two elements are not required at the same time, such as when two multiplications are performed on separate cycles, they can share the same resource. Binding can be performed by **clique partitioning** [Tse86]. The problem is modeled as a graph, as shown in Figure 10-54: the elements to be allocated (variables, operations, etc.) are represented by nodes; an edge is placed between two nodes if the elements are not needed at the same time (e.g., two variables are not simultaneously alive). A clique is a set of nodes which form a complete graph—there is an edge between every pair of nodes in the clique. All the elements (such as variables) contained in a clique can be assigned to the same resource (such as a register) because no two of them are needed at the same time. A complete binding of elements onto resources corresponds to a clique covering of the graph—every node is an element of one clique. There may be many clique coverings; by assigning costs, we can optimize the binding to minimize those costs.

10.11 Hardware/Software Co-Design

Embedded CPUs—programmable processors used to implement some of the functions of a system—are becoming increasingly important in VLSI design due to the explosion in fabrication capability. The huge chips that can be fabricated today often must use pre-designed components such as CPUs to be able to complete the design in any feasible time span. In addition, because a CPU can be time-shared across several different algorithms, embedded CPUs often give higher hardware utilization than architectures consisting entirely of special-purpose function units. **Hardware/software co-design** is a methodology for the simultaneous design of the hardware architecture and the software executing on that hardware. The goal of co-design is to develop a custom multiprocessor, which may consist of one or more CPUs as well as hardwired, special-purpose function units, as well as custom software which efficiently uses that multiprocessor.

Hardware/software co-simulation uses specialized techniques for simulating systems which include software executing on CPUs and hardwired logic. Co-simulation is challenging because different parts of the simulation run at very different rates; high-level models must be applied to the software to achieve reasonable simulation speeds, while more detailed simulation algorithms must often be applied to special-purpose function units to give the accuracy required to ensure that the system operates properly. Ptolemy [Buc91] is one well-known co-simulation system. It uses an object-oriented execution model to support simulation of function units in different universes which correspond to differing implementation domains.

Hardware/software co-synthesis synthesizes the hardware engine topology and the software that executes on that hardware from a high-level description. The goal of co-synthesis is to synthesize an implementation which meets performance (and perhaps other) goals while minimizing manufacturing costs; simultaneously considering the hardware and software architectures allows more robust design trade-offs to be made. There are several different styles of co-synthesis. **Hardware/software partitioning**

targets a hardware template of a CPU and one or more ASICs which act as accelerators; the central processing unit and ASICs communicate via the CPU bus. The Vulcan system [Gup96] starts with all functions allocated to ASICs and moves selected functions to the processor to reduce implementation cost. The COSYMA system [Hen94] uses the opposite approach—it starts with all functions implemented in software on the CPU and moves selected functions in the inner loop of the program specification to meet performance requirements. **Distributed system co-synthesis** generates systems with arbitrary hardware architectures. Wolf [Wol97] used heuristic algorithms to synthesize a hardware architecture while allocating software processes to the hardware units.

10.12 References

The best-known logic optimization programs are Espresso [Bra84] for two-level simplification and *misII* [Bra87] for multilevel networks. The article by Brayton, Hachtel, and Sangiovanni-Vincentelli [Bra90] gives a good overview of the state-of-the-art in multilevel logic optimization, while Hachtel and Somenzi's book [Hac96] provides more details. DeMicheli's book [DeM94] describes CAD algorithms for logical, sequential, and behavioral design. Michel et al. [Mic92] describe synthesis for systems as well as related topics in hardware description, simulation, and testing. Gajski et al. [Gaj92] survey system-level synthesis algorithms. *Digital System Testing and Testable Design* [Abr90] provides excellent descriptions of both test generation algorithms and design for testability. Several books [DeM96, Sta98] describe hardware/software co-design algorithms.

10.13 Problems

10-1. Animate the simulation of a domino logic gate, showing how values are propagated at each simulation step. Simulate the complete precharge-evaluate cycle, assuming that one pulldown is turned on. Assume that all non-primary input nodes are X at the start of simulation.

10-2. Use the Rivest-Fiduccia method to route this channel:

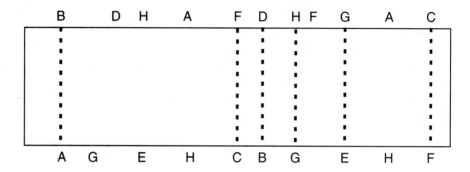

10-3. Use the current direction rules of Section 10.6 to determine the direction of current flow in each transistor in a Manchester carry-chain adder cell. Show your steps in deriving the flows.

10-4. Is it reasonable to assume that all nodes in a Boolean network have equal delay? If so, justify your answer; if not, propose a simple delay model for a node.

10-5. Find the minimum-cost mapping of the Boolean network given below onto these two gate libraries:

    a) inverter (cost = 1), 2-input NAND (cost = 3), 2-input NOR (cost = 3);

    b) inverter (cost = 1), 2-input NOR (cost = 3), 3-input NOR (cost = 4), 2-input NAND (cost = 3), 3-input NAND (cost = 4).

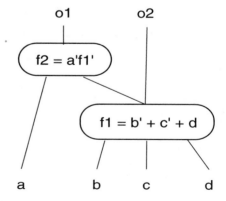

10-6. Encode these six states, using as few bits as possible, while satisfying all these coding constraints: $\{s1, s2, s3\}$, $\{s2, s5\}$, $\{s1, s6\}$, $\{s5, s6\}$.

10-7. For this code fragment, assuming that a variable is no longer needed after its last use:

```
x1 <= a + b + c;
x2 <= b - d;
x3 <= x1 - d;
x4 <= x1 * x2;
```

a) find the as-soon-as-possible schedule;

b) construct an allocation graph;

c) find an assignment of variables to registers requiring the minimum number of registers.

Appendices

A

Chip Designer's Lexicon

Thanks to John Redford and Derek Beatty for many colorful terms.

ALU   Arithmetic logic unit, which can perform several different arithmetic and logic operations as determined by control signals. (See Chapter 6.)

AOI   An and-or-invert gate. (See Chapter 3.)

ALAP   As-late-as-possible, a schedule that performs operations at the last possible time. (See Chapter 8.)

ASAP   As-soon-as-possible, a schedule that performs operations at the earliest possible time. (See Chapter 8.)

ASIC   Application-specific integrated circuit.

ASM chart   A technique for register-transfer design. Elements of the chart correspond to states and transitions in the register-transfer machine. Data path operations can be specified as annotations to those states and transitions. (See Chapter 10.)

ATPG   See *automatic test pattern generation*.

AWE   Asymptotic waveform evaluation, an algorithm used for circuit analysis. (See Chapter 4.)

abutment   A connection between two layout blocks formed without additional wiring. (See Chapter 7.)

allocation   The assignment of operations to function units. (See Chapter 8.)

aggressor net   In crosstalk, the net that generates the noise. (See Chapter 3.)

AND plane   The block of logic in a PLA that computes the AND part of the required sum-of-products. (See Chapter 6.)

area router   A router that can operate in non-rectangular, arbitrary-shaped regions. (See Chapter 7.)

architecture-driven voltage scaling

A technique for reducing power consumption in which the power supply voltage is reduced and logic operating in parallel is increased to make up for the performance deficiency. (See Chapter 8.)

array multiplier

A multiplier built from a two-dimensional array of adders and additional logic. (See Chapter 6.)

arrival time   The time at which a signal transition arrives at a given point in a logic network. (See Chapter 5.)

aspect ratio   The width/height ratio of a layout block. (See Chapter 7.)

automatic test pattern generation

Use of a program to generate a set of manufacturing tests. (See Chapter 10.)

BIST   See built-in self-test.

BGA   Ball-grid array, a type of package. (See Chapter 7.)

balanced tree   In clock distribution, a wiring tree synthesized with RC loads that are balanced across each set of branches of the tree. (See Chapter 7.)

Baugh-Wooley multiplier   A multiplication algorithm for two's-complement signed numbers. (See Chapter 6.)

bed of nails   A set of probes used to test a printed circuit board.

behavioral synthesis See *high-level synthesis*.

belt buckle   An extremely large chip. See *lots per die*.

binding   In high-level synthesis, synonym for *allocation*.

bit-slice   One bit of a regular, n-bit design. Refers to design styles in both logic and layout. (See Chapter 4.)

body effect   Variation of threshold voltage with source/drain voltage. (See Chapter 2 for the definition of body effect and Chapter 3 for its effect on logic gate design.)

Booth encoding A technique for reducing the number of stages in array multipliers.

bottle   CRT in a terminal or workstation (West Coast USA).

bottomwall capacitance

    Junction capacitance from the bottom of a diffusion region to the substrate. (See Chapter 2.)

buffer   An amplifier inserted in a wiring network to improve performance. (See Chapter 3.)

built-in self-testA testing scheme that uses logic built into the chip to test the remainder of the chip. (See Chapter 8.)

burn-in   The initial operation of a part before it leaves the factory. See *infant mortality*.

bus   A common connection.

CPU   Central processing unit.

carry-lookahead adder An adder that evaluates propagate and generate signals in a carry-lookahead network which directly computes the carry out of a group of bits. (See Chapter 6.)

carry-select adder   An adder that first generates alternate results for different possible carry-ins, then selects the proper result based on the
actually carry-in. (See Chapter 6.)

carry-skip adder   An adder that recognizes certain conditions for which
the carry into a group of bits may be propagated directly to the
next group of bits. (See Chapter 6.)

ceramic package

A package for an integrated circuit made from ceramics, which
offers better thermal conductivity and isolation from the elements
than a plastic package. (See Chapter 7.)

chaining   Performing two data operations, such as two additions, in the
same clock cycle. (See Chapter 8.)

channel   A rectangular routing region. (See Chapter 4.)

channel graph   A graph that describes the connections between channels in a
floorplan. (See Chapter 7.)

channel router

A routing program designed to route within a rectangular routing
region. (See Chapter 4 for an introduction to channel routing and
Chapter 10 for more details on channel routing algorithms.)

channel utilization   The number of wires that flow through a channel. (See
Chapter 7.)

charge sharing   Storing charge in parasitic capacitances such that the circuit produces erroneous results. (See Chapter 4.)

circuit under test, CUT

Testing terminology for the logic undergoing testing.

clock   A signal used to load data into a memory element.

clocked inverter

An inverter with additional transistors which cause its output to be
in a high-impedance state when the clock is not active. (See Chapter
4.)

color plan   A sketch of wiring over a chip or a large section of the chip which has been drawn in color to emphasize relationships between the layers and which emphasizes the decisions on how layers are to be used in the layout design. (See Chapter 7.)

control dependency   A set of control decisions, one of which depends on the other. (See Chapter 8.)

controllability

The ability to set (directly or indirectly) the value of a node on chip. See also *observability*. (See Chapter 4.)

controller   A state machine designed primarily to generate control signals. (See Chapter 8.)

core -limited   A chip whose size is determined by its core logic, not its pad frame.

clock distribution

The problem of distributing a clock signal to all points within a chip with acceptable delay, skew, and signal integrity. (See Chapter 7.)

clocked inverter An inverter with extra transistors that cause the inverter's output to be an open circuit when the clock input is disabled. (See Chapter 5.)

clocking discipline   A set of rules that, when followed, ensure that a sequential system will operate correctly across a broad range of clock frequencies. (See Chapter 5.)

crosstalk   Noise generated by one line interfering with another. (See Chapter 3.)

DCSL   A low-power variant of DCVSL. (See Chapter 3.)

DCVSL   A logic family which uses a latched pullup stage. (See Chapter 3.)

DIP   Dual in-line package, a type of package. (See Chapter 7.)

DRAM   Dynamic random-access memory. A three-transistor cell was an early form; the one-transistor cell is universal in commodity DRAM and increasingly used in logic chips. See *embedded RAM*.

data dependency   A relationship between two data computations in which the result of one is needed to compute the other. (See Chapter 8.)

data path   A unit designed primarily for data-oriented operations. Often designed in *bit-slice* style. (See Chapter 6.)

data path-controller architecture A sequential machine built from a data path plus a controller that responds to the data path's outputs and provides the data path's control inputs. (See Chapter 8.)

database   A program that provides access to and maintains the consistency of data. (See Chapter 10 for a discussion of CAD databases.)

decoupling capacitors   Capacitors added either on-chip or off-chip to reduce power/ground noise. (See Chapter 7.)

delay   In logic gate design, particularly measured between 50% points in the waveform. (See Chapter 3.)

departure time The time at which a signal transition leaves a given point in a logic network. (See Chapter 5.)

design flow   A series of steps used to design a chip. (See Chapter 9.)

design methodology   Generally similar to a design flow, though this is perhaps a more general term. (See Chapter 9.)

design rule   In general, a rule that governs design procedures. Most frequently applied to layout rules. (See Chapter 2.)

detailed routing The determination of the exact layout of a set of wires; compare to *global routing*. (See Chapter 7.)

dice   To cut a wafer into die; singular form of *die*.

diffusion   Generic term for any n-type or p-type region that is used to form transistors or wires. (See Chapter 2.)

die   Chips after slicing from the wafer but before packaging.

direct write   Exposing photoresist by writing directly on the wafer without masks using an electron beam or x-ray lithography system.

distributed control A controller built from several communicating machines. (See Chapter 8.)

dog and pony show
     A presentation to management.

dogleg   A style of channel routing that allows multiple horizontal segments. (See Chapter 4.)

domino   A common form of dynamic logic gate. (See Chapter 3.)

dot-com   An extinct form of company to which many CAD engineers went to seek their fortunes, only to return empty-handed.

drain   One of the transistor terminals connected to the channel. (See Chapter 2.)

drawn length The length of the transistor channel as drawn in the layout sent to manufacturing. The masks are often post-processed after tapeout and before manufacturing, making the fabricated and drawn lengths of the gate different. (See Chapter 2.)

drop-in   See *test structure*.

dynamic latch A latch that uses gate capacitance as a storage element and is volatile. (See Chapter 5.)

dynamic logic Logic that relies on charge stored on a transistor's gate capacitance. (See Chapter 4.)

e beam   An electron beam lithography machine. See also *direct write*.

effective capacitance A capacitance value chosen to estimate the gate delay induced by a wiring load. (See Chapter 4.)

Elmore delay A wiring delay model for RC transmission lines. (See Chapter 3.)

embedded CPU A CPU used in a larger system design.

embedded RAM Memory fabricated on the same die as logic components.

emulator   An FPGA-based machine into which a logic design can be compiled to be executed at relatively high speeds for prototyping and debugging. (See Chapter 9.)

FPGA   Field-programmable gate array. (See Chapter 6.)

fanin   All the gates which drive a given input of a logic gate.

fanout   All the gates driven by a given gate.

flash memory   An EEPROM memory that can be erased and reprogrammed using typical digital voltages. (See Chapter 6.)

flip-flop   A type of memory element not normally transparent during clocking. (See Chapter 4.)

floorplan   A sketch used to plan a layout design. (See Chapter 7.)

framework   A style of CAD database which provides utilities used by a variety of CAD tools.

functional testing
                      Testing of a component at low speed.

fringe capacitance
                      Capacitance around the edges of a pair of parallel plates. (See Chapter 2.)

full adder   An adder that generates both a sum and a carry. (See Chapter 6.)

GDS2   A common data format used to deliver mask information.

gate   The transistor terminal that controls the source-drain current. (See Chapter 2.)

global routing Determining the paths of wires through channels or other routing areas without determining the exact layout of those wires; compare to *detailed routing*. (See Chapter 7.)

ground bounce, ground noise

>Variations in ground voltage due to impedance on the ground wires. (See Chapter 7.)

ground plane　A large section of metallization used to provide coupling to ground and reduce the effect of other signal coupling. (See Chapter 7.)

H tree　A style of clock distribution network in which wires are organized as a hierarchy of Hs. (See Chapter 7.)

half adder　An adder that puts out only a sum. (See Chapter 6.)

hardware/software co-design

>The simultaneous design of an embedded CPU system and the software that will execute on it.

hardwired controller　A controller that is designed using random logic; compare to *microcoded controller*. (See Chapter 8.)

high-level synthesis

>CAD techniques for allocation, scheduling, and related tasks.

Hightower routing

>A common algorithm for area routing. (See Chapter 7.)

hit by a truck　The canonical means of losing a key technical person at a critical point in a project.

hold time　The interval for which a memory element data input must remain stable after the clock transition. (See Chapter 5.)

infant mortality

>The failure of chips during their first few hours of operation. See *burn-in*.

LFSR　Linear feedback shift register, a sequential machine used to generate pseudo-random sequences. (See Chapter 8.)

LSSD　Level-sensitive scan design, a method by which registers are operated in a shift mode during testing to observe and set state internal to the chip. (See Chapter 5.)

latch   A type of memory element that is transparent when the clock is active. (See Chapter 5.)

Lee/Moore router
                             A common algorithm for area routing. (See Chapter 7.)

linear region   The region of transistor operation in which the drain current is a strong function of the source/drain voltage. (See Chapter 2.)

logic synthesis   The automatic design of a logic network implementation. (See Chapter 10.)

lot   A set of wafers run through fabrication simultaneously; the basic unit of production.

lots per die   A yield measure for extremely large chips. See *belt buckle*.

MTCMOS   Multiple threshold CMOS, a low-power logic family. (See Chapter 3.)

Manchester carry chain A form of precharged carry chain that uses pass transistors. (See Chapter 6.)

Manhattan geometry
                             Masks which use only 90-degree angles.

memory elementA generic term for any storage element: flip-flop, latch, RAM, etc. (See Chapter 5.)

metal migration
                             A failure mode of metal wires caused by excessive current relative to the size of the wire. (See Chapter 2.)

microcoded controller   A controller that is designed using a microsequencer; compare to *hardwired controller*. (See Chapter 8.)

multiplexer   A combinational logic unit that selects one out of n inputs based on a control signal.

n-type diffusion An n-doped region. (See Chapter 2.)

NORA   A style of precharged logic.

no-op   A useless person.

OAI   An or-and-invert gate. (See Chapter 3.)

observability   The ability to determine (directly or indirectly) the value of a node on a chip. See also *controllability.*

one-hot code   A unary code used for state assignment or other codes in which each symbol is represented by a single true bit.

one-transistor DRAM   A dynamic RAM circuit that uses one capacitor to store the value and one transistor to access the value. Also called *one-T DRAM.* (See Chapter 6.)

OR plane   The block of logic in a PLA that computes the OR part of the required sum-of-products. (See Chapter 6.)

overdamped   An RLC circuit that does not oscillate. (See Chapter 3.)

π model   A model for the load on a gate that uses two capacitors bridged by a resistor. (See Chapter 4.)

p-type diffusion A p-doped region. (See Chapter 2.)

PCB   Printed circuit board.

PGA   Pin grid array, a type of package. (See Chapter 7.)

PLA   Programmable logic array. (See Chapter 6.)

PLCC   Plastic leadless chip carrier, a type of package. (See Chapter 7.)

PLL   See *phase-locked loop.*

PODEM   A test generation algorithm. (See Chapter 10.)

PG   See *pattern generator.*

package   Any carrier for an integrated circuit. (See Chapter 7 for examples.)

pad   A large metal region used to make off-chip connections. (See Chapter 7.)

pad frame   A set of pads and associated circuitry arranged around the edges of a rectangle, with room for logic in the middle. (See Chapter 7.)

pad-limited   A chip whose size is limited by its pad frame, not its core logic.

parametric testing
                       Testing for process-determined parameters: k', V_T, etc.

pass transistor A single transistor (usually n-type) used for switch logic. (See Chapter 3.)

pattern generator
                       A machine that makes masks for fabrication. Pattern generator machines are replaced by electron beam machines for fine-line masks—see *e beam*.

performance testing
                       Testing the speed at which a component runs.

phase   A clock signal that has a specified relationship to other clock phases. (See Chapter 5.)

phase-locked loop
                       A circuit that is often used to generate an internal clock from a slower external clock source. (See Chapter 7.)

pin   The connection between a package and a board. (See Chapter 7.)

pipelining   A logic design technique that adds ranks of memory elements to reduce clock cycle time at the cost of added latency.

placement   The physical arrangement of elements. (See Chapter 4 for gate placement and Chapter 7 for more global placement considerations.)

plastic package
                       A package made from plastic with metal leads for electrical connections. Is cheaper than a ceramic package but provides lower thermal conductivity. (See Chapter 7.)

plate capacitance
A capacitance between two parallel plates. The capacitance mechanism for transistor gates and metal capacitance. (See Chapter 2.)

polysilicon   Material used for transistor gates and wires. (See Chapter 2.)

power-down mode
An operating mode of a digital system in which large sections are turned off.

precharging   Charging a storage node for possible later discharge. (See Chapter 3.)

primary input An input to the complete system, as opposed to an input to a logic gate in the system.

primary output An output of the complete system, as opposed to an output of a logic gate in the system.

probe card   Used to connect a tester to an unpackaged integrated circuit.

propagation time   The time required for a signal to travel through combinational logic. (See Chapter 5.)

pseudo-nMOS A circuit family that uses a p-type resistive load. (See Chapter 3.)

pulldown   Any transistor used to pull a gate output toward V_{SS}. (See Chapter 3.)

pulldown network
The network of transistors in a logic gate responsible for pulling the gate output toward V_{SS}. (See Chapter 3.)

pullup   Any transistor used to pull a gate output toward V_{DD}. (See Chapter 3.)

pullup network
The network of transistors in a logic gate responsible for pulling the gate output toward V_{DD}. (See Chapter 3.)

RAM   Random-access memory. May be dynamic or static. (See Chapter 6.)

ROM   Read-only memory. (See Chapter 6.)

real estate   Chip area.

recirculating latch   A latch with cross-coupled inverters to provide non-volatile storage. (See Chapter 5.)

redundant   In combinational logic, an expression that is not minimal. (See Chapter 3.)

refresh   Restoring the dynamically-stored value in a memory. (See Chapter 6.)

register   Typically a memory element used in a data path, but may be used as synonymous with memory element.

register graph   A graph used in test generation that describes the connections between registers. (See Chapter 8.)

reticle   An alternate form of mask, which covers only a small part of the wafer and is repeated across the wafer surface.

retiming   Moving memory elements through combinational logic to change the clock period. (See Chapter 5.)

river routing   Routing in which wires form meandering paths but do not cross one another. (See Chapter 7.)

routing   The physical design of wiring. (See Chapters 4 and 7.)

rubylith   Early material for generating masks—a red sheet of plastic over a clear plastic base sheet which could be cut and peeled away to produce artwork for photographic reduction.

SCR   See *silicon-controlled rectifier.*

SRAM   Static read-only memory. (See Chapter 6.)

saturation region
                           The region of transistor operation that is roughly independent of the source/drain voltage. (See Chapter 2.)

scan chain, scan path

A set of registers that can be operated as a shift register for reading and writing during testing. See *LSSD*.

scheduling   The assignment of operations to clock cycles. (See Chapter 8.)

sense amplifier A differential amplifier used to sense the state of bit lines in memories. (See Chapter 6.)

sequential depth   The number of intervening registers between a selected register and a primary input. (See Chapter 8.)

setup time   The time by which a memory element's data input must arrive for it to be properly stored by the memory element. (See Chapter 5.)

shifter   A logic unit designed for shift operations. (See Chapter 6.)

short circuit power

The power consumed by a logic gate or network when both pullup and pulldown transistors are on. (See Chapter 3.)

sidewall capacitance

Junction capacitance from the side of a diffusion region to the substrate. (See Chapter 2.)

signal probability   The probability that a signal will switch, used in power analysis. (See Chapter 4.)

silicide   An improved gate material.

silicon-controlled rectifier In VLSI circuits, a parasitic device that can cause the chip to latch up. (See Chapter 2.)

sign-off   The approval of a design for manufacturing (or possibly some intermediate point in the design.)

signature analysis   A built-in self-test technique. (See Chapter 8.)

skin effect   The result of electromagnetic fields in low-resistance conductors that causes current to be carried primarily along the conductor's skin. (see Chapter 2.)

slicing structure   A floorplan that can be sliced into two sections without cutting any block, making it easier to route. (See Chapter 7.)

solder bump   A technique for making connections to a chip across its entire surface, not just at the periphery.

source   One of the transistor terminals connected to the gate. (See Chapter 2.)

spin   A workaholic's term for a *turn*.

state   The current values of the memory elements. (See Chapter 5.)

state assignment   The selection of binary codes for symbolic states. (See Chapter 5.)

state transition graph   A specification of a sequential machine, equivalent to a *state transition table*. (See Chapter 5.)

state transition table   A specification of a sequential machine, equivalent to a *state transition graph*. (See Chapter 5.)

static logic   Logic that does not rely on dynamically-stored charge.

step-and-repeat   The process of patterning a wafer with a reticle.

stuck-at-0/1   A fault model that assumes that a faulty gate's output is always either 0 or 1. (See Chapter 4.)

stuck-at-open   A fault model that assumes that a faulty gate's output is always either electrically open.

suit   A manager. See *no-op.*

switchbox   A rectangular routing region with pins on all four sides. (See Chapter 7.)

synthesis subset   A subset of a hardware description language that can be synthesized into hardware. (See Chapter 8.)

tape out   Generate a tape for PG. When working for a munificent employer, the precondition for a major party.

tapered wire A wire whose width varies along its width, usually to reduce the wire delay. (See Chapter 3.)

test structure Features added to the wafer for measuring processing parameters.

test synthesis The creation of test vectors from a state transition diagram or other non-gate description of the logic.

tester A machine that applies test vectors to chips on the manufacturing line.

threshold voltage
The gate voltage at which a transistor's drain current is deemed to be significant. (See Chapter 2.)

toaster 1) An extremely cost-sensitive application. 2) A chip that greatly exceeds its power budget.

transistor sizing
The determination of the appropriate W/Ls for transistors for performance or other design goals. (See Chapter 3.)

transmission gate A pair of n-type and p-type transistors connected in parallel and used to build switch logic. (See Chapter 3.)

transition time The time it takes a gate to rise or fall, often measured from 10% to 90% for rise time and visa versa for fall time. (See Chapter 3.)

tube CRT in a terminal or workstation (East Coast USA).

turn One iteration of the complete design cycle.

underdamped An RLC circuit that oscillates. (See Chapter 3.)

unknown voltage A voltage that represents neither logic 0 nor logic 1. (See Chapter 3.)

VHDL A hardware description language. (See Chapter 8.)

VTCMOS Variable threshold CMOS, a low-power logic family. (See Chapter 3.)

vector   Inputs applied to a chip.

Verilog   A hardware description language. (See Chapter 8.)

via   A hole in the chip's insulating layer that allows connections between different layers of interconnect. (See Chapter 2.)

victim net   In crosstalk, the net that receives the noise. (See Chapter 3.)

voltage contrast
                      A technique for reading voltages on an operating chip by scanning the chip with an electron beam and measuring the deflected current. (See Chapter 9.)

voltage scaling
                      Any one of several techniques for reducing the power supply voltage of a chip to lower its power consumption. (See Chapters 3 and 4.)

wafer start   A unit of production—the start of one wafer through the fabrication line. Both fab line capacity and chip production are measured in units of wafer starts.

Wallace tree   A design for high-speed multiplication. (See Chapter 6.)

wave pipelining
                      An advanced logic design methodology in which more than one signal is traveling through the logic between successive ranks of memory elements.

win the lottery   To get a much higher salary from a competitor.

windmill   A configuration of routing channels for which there is no unique routing order of the channels. (See Chapter 7.)

xter, xstr   Synonyms for transistor.

zipper   A logic design family similar to domino logic but without the output-stage inverter.

B

Chip Design Projects

B.1 Class Project Ideas

Below are some ideas for chip projects. The ideas vary greatly in complexity, and you can vary the complexity of a project by changing details of the specification. Remember the First Rule of Projects: *No matter how long you think your project will take, it will take longer.*

- **Alarm clock**. The user can set a time for the alarm, turn the alarm on and off, and change the current time. The alarm sounds when the clock is on and the alarm time has been reached.

- **Queue manager**. Enqueues data words from a source and dequeues them to a sink. For added challenge, pool n queues in a common memory, as described below.

- **Taxi meter**. The meter is turned on when a fare enters the cab. It also keeps track of mileage and waiting time. It computes the fare continuously during the trip.

- **Digital filter**. A biquadratic filter with programmable coefficients, for example.

- **Multiplexer**. Several data ports send data words to the chip; a one-bit output transmits the data as quickly as possible. An on-chip buffer matches input and output rates.

- **Programmable house controller**. It can be programmed to initiate a number of simple events initiated by a clock.

- **Systolic array element**. A computing element for some simple systolic array to do signal processing tasks.

- **BCD arithmetic unit**. It takes number strings in binary-coded decimal (BCD) and performs arithmetic operations on them. Handy for building calculators.

- **Very simple microprocessor**. With the accent on *very simple*. Even a PDP-8 is a major project.

- **Programmable metronome**. Plays a programmable rhythm at a programmable tempo.

- **Universal synchronous/asynchronous receiver/transmitter (USART)**. A USART can independently send and receive serial data streams, using either synchronous or asynchronous communication.

- **Display processor**. It produces a raster-order of a screen, filling the windows in real time. Window contents are stored in off-chip memory and window sizes/locations can be changed.

- **Simple Ethernet[1] transmitter or receiver**. Sends data on serial link, retransmitting on collision; listens for data addressed to chip's address.

- **Floating-point multiplier**. Multiplies two numbers in some floating-point format.

- **Regular expression comparator**. The chip input is a regular expression; looks for regular expression occurrences in a serial character stream.

B.2 Project Proposal and Specification

The first step in design is preparing a specification that describes what your chip will do but not the details of how the chip will do it. The proposal must, however, be a concrete and complete description of your chip's function. When defining your specification you should think as thoroughly as you can through the implementation costs of design features. Try to consider how you would build what you are proposing to be sure that your project is not too complex.

---

1. Ethernet is a trademark of Xerox.

The kitchen timer description in Chapter 9 should help you understand what goes into a good specification. You probably want to develop your specification in stages. For class purposes, you want to make sure your project is feasible before you develop a full-blown specification; in the real world, you want to make sure that you and the customer for the chip agree on the basics of what must be done before work progresses too far. An initial proposal should include: a brief description of the environment in which the chip is to operate; a plain-language description of what the chip is to do; a list of its major inputs and outputs; any performance or power requirements.

A more complete specification should include:

- A summary description of what the chip does and its operating environment.

- A list of the chip's input and output pins, along with any constraints on pin order in the package.

- Required clock rate and power consumption.

- A list of major modes in which the chip operates.

- For each mode:

  - pins that control the mode;

  - function executed in that mode;

  - performance constraints on execution, such as minimum and maximum times between inputs and outputs.

B.3 Design Plan

Your design plan is a first cut at your chip design down to the logic and layout. Let your motto be *I Think Before I Build*—a little thought early in the design process will save much grief later. It is completed before you start entering your design into tools. The design plan includes first designs of the chip architecture and all the major chip components. The report on your design plan should include:

✓ A recap of the chip's specifications, plus a list changes made to the initial specification.

✓ For the entire chip:

• A list of the chip's inputs and outputs, including their clocking types.

• A detailed block diagram, showing the chip's architecture and a description of the chip's internal operation. The operational   description   may include program fragments but should include some connecting English description.

• The behavior simulator:
   - code;
   - sample execution.

• Wiring plans:
   - V_{DD} and V_{SS};
   - clocks;
   - signals.

• A clocking plan for the chip, showing the clocking types of all major signals and, if a two-phase system, a two-coloring of the block diagram.

• A rough estimate of the critical timing paths through the chip.

• A validation plan: a pseudo-code description of a program that generates inputs to the chip and tests the chip's results. The tests should be thorough.

✓ For each component:

✓ For non-standard components, a brief description of their operation. (You don't need to explain how a counter works, just the basics, like how far it counts.)

✓ A list of all inputs and outputs, including their clocking types.

✓ For a custom-designed component, stick diagrams of all its cells and a floor plan of the component showing how it is assembled; for synthesized components, state transition tables, PDL++ programs, or other required generator files.

✓ For a custom component, a block or logic diagram and a two-coloring to show clocking correctness.

✓ A size estimate for the component.

The cell data sheet on the next page is an example of how you can organize the cell descriptions.

Cell Data Sheet

Name:

Function:

Size:

Delay:

Pinout:

name	direction	clocking	side/position	layer

Logic + clocking:

Sticks:

B.4 Design Checkpoints and Documentation

Once you start entering your design, you should check your progress regularly to ensure that you will finish in a timely manner and that you catch potential problems as soon as possible.

B.4.1 Subsystems Check

You should have completed:

✓ Layout of all the subsystems enumerated in your design plan.

✓ Simulation to validate the subsystems.

Warning: Most system errors are interface problems—components that are believed to work properly in isolation turn out to have incompatible interfaces. Though you cannot simulate components against each other until you wire them together in a layout, you should try during your simulation to make sure that your components are in fact compatible.

B.4.2 First Layout Check

You should turn a first complete layout for your chip that includes:

✓ all components placed and all signals routed;

✓ all signals to pins routed to the edge of the layout.

B.4.3 Project Completion

When you are finished, you should be able to show:

✓ The mask data ready for shipment to production.

✓ A check plot of the final layout.

✓ Descriptions of the major cells: size, speed.

✓ Floorplan showing the chip layout organization and the global routing of signals, power, and clocks.

✓ The results of functional simulation, showing that the chip meets its functional requirements.

✓ Performance analysis of the complete chip.

✓ Testability analysis of the complete chip.

C

Kitchen Timer Model

C.1 Hardware Modeling in C

A functional simulator can be written as an ordinary C [Ker70] language program; while some of the simulator's functions will be harder to program in C, the advantages of working in a familiar language generally pay off for a novice hardware designer. (Of course, such a simulator could be written in Pascal, Lisp, or other ordinary programming language, though details of the programming may vary.) We will use as an example a functional model program of the kitchen timer of Chapter 9.

It is possible to use C to create a register-transfer simulator that mimics the chip's behavior on a cycle-by-cycle basis. However, it is often useful to relax that requirement and simulate the function without strict concern for the clock cycle on which each input and output occurs. We may not care about the exact time behavior of the chip, and the functional simulator should not overly constrain the implementation choices; building a more abstract model also helps us think more clearly without worrying about excessive detail.

The main body of the simulator program is a simple read/execute/write loop, with the general form:

```
initialize_chip(&chip_state);
while (TRUE) {
    read_chip_inputs(&chip_inputs);
    run_one_cycle(chip_inputs,&chip_outputs,&chip_state);
```

```
write_chip_outputs(chip_outputs);
}
```

The timer chip model consists of three components—the two timers and the controller—which all execute in parallel. In a parallel programming language, we could
write one process for each component, describe the communication (conducted on the
chip through wires) between them, and let the programming language handle the
details. Since C does not directly support parallelism, we must hand-craft our code to
achieve the desired effect. We use C functions to model components for modularity
and clarity.

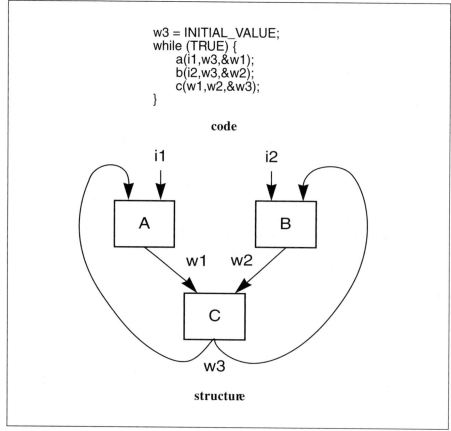

```
w3 = INITIAL_VALUE;
while (TRUE) {
    a(i1,w3,&w1);
    b(i2,w3,&w2);
    c(w1,w2,&w3);
}
```

code

structure

Figure C-1:
Ordering model
execution to sat-
isfy data depen-
dencies.

One important thing to remember when building a simulation program is that the functions which model components must be called in the order required by data dependencies. As illustrated in Figure C-1, we must call the function which drives a wire before we call any function which examines the value of that wire. In this example, it doesn't matter whether the A() or B() function is executed first, since there is no data dependency between them. In the simulator program's main body, we first called the input routine, then the controller routine to work on the button values, and finally the timer function to process the commands from the controller.

It is also important to remember that a function's internal variables should not be used to store a component's current state. Normal function variables are destroyed when the function exits; when they go away, the component's state goes with them. If static variables (which in C retain their values between function calls) are used, the state will not be destroyed but we cannot use the same function to simulate two instances of that component type. We pass into a function not only the values on its inputs, but also its present state value. The function produces the outputs and modifies the state. Under this scheme, a function corresponds to a component type and a function call with a particular state corresponds to the execution of a particular component instance.

This scheme is required for two reasons. First, we may have several copies of a component type, such as the two timers, and we don't want to write two functions for the same basic design. Second, the state of a component must be stored outside the function body since the chip is modeled by a sequence of function calls. If the state were stored in a function's local variables, the state value would disappear when the function returned to its caller. While we could use global variables to store the state, passing it in as an argument to the function lets us reuse the function for several instances of a component.

There are some parts of the chip's function that we can leave indeterminate in this functional model. For example, we don't care exactly how long it takes the chip to respond to a button push, as long as the response time is reasonable and the chip always responds to a button event. In other designs, the details of chip behavior may become more important: for example, a chip may be required to respond to an input within two clock cycles or 100 *ns*.

C.1.1 Simulator

```
#include <stdio.h>
/* Written by Wayne Wolf.
 A functional model for the kitchen timer chip.
 */
```

```
/*
 Some types and constants.
*/

typedef int boolean;
#define TRUE 1
#define FALSE 0

#define NTIMERS 2 /* number of timers implemented in machine */
#define NDIGITS 4 /* number of digits in the display */

/*
 The keyboard.
*/

typedef struct {
     boolean select[NTIMERS]; /* timer select buttons */
     boolean seconds, minutes; /* time increment buttons */
     boolean go, clear; /* start the current timer, clear it and stop bell */
     boolean clock; /* seconds clock */
} button_inputs;

void read_buttons(bin)
     button_inputs *bin;
{
int i;

printf("select buttons? "); fflush(stdout);
for (i=0; i<NTIMERS; i++)
scanf("%d",&(bin->select[i]));
printf("second/minute buttons? "); fflush(stdout);
scanf("%d %d", &(bin->seconds), &(bin->minutes));
printf("go, clear buttons? "); fflush(stdout);
scanf("%d %d", &(bin->go), &(bin->clear));
printf("seconds clock? "); fflush(stdout);
scanf("%d",&(bin->clock));
}

/*
 One timer.
*/

typedef struct {
     boolean go; /* TRUE when this timer is running */
     boolean reset; /* TRUE to turn off timer and stop bell */
     boolean tick; /* seconds */
     boolean incr_seconds, /* increment seconds */
     incr_minutes; /* increment minutes */
} timer_inputs;

typedef struct {
```

```
    int value; /* timer value */
    boolean done; /* set TRUE when counter = 0 */
} timer_outputs;

typedef struct {
    int timer; /* current time */
    boolean running; /* TRUE when timer is running */
} timer_state;

void timer(in,out,state)
    timer_inputs *in;
    timer_outputs *out;
    timer_state *state;
{
 /* Set timer running if necessary. */
state->running = (state->running || in->go) && !in->reset;
 /* Increment time if requested and not running. */
if (!state->running) {
    if (in->incr_seconds)
    state->timer = state->timer + 1;
    if (in->incr_minutes)
    state->timer = state->timer + 60 /* 60 seconds/minute */;
    }
 /* decrement time if clock ticked */
if (state->running && in->tick && state->timer > 0)
    state->timer = state->timer - 1;
/* determine outputs -- timer value and bell */
out->value = state->timer;
out->done = state->running && (state->timer == 0);
}

/*
 The controller.
*/

typedef struct {
    /* select one of NTIMER timers to display */
    boolean timer[NTIMERS];
    /* turn on currently selected timer, turn it off */
    boolean go, clear;
    /* increment seconds/minutes on selected timer */
    boolean incr_seconds, incr_minutes;
} controller_inputs;

typedef struct {
    /* control signals to each of the timers */
    boolean incr_seconds[NTIMERS],
    incr_minutes[NTIMERS],
    go[NTIMERS], /* set TRUE to start timer running */
    reset[NTIMERS] /* set TRUE to reset timer */;
    /* control signals to display/bell */
    int selected; /* number of timer selected */
```

```
        boolean display; /* TRUE when display should be on */
} controller_outputs;

typedef struct {
    /* state of buttons for edge detection */
    boolean incr_seconds, incr_minutes, go, clear;
    /* currently selected timer */
    int selected;
} controller_state;

void controller(in,out,state)
    controller_inputs *in;
    controller_outputs *out;
    controller_state *state;
{
int i;
boolean some_selection;

 /* Initialize any outputs that are not always set. */
for (i=0; i<NTIMERS; i++) {
    out->incr_minutes[i] = FALSE;
    out->incr_seconds[i] = FALSE;
    out->go[i] = FALSE;
    out->reset[i] = FALSE;
    }
 /* Did the user select a new timer? */
for (i=0; i<NTIMERS; i++)
    state->selected = (i == in->timer[i]);
 /* Do we need to increment seconds or minutes? Perform
    edge detection on button inputs as well. */
if (in->incr_seconds && !state->incr_seconds)
    out->incr_seconds[state->selected] = TRUE;
state->incr_seconds = in->incr_seconds;
if (in->incr_minutes && !state->incr_minutes)
    out->incr_minutes[state->selected] = TRUE;
state->incr_minutes = in->incr_minutes;
/* Do we need to turn on or clear the currently selected timer?
    Perform edge detection as well. */
if (in->go && !state->go)
    out->go[state->selected] = TRUE;
state->go = in->go;
if (in->clear && !state->clear)
    out->reset[state->selected] = TRUE;
state->clear = in->clear;
/* Tell display which timer is selected. */
out->selected = state->selected;
}

/*
 The display.
*/
```

```c
typedef struct {
    int values[NTIMERS]; /* NTIMER timer values */
    int selected; /* timer to display */
    boolean on; /* TRUE when display is on */
    boolean done[NTIMERS]; /* says when a timer is finished */
} display_inputs;

typedef struct {
    int digit; /* the digit to display: 0 => 10s minutes, 1 => 1s minutes,
                    2 => 10s seconds, 3 => 10s seconds */
} display_state;

typedef struct {
    int timer; /* number of timer selected */
    int minutes, seconds; /* the total time, kept here for convenience */
    int onedigit, digitselect[NDIGITS]; /* the current digit being
                    displayed and the select signal for that digit */
    boolean bell; /* rings the bell */
} display_outputs;

void display(in,out,state)
    display_inputs *in;
    display_outputs *out;
    display_state *state;
{
int mytimer; /* temporary -- holds value of currently selected timer */
int i;

/* tell display which timer is on show */
out->timer = in->selected;
 /* select the right time value */
mytimer = in->values[in->selected];
/* decode into minutes/seconds */
out->minutes = mytimer / 60;
out->seconds = mytimer - (out->minutes * 60);
/* decode a digit and send the select to the display */
for (i=0; i<NDIGITS; i++)
    out->digitselect[i] = 0;
switch (state->digit) {
    case 0:
        out->onedigit = out->minutes / 10;
        out->digitselect[0] = 1;
        break;
    case 1:
        out->onedigit = out->minutes - (out->minutes / 10)*10;
        out->digitselect[1] = 1;
        break;
    case 2:
        out->onedigit = out->seconds / 10;
        out->digitselect[2] = 1;
        break;
    case 3:
```

```
                out->onedigit = out->seconds - (out->seconds / 10)*10;
                out->digitselect[3] = 1;
                break;
        default:
                fprintf(stderr,"unknown digit select %d\n",in->selected);
                break;
        }
/* select a digit */
if (state->digit == 3)
        state->digit = 0;
else
        state->digit++;
/* ring the bell? */
out->bell = FALSE;
for (i=0; i<NTIMERS; i++)
        out->bell = out->bell || in->done[i];
}


void write_display(out)
        display_outputs *out;
{
 /* print the whole time for ease of debugging */
printf("timer %1d\n\t%2d:%2d",out->timer,out->minutes,out->seconds);
printf("\tcurrent digit(%1d%1d%1d%1d) = %d\n",out->digitselect[0],
        out->digitselect[1],out->digitselect[2],out->digitselect[3],
        out->onedigit);
if (out->bell)
        printf(" DING!\n\n");
else
        printf("\n\n");
}


/*
 The main simulation program.
*/


main()
{
controller_inputs cin;
controller_outputs cout;
controller_state cstate;
timer_inputs tin[NTIMERS];
timer_outputs tout[NTIMERS];
timer_state tstate[NTIMERS];
button_inputs bin;
display_inputs din;
display_state dstate;
display_outputs dout;
int i;
int bell;

/* initialize state */
```

```
for (i=0; i<NTIMERS; i++) {
    tstate[i].timer = 0;
}
dstate.digit = 0;
/* run indefinitely */
while (!feof(stdin)) {
    /* read inputs and distribute to components */
    read_buttons(&bin);
    for (i=0; i<NTIMERS; i++)
        cin.timer[i] = bin.select[i];
    cin.incr_seconds = bin.seconds;
    cin.incr_minutes = bin.minutes;
    cin.go = bin.go;
    cin.clear = bin.clear;
    /* compute for one cycle: first controller, then the timers
        themselves */
    controller(&cin,&cout,&cstate);
    for (i=0; i<NTIMERS; i++) {
        tin[i].go = cout.go[i];
        tin[i].incr_seconds = cout.incr_seconds[i];
        tin[i].incr_minutes = cout.incr_minutes[i];
        tin[i].reset = cout.reset[i];
        tin[i].tick = bin.clock;
        timer(&(tin[i]),&(tout[i]),&(tstate[i]));
        din.values[i] = tout[i].value;
        din.done[i] = tout[i].done;
        }
    din.on = cout.display;
    din.selected = cout.selected;
    display(&din,&dout,&dstate);
    /* write outputs and ring bell */
    write_display(&dout);
    }
}
```

C.1.2 Sample Execution

```
% timer
select buttons? 1 0
second/minute buttons? 0 0
go, clear buttons? 0 0
seconds clock? 0
timer 0
    0: 0     current digit(1000) = 0


select buttons? 0 0
second/minute buttons? 1 0
go, clear buttons? 0 0
seconds clock? 0
timer 0
```

0: 1   current digit(0100) = 0

select buttons? 0 0
second/minute buttons? 0 0
go, clear buttons? 0 0
seconds clock? 0
timer 0
    0: 1   current digit(0010) = 0

select buttons? 0 0
second/minute buttons? 1 0
go, clear buttons? 0 0
seconds clock? 0
timer 0
    0: 2   current digit(0001) = 2

select buttons? 0 0
second/minute buttons? 1 0
go, clear buttons? 0 0
seconds clock? 0
timer 0
    0: 2   current digit(1000) = 0

select buttons? 0 0
second/minute buttons? 0 0
go, clear buttons? 0 0
seconds clock? 0
timer 0
    0: 2   current digit(0100) = 0

select buttons? 0 0
second/minute buttons? 0 0
go, clear buttons? 1 0
seconds clock? 0
timer 0
    0: 2   current digit(0010) = 0

select buttons? 0 0
second/minute buttons? 0 0
go, clear buttons? 0 0
seconds clock? 1
timer 0
    0: 1   current digit(0001) = 1

select buttons? 0 0

```
second/minute buttons? 0 0
go, clear buttons? 0 0
seconds clock? 1
timer 0
    0: 0      current digit(1000) = 0
 DING!

select buttons? 0 0
second/minute buttons? 0 0
go, clear buttons? 0 0
seconds clock? 1
timer 0
    0: 0      current digit(0100) = 0
 DING!

select buttons? 0 0
second/minute buttons? 0 0
go, clear buttons? 0 1
seconds clock? 0
timer 0
    0: 0      current digit(0010) = 0
```

References

[Abr90]   Miron Abramovici, Melvin A. Breuer, and Arthur D. Friedman, *Digital System Testing and Testable Design*, Computer Science Press, Rockville, MD, 1990.

[Ake67]   S. B. Akers, "A modification of Lee's path connection algorithm," *IEEE Transactions on Electronic Computers*, February, 1967, pp. 97-98.

[Bak90]   H. B. Bakoglu, *Circuits, Interconnections, and Packaging for VLSI*, Addison-Wesley, 1990.

[Bas95]   Mick Bass, Terry W. Blanchard, D. Douglas Josephson, Duncan Weir, and Daniel L. Halperin, "Design methodologies for the PA 7100LC microprocessor," Hewlett-Packard Journal, 46(2), April, 1995, pp. 23-35.

[Bau73]   Charles R. Baugh and Bruce A. Wooley, "A two's complement parallel array multiplication algorithm," *IEEE Transactions on Computers*, C-22(12), December, 1973, pp. 1045-1047.

[Ben95]   Jack D. Benzel, "Bugs in black and white: imaging IC logic levels with voltage contrast," Hewlett-Packard Journal, 46(2), April, 1995, pp. 102-106.

[Ber95]   R. A. Bergamaschi, R. A. O'Connor, L. Stok, M. Z. Moricz, S. Prakash, A. Kuehlmann, and D. S. Rao, "High-level synthesis in an industrial environment," *IBM Journal of Research and Development*, 39(1/2), January/March, 1995, pp. 131-148.

[Bha95]   J. Bhasker, A VHDL Primer, revised edition, Englewood Cliffs NJ: Prentice Hall, 1995.

[Bis90]   Philip E. Bishop, Grady L. Giles, Sudarshan N. Iyengar, C. Thomas Glover, and Wai-on Law, "Testability considerations in the design of the MC68340 integrated processor unit," *Proceedings, 1990 International Test Conference*, IEEE Computer Society Press, 1990, pp. 337-346.

[Boe93]   K. D. Boese, A. B. Kahng, B. A. McCoy, and G. Robins, "Fidelity and near-optimality of Elmore-based routing constructions," in *Proceedings, ICCD '93*, IEEE Comptuer Society Press, 1993, pp. 81-84.

[Boo51]   Andrew D. Booth, "A signed binary multiplication technique," *Quart. Journal of Mech. and Appl. Math.*, Vol. IV, Pt. 2, 1951, pp. 236-240.

[Bow95]   William J. Bowhill, Shane L. Bell, Bradley J. Benschneider, Andrew J. Black, Sharon M. Britton, Ruben W. Castelino, Dale R. Donchin, John H.

Edmondson, Harry R. Fair III, Paul E. Gronowski, Anil K. Jain, Patricia L. Kroesen, Marc E. Lamere, Bruce J. Loughlin, Shekhar Mehta, Robert O. Mueller, Robert P. Preston, Sribalan Santhanam, Timothy A. Shedd, Michael J. Smith, and Stephen C. Thierauf, "Circuit implementation of a 300-MHz 64-bit second-generation CMOS Alpha CPU," Digital Technical Journal, 7(1), 1995, pp. 100-118.

[Bra84]   R. K. Brayton, C. McMullen, G. D. Hachtel, and A. Sangiovanni-Vincentelli, Logic Minimization Algorithms for VLSI Synthesis, Kluwer Academic Publishers, Norwell, MA, 1984.

[Bra87]   R. K. Brayton, R. Rudell, A. Sangiovanni-Vincentelli, and A. Wang, "MIS: A Multiple-Level Logic Optimization System," *IEEE Transactions on CAD/ICAS,* CAD-6, November, 1987, pp. 1062-1081.

[Bra90]   R. K. Brayton, G. D. Hachtel, and A. L. Sangiovanni-Vincentelli, "Multilevel logic synthesis," *Proceedings of the IEEE*, 78(2), February, 1990, pp. 264-300.

[Bro93]   T. Brodnax, M. Schiffli, and F. Watson, "The PowerPC 601 design methodology," in *Proceedings, ICCD '93*, IEEE Computer Society Press, 1993, pp. 248-252.

[Bre77]   Melvin A. Breuer, "A class of min-cut placement algorithms," *Proceedings, 14th Design Automation Conference*, ACM/IEEE, 1977, pp. 284-290.

[Bry86]   Randal E. Bryant , Derek Beatty , Karl Brace , Kyeongsoon Cho, and Thomas Scheffler, "COSMOS: a compiled simulator for MOS circuits," *Proceedings, 23rd Design Automation Conference*, ACM/IEEE, 1986, pp. 9-16.

[Bry87A]   Randal E. Bryant, "A survey of switch-level simulation," *IEEE Design & Test*, August, 1987, pp.26-40.

[Bry87B]   Randal E. Bryant, "Algorithmic aspects of symbolic switch network analysis," *IEEE Transactions on CAD/ICAS*, CAD-6(4), July, 1987, pp. 618-633.

[Bry87C]   Randal E. Bryant, "Boolean analysis of MOS circuits," *IEEE Transactions on CAD/ICAS*, CAD-6(4), July, 1987, pp. 634-649.

[Bre90]   J. R. Brews, "The submicron MOSFET." Chapter 3 in S. M. Sze, ed., *High-Speed Semiconductor Devices*, John Wiley & Sons, 1990.

[Buc91]   Joseph Buck, Soonhoi Ha, Edward A. Lee, and David G. Messerschmitt, "Ptolemy: a platform for heterogeneous simulation and prototyping," *Proceedings, 1991 European Simulation Conference*, June, 1991.

[Cal90]   Richard E. Calcagni and Will Sherwood, "VAX 6000 Model 400 CPU chip set functional design verification," *Digital Technical Journal*, 2(2),

Spring, 1990, pp. 64-72.

[Cal96] Thomas K. Callaway and Earl E. Schwartzlander, Jr., "Low Power Arithmetic Components," Chapter 7 in Jan M. Rabaey and Massoud Pedram, eds., *Low Power Design Methodologies*, Kluwer Academic Publishers, Norwell MA, 1996.

[Cam91] Raul Camposano, "Path scheduling for synthesis," *IEEE Transactions on CAD/ICAS*, 10(1), January, 1991, pp. 85-93.

[Cha92] Anantha P. Chandrakasan, Samuel Sheng, and Robert W. Brodersen, "Low-power CMOS digital design," *IEEE Journal of Solid-State Circuits*, 27(4), April, 1992, pp. 473-484.

[Cha01] Anantha Chandrakasan, William J. Bowhill, and Frank Fox, eds., *Design of High-Performance Microprocessor Circuits*, New York: IEEE Press, 2001.

[Cha93] K. Chaudhary, A. Onozawa, and E. S. Kuh, "A spacing algorithm for performance enhancement and cross-talk reduction," in *ICCAD-93 Digest of Technical Papers*, IEEE Computer Society Press, 1993, pp. 697-702.

[Che84] C.F.Chen and P. Subramaniam, "The Second Generation MOTIS Timing Simulator—An Efficient and Accurate Approach for General MOS Circuits," *Proceedings of the IEEE International Symposium on Circuits and Systems*, 1984, pp. 538-542.

[Che00] Chung-Kuan Cheng, John Lillis, Shen Lin, and Norman Chang, *Interconnect Analysis and Synthesis*, New York: Wiley Interscience, 2000.

[Che89] K.-T. Cheng and V. D. Agrawal, "An economical scan design for sequential logic test generation,," *Digest of Papers, 19th International Symposium on Fault-Tolerant Computing*, IEEE Computer Society Press, Los Alamitos, CA, 1989, pp. 28-35.

[Cho89] Paul Chow, ed., *The MIPS-X Microprocessor*, Kluwer Academic Publishers, Norwell MA, 1989.

[Cla73] Christopher R. Clare, *Designing Logic Using State Machines*, McGraw-Hill, 1973.

[Con93] Jason Cong and Kwok-Shing Leung, "Optimal wiresizing under the distributed Elmore delay model," in *Proceedings, ICCAD-93*, IEEE Computer Society Press, 1993, pp. 634-939.

[Cor90] Thomas H. Cormen , Charles E. Leiserson, and Ronald L. Rivest, *Introduction to Algorithms*, McGraw-Hill/MIT Press, 1990.

[De01] Vivek De, Yibin Ye, Ali Keshavarzi, Siva Narendra, James Kao, Dinesh Somasekhar, Raj Nair, Shekhar Borkar, "Techniques for leakage power reduction," Chapter 3 in Anantha Chanddrakasan, William J. Bowhill, and Frank Fox, eds., *Design of High-Performance Microprocessor Circuits*,

New York: IEEE Press, 2000.

[DeM85] Giovanni De Micheli, Robert K. Brayton, Alberto Sangiovanni-Vincentelli, "Optimal State Assignment for Finite State Machines," *IEEE Transactions on CAD/ICAS*, CAD-4, 1985, pp. 269-285.

[DeM94] Giovanni De Micheli, *Synthesis of Digital Circuits*, McGraw-Hill, 1994.

[DeM96] G. De Micheli and M. Sami, eds., Hardware/Software Co-Design,, Norwell MA: Kluwer Academic Publishers, NATO ASI Series, 1996.

[DeM01] Giovanni De Micheli, Rolf Ernst, and Wayne Wolf, *Readings in Hardware/Software Co-Design*, San Francisco: Morgan Kaufman, 2001.

[Den74] Robert H. Dennard, Fritz H. Gaensslen, Hwa-Nien Yu, V. Leo Rideout, Ernest Bassous, and Andre R. LeBlanc, "Design of ion-implanted MOSFET's with very small physical dimensions," *IEEE Journal of Solid-State Circuits*, SC-9(5), October, 1974, pp. 256-268.

[Den85] Peter Denyer and David Renshaw, *VLSI Signal Processing: A Bit-Serial Approach*, Addison-Wesley, 1985.

[Deu76] David N. Deutsch, "A 'dogleg' channel router," *Proceedings, 13th Design Automation Conference*, ACM/IEEE , 1976 , pp. 425-433.

[Dev91] Srinivas Devadas, Kurt Keutzer, and Sharad Malik, "Delay computation in combinational logic circuits: theory and algorithms," *Proceedings, ICCAD-91*, IEEE Computer Society, 1991, pp. 176-179.

[Dev88] Srinivas Devadas, Hi-Keung Ma, A. Richard Newton, and A. Sangiovanni-Vincentelli, "MUSTANG: state assignment of finite state machines targeting multilevel logic implementations," *IEEE Transactions on CAD/ICAS*, CAD-7(12), December, 1988, pp. 1290-1300.

[Die78] Donald L. Dietmeyer, *Logic Design of Digital Systems*, second edition, Allyn and Bacon, 1978.

[Dob92] D. W. Dobberpuhl, R. T. Witek, R. Allmon , R. Anglin, D. Bertucci, S. Brittoni, L. Chao, R. A. Conrad, D. E. Dever, B. Gieseke, S. M. N. Hassoun, G. W. Hoeppner, K. Kuchler, M. Ladd, B. M. Leary, L. Madden, E. J. McLellan, D. R. Meyer, J. Montanaro, D. A. Priore, V. Rajagopalan, S. Samudral , and S. Santhanam, "A 200-MHz 64-b Dual-Issue CMOS Microprocessor", *IEEE Transactions on Solid-State Circuits*, 27(11), November, 1992, pp. 1555-1568.

[Du91] X. Du, G. Hachtel, B. Lin, and A. R. Newton, "MUSE: A MUltilevel Symbolic Encoding Algorithm for state assignment," 10(1), January, 1991, pp. 28-38.

[Dun85] Alfred E. Dunlop and Brian W. Kernighan, "A procedure for placement of standard-cell VLSI circuits," *IEEE Transactions on CAD/ICAS*, CAD-4(1), January, 1985, pp. 92-98.

[Dut01]   Santanu Dutta, Rune Jensen, and Alf Rieckmann, "Architecture and implementation of VIPER: a multiprocessor SOC for ASTB and DTV systems," *IEEE Design & Test of Computers*, July-August 2001.

[Elm48]   W. C. Elmore, "The transient response of damped linear networks with particular regard to wideband amplifiers," Journal of Applied Physics, 19, January, 1948, pp. 55-63.

[Eng96]   J. J. Engel, T. S. Guzowski, A. Hunt, D. E. Lackey, L. D. Pickup, R. A. Proctor, K. Reynolds, A. M. Rincon, and D. R. Stauffer, "Design methodology for IBM ASIC products," *IBM Journal of Research and Development*, 40(4), July, 1996, pp. 387-406.

[Fis85]   J. P. Fishburn and A. E. Dunlop, "TILOS: A Posynomial Programming Approach to Transistor Sizing," *Proceedings, ICCAD-85* , IEEE Computer Society, 1985, pp. 326-328.

[Fis90]   John P. Fishburn, "Clock skew optimization," *IEEE Transactions on Computers*, 39(7), July, 1990, pp. 945-951.

[Fis92]   J. P. Fishburn, private communication, December, 1992.

[Fis95]   J. P. Fishburn and C. A. Schevon, "Shaping a distributed-RC line to minimize Elmore delay," *IEEE Transactions on CAS-I*, 42, December, 1995, pp. 1020-1022.

[Fri75]   A. D. Friedman and P. R. Menon, *Theory and Design of Switching Circuits*, Computer Science Press, Rockville, MD, 1975.

[Fri86]   Eby G. Friedman and Scott Powell, "Design and analysis of a hierarchical clock distribution system for synchronous standard cell/macrocell VLSI," *IEEE Journal of Solid-State Circuits*, SC-21(2), April, 1986, pp. 240-246.

[Fri95]   Eby G. Friedman, ed., *Clock Distribution Networks in VLSI Circuits and Systems*, IEEE Press, 1995.

[For87]   Mark Forsyth, William S. Jaffe, Darius Tanksalvala, John Wheeler, and Jeff Yetter, "A 32-bit VLSI CPU with 15-MIPS peak performance," *IEEE Journal of Solid-State Circuits*, SC-22(5), October, 1987, pp. 768-775.

[Gal80]   J. Gality, Y. Crouze, and M. Verginault, "Physical *vs.* logical fault models for MOS LSI circuits: impact on their testability," *IEEE Transactions on Computers*, C-29(6), June, 1980, pp. 527-531.

[Gaj92]   Daniel Gajski, Nikil Dutt, Allen Wu, and Steve Lin, *High-Level Synthesis: Introduction to Chip and System Design*, Kluwer Academic Publishers, Norwell, MA, 1992.

[Gao94]   T. Gao and C. L. Liu, "Minimum crosstalk switchbox routing," in *ICCAD-94 Digest of Technical Papers*, IEEE Computer Society Press, 1994, pp. 610-615.

[Gar79]   Michael R. Garey and David S. Johnson, *Computers and Intractability: A Guide to the Theory of NP-Completeness*, W. H. Freeman and Company, 1979.

[Gei90]   Randall L. Geiger, Phillip E. Allen, and Noel R. Strader, *VLSI: Design Techniques for Analog and Digital Circuits*, McGraw-Hill, 1990.

[Gel86]   Patrick P. Gelsinger, "Built in self test of the 80386," *Proceedings, ICCD-86*, IEEE Computer Society, 1986, pp. 169-173.

[Gel87]   Patrick P. Gelsinger, "Design and test of the 80386," IEEE Design & Test, June, 1987, pp. 42-50.

[Gha94]   Sorab K. Ghandhi, *VLSI Fabrication Principles: Silicon and Gallium Arsenide*, second edition, John Wiley and Sons, 1994.

[Gho91]   A. Ghosh and S. Devadas and A. R. Newton, "Test generation and verification for highly sequential circuits," *IEEE Transactions on CAD/ICAS*, 10(5), May, 1991, pp. 652-667.

[Gie97]   Bruce A. Gieseke, Randy L. Allmon, Daniel W. Bailey, Bradley J. Benschneider, Sharon M. Birtton, John D. Clouser, Harry R. Fair III, James A. Farrell, Michael K. Gowan, Christopher L. Houghton, James B. Keller, Thomas H. Lee, Daniel L. Liebholz, Susan C. Lowell, Mark D. Matson, Richard J. Matthew, Victor Peng, Michael D. Quinn, Donald A. Priore, Michael J. Smith, and Kathryn E. Wilcox, "A 600 MHz superscalar RISC microprocessor with out-of-order execution," in *Digest of Technical Papers, 1997 IEEE International Solid State Circuits Conference*, Castine ME: John H. Wuorinen, 1997, pp. 176-177.

[Gla85]   Lancer A. Glasser and Daniel W. Dobberpuhl, *The Design and Analysis of VLSI Circuits*, Addison-Wesley, 1985.

[Goe81]   P. Goel, "An implicit enumeration algorithm to generate tests for combinational logic circuits," *IEEE Transactions on Computers*, C-30(3), March, 1981, pp. 215-222.

[Gup96]   Rajesh K. Gupta and Giovanni de Micheli, "A co-synthesis approach to embedded system design automation," Design Automation for Embedded Systems, 1(1-2), January, 1996, pp. 69-120.

[Hac96]   Gary D. Hachtel and Fabio Somenzi, *Logic Synthesis and Verification Algorithms*, Kluwer Academic Publishers, 1996.

[Has71]   Akihiro Hashimoto and James Stevens, "Wire routing by optimizing channel assignment within large apertures," *Proceedings, 8th Design Automation Workshop*, SHARE Committee, 1971, pp. 155-169.

[Hat87]   M. Hatamian and G. Cash, "Parllel bit-level pipelined VLSI designs for high-speed signal processing," *Proceedings of the IEEE*, 75(9), September, 1987, pp. 1192-1202.

[Hen94]   Jorg Henkel, Rolf Ernst, Ullrich Holtmann, and Thomas Benner, "Adaptation of partitioning and high-level synthesis in hardware/software co-synthesis," in *Proceedings, ICCAD-94*, IEEE Computer Society Press, 1994, pp. 96-100.

[Hig69]   D. W. Hightower, "A solution to line-routing problems on the continuous plane," *Proceedings, 6th Design Automation Workshop*, SHARE Committee, 1969, pp. 1-24.

[Hil89]   Dwight Hill , Don Shugard , John Fishburn, and Kurt Keutzer, *Algorithms and Techniques for VLSI Layout Synthesis*, Kluwer Academic Publishers, Norwell, MA, 1989.

[Hil80]   Frederick S. Hillier and Gerald J. Lieberman, Introduction to Operations Research, third edition, Holden-Day, 1980.

[Hod83]   David A. Hodges and Horace G. Jackson, *Analysis and Design of Digital Integrated Circuits*, McGraw-Hill, 1983.

[Hoo90]   Donald F. Hooper and John C. Eck, "Synthesis in the CAD system used to design the VAX 9000 system," *Digital Technical Journal*, 2(4), Fall, 1009, pp. 118-129.

[Hor83A]   Mark Horowitz, *Timing Models for MOS Circuits*, Ph.D. Thesis, Stanford University, December, 1983.

[Hor83B]   M. Horowitz and R. W. Dutton, "Resistance Extraction from Mask Layout Data," *IEEE Transactions on CAD/ICAS*, CAD-2(3), 1983, pp. 145-150.

[Hwa91]   C.-T. Hwang , J.-H. Lee, and Y.-C. Hsu, "A formal approach to the scheduling problem in high-level synthesis," *IEEE Transactions on CAD/ICAS*, 10(4), April, 1991, pp. 464-475.

[IEE93]   IEEE, *IEEE Standard VHDL Language Reference Manual*, Std 1076-1993, New York: IEEE, 1993.

[Int94]   Intel Corporation, *Microprocessors: Vol. III*, 1994.

[Ism00]   Yehea I. Ismail and Eby G. Friedman, "Effects of inductance on the propagation delay and repeater insertion in VLSI circuits," *IEEE Transactions on VLSI Systems*, 8(2), April 2000, pp. 195-206.

[Jae75]   Richard C. Jaeger, "Comments on 'An optimized output stage for MOS integrated circuits," *IEEE Journal of Solid-State Circuits*, SC10(3), June, 1975, pp. 185-186.

[Jai83]   S. K. Jain and V. D. Agrawal, "Test generation for MOS circuits using the D algorithm," *Proceedings, 20th Design Automation Conference*, IEEE Computer Society, 1983, pp. 64-70.

[Jha90]   Niraj K. Jha and Sandip Kundu, *Testing and Reliable Design of CMOS Circuits*, Kluwer Academic Publishers, Norwell, MA, 1990.

[Jou84]   Norman P. Jouppi, *Timing Verification and Performance Improvement of MOS VLSI Designs*, Ph.D. Thesis, Stanford University, October, 1984.

[Kei01]   Doris Keitel-Schulz and Norbert Wehn, "Embedded DRAM development: technology, physical design, and application issues," *IEEE Design & Test of Computers*, May-June 2001, pp. 7-15.

[Ker70]   B. W. Kernighan and S. Lin, "An efficient heuristic procedure for partitioning graphs," *Bell System Technical Journal*, 49(2), 1970, pp. 291-308.

[Ker78]   Brian W. Kernighan and Dennis M. Ritchie, *The C Programming Language*, Prentice-Hall, 1978.

[Keu87]   Kurt Keutzer, "DAGON: Technology Binding and Local Optimization by DAG Matching," *Proceedings, 24th Design Automation Conference*, ACM/IEEE, 1987, pp. 341-347.

[Keu91]   Kurt Keutzer, Sharad Malik, and Alexander Saldanha, "Is redundancy necessary to reduce delay?," *IEEE Transactions on CAD/ICAS*, 10(4), April, 1991, pp. 427-435.

[Kir94]   D. Kirkpatrick and A. Sangiovanni-Vincentelli, "Techniques for crosstalk avoidance in the physical design of high-performance digital systems," in *ICCAD-94 Digest of Technical Papers*, IEEE Computer Society Press, 1994, pp. 616-619.

[Kog95]   Peter M. Kogge, Toshio Sunaga, Hisatada Miyataka, Koji Kitamura, and Eric Retter, "Combined DRAM and logic chip for massively parallel systems," in W. J. Dally, J. W. Poulton, and A. T. Ishii, eds., *16th Conference on Advanced Reserach in VLSI*, IEEE Computer Society Press, 1995, pp. 4-16.

[Kor93]   Israel Koren, *Computer Arithmetic Algorithms*, Prentice Hall, New York, 1993.

[Kra82]   R. H. Krambeck, C. M. Lee, H. F. S. Law, "High-Speed Compact Circuits with CMOS," *IEEE Journal of Solid-State Circuits*, SC-17(3), June, 1982, pp. 614-619.

[Kur96]   T. Kuroda, T. Fujita, S. Mita, T. Nagamatsu, S. Yoshioka, K. Suzuki, F. Sano, M. Norishima, M. Murota, M. Kato, M. Kinugawa, M. Kakumu, and T. Sakurai, "A 0.9V, 150 MHz, 10-mW, $4mm^2$, 2-D discrete cosine transform core processor with variable threshold-voltage (VT) scheme," *IEEE Journal of Solid-State Circuits*, 31(11), November 1996, pp. 1770-1779.

[Lan96]   Paul Landman, Renu Mehra, and Jan M. Rabaey, "An integrated CAD environment for low-power design," *IEEE Design & Test of Computers*, Summer, 1996, pp. 72-82.

[Lan91]   Dirk Lanneer, Stefaan Note, Francis Depuydt, Marc Pauwels, Francky
           Catthoor, Gert Goossens, and Hugo De Man, "Architectural synthesis for
           medium and high throughput signal processing with the new CATHE-
           DRAL environment," in Raul Camposano and Wayne Wolf, eds., *High-
           Level VLSI Synthesis*, Kluwer Academic Publishers, Norwell MA, 1991.

[Lee61]   C. Y. Lee, "An algorithm for path connections and its applications," *IRE
           Transactions on Electronic Computers*, EC-10, September, 1961, pp.
           346-365.

[Lee92]   Tien-Chien Lee, Wayne H. Wolf, Niraj K. Jha, and John M. Acken,
           "Behavioral synthesis for easy testability in data path allocation," *Pro-
           ceedings, ICCD-92*, IEEE Computer Society Press, Los Alamitos, CA,
           1992, pp. 29-32.

[Leh61]   M. Lehman and N. Burla, "Skip techniques for high-speed carry-propaga-
           tion in binary arithmetic units," *IRE Transactions on Electronic Com-
           puters*, December, 1961, pp. 691-698.

[Lei83]   Charles E. Leiserson, Flavio M. Rose, and James B. Saxe, "Optimizing
           synchronous circuitry by retiming," *Proceedings, Third Caltech Confer-
           ence on VLSI* , Randal Bryant, ed., Computer Science Press, Rockville,
           MD, 1983, pp. 87-116.

[Lig88]   Michael Lightner and Wayne Wolf, "Experiments in logic optimization,"
           Proceedings, ICCAD-88, ACM/IEEE, 1988, pp. 286-289.

[Lin87]   Chin Jen Lin and Sudhakar M. Reddy, "On delay fault testing in logic cir-
           cuits," *IEEE Transactions on CAD/ICAS*, CAD-6(5), September, 1987,
           pp. 694-703.

[Lin91]   B. Lin and A. R. Newton, "Jedi: a state assignment algorithm for multi-
           level logic," *IEEE Transactions on CAD/ICAS*, 1991.

[Lip90A]   R. J. Lipton, D. N. Serpanos, and W. H. Wolf, "PDL++: an optimizing
           generator language for register-transfer design," *Proceedings, ISCAS-90*,
           IEEE Circuits and Systems Society, May, 1990, pp. 1135-1138.

[Lip90B]   Roger Lipsett , Carl Schaefer, and Cary Ussery, *VHDL: Hardware
           Description and Design*, Kluwer Academic Publishers, Norwell, MA,
           1990.

[Liu93]   Dake Liu and Christer Svensson, "Trading speed for low power by choice
           of supply and threshold voltages," *IEEE Journal of Solid-State Circuits*,
           January 1993, pp. 10-17.

[Lyo91]   Jose A. Lyon, Mike Gladden, Eytan Hartung, Eric Hoang, and K. Raghu-
           nathan, "Testability features of the 68HC16Z1," *Proceedings, 1991
           International Test Conference*, IEEE Computer Society Press, 1991, pp.
           122-130.

[Mai91]   Frederic Mailhot and Giovanni De Micheli, "Algorithms for technology mapping based on binary decision diagrams and on Boolean operations," Stanford University Computer Systems Laboratory Technical Report No. CSL-TR-91-486, August, 1991.

[Mal87]   W. Maly, *Atlas of IC Technologies: An Introduction to VLSI Processes*, Benjamin-Cummings, 1987.

[Mal90]   Sharad Malik, *Combinational Logic Optimization Techniques in Sequential Logic Synthesis*, Ph.D. Thesis, University of California, Berkeley, November, 1990.

[Man96]   John G. Maneatis, "Low-jitter and process-independent DLL and PLL based on self-biased techniques," in *Digest of Technical Papers, 1996 IEEE International Solid-State Circuits Conference*, Castine ME: John Wuorinen, 1996, pp. 130-121.

[Mar92]   John Markoff, "Rethinking the national chip policy," *New York Times*, July 14, 1992.

[Mau88]   Peter M. Maurer, "Design verification of the WE 32106 math accelerator unit," *IEEE Design & Test of Computers*, June, 1988, pp. 11-21.

[Maz92]   Stanley Mazor and Patricia Langstraat, *A Guide to VHDL*, Kluwer Academic Publishers, Norwell, MA, 1992.

[Mic92]   Petra Michel, Ulrich Lauther, and Peter Duzy, eds., T*he Synthesis Approach to Digital System Design*, Kluwer Academic Publishers, Norwell, MA, 1992.

[McC86]   Edward J. McCluskey, *Logic Design Principles with Emphasis on Testable Semicustom Circuits*, Prentice-Hall, 1986.

[McW80]   T. M. McWilliams, *Verification of Timing Constraints on Large Digital Systems*, Ph.D. Thesis, Stanford University, May , 1980,

[Mea80]   Carver Mead and Lynn Conway, *Introduction to VLSI Systems*, Addison-Wesley, 1980.

[Mik68]   K. Mikami and K. Tabuchi, "A computer program for optimal routing of printed circuit connectors," *IFIPS Proceedings*, H47, 1968, pp. 1745-1478.

[MSU89]   Center for Integrated Systems, Mississippi State University, SCMOS Standard Cell Library, 1989.

[Mon93]   Jose' Montiero, Srinivas Devadas, and Abhijit Ghosh, "Retiming sequential circuits for low power," in *Proceedings, ICCAD-93*, IEEE Computer Society Press, 1993, pp. 398-402.

[Mul77]   Richard S. Muller and Theodore I. Kamins, *Device Electronics for Integrated Circuits*, John Wiley and Sons, 1977.

[Mur93]   Shyam P. Murarka, *Metallization: Theory and Practice for VLSI and ULSI*, Butterworth-Heinemann, 1993.

[Mut95]   S. Mutoh, T. Douseki, Y. Matsuya, T. Aoki, S. Shigematsu, and J. Yamada, "1-V power supply high-speed digital circuitry with multithreshold CMOS," *IEEE Journal of Solid-State Circuits*, 30(8), August 1995, pp. 147-854.

[Nag75]   L. W. Nagel, "SPICE 2: A Computer Program to Simulate Semiconductor Circuits," UCB/ERL M520, University of California, Berkeley, May 1975.

[Nie91]   Thomas Niermann and Janak H. Patel, "HITEC: a test generation package for sequential circuits," *Proceedings of the European Conference on Design Automation*, 1991, IEEE Computer Society Press, Los Alamitos, CA, pp. 214-218.

[Noi82]   David Noice, Rob Mathews, and John Newkirk , "A Clocking Discipline for Two-Phase Digital Systems," *Proceedings, International Conference on Circuits and Computers* ," IEEE Computer Society, 1982, pp. 108-111.

[NYT93]   Lawrence M. Fisher, "Intel raising capacity of chip factory," *New York Times*, April 2, 1993, p. D2.

[Ost83]   John K. Osterhout, "Crystal: A Timing Analyzer for nMOS VLSI Circuits," *Proceedings, Third Caltech Conference on VLSI*, Randal Bryant, ed. Computer Science Press, Rockville, MD, 1983, pp. 57-69.

[Ost84]   John K. Osterhout , Gorton T. Hamachi, Robert N. Mayo, Walter S. Scott , and George S. Taylor, "Magic: A VLSI Layout System," *Proceedings, 21st Design Automation Conference*, ACM/IEEE, 1984, pp. 152-159.

[Ott80]   R. H. J. M. Otten, "Complexity and diversity in IC layout design," *Proceedings of the International Conference on Circuits and Computers*, IEEE Computer Society, 1980, pp. 464-467.

[Pan87]   Barry Michael Pangrle and Daniel D. Gajski, "Design tools for intelligent silicon compilation," *IEEE Transactions on CAD/ICAS*, CAD-6(6), November, 1987, pp. 1098-1111.

[Pap90]   Chris Papachristou and Scott Chiu, "High level synthesis with self testability," *Proceedings, Techcon '90*, Semiconductor Research Corporation, October, 1990, pp. 407-410.

[Par92]   Kenneth P. Parker, The Boundary-Scan Handbook, *Kluwer Academic Publishers*, Norwell, MA, 1992.

[Pat98]   David A. Patterson and John L. Hennessy, *Computer organization and design: the hardware/software interface*, San Francisco: Morgan Kaufman, 1998.

602

[Pau89]   P. G. Paulin and J. P. Knight, "Force-directed scheduling for the behavioral synthesis of ASICs," *IEEE Transactions on CAD/ICAS*, 8(6), June, 1989, pp. 661-679.

[Ped96]   Massoud Pedram, "Logic Synthesis for Low Power," Chapter 6 in Jan M. Rabaey and Massoud Pedram, eds., *Low Power Design Methodologies*, Kluwer Academic Publishers, 1996.

[Pet67]   Richard L. Petritz, "Current status of large scale integration technology," *IEEE Journal of Solid-State Circuits*, SC-2(4), December, 1967, pp. 130-147.

[Pil90]   Lawrence T. Pillage and Ronald A. Rohrer, "Asymptotic waveform evaluation for timing analysis," *IEEE Transactions on CAD/ICAS*, 9(4), 352-366, April 1990.

[Pre79]   Bryan T. Preas, *Placement and Routing Algorithms for Hierarchical Integrated Circuit Layout*, Ph.D. Thesis, Stanford University, August, 1979.

[Pre88]   Bryan T. Preas and Michael J. Lorenzetti, eds. *Physical Design Automation of VLSI Systems*, Benjamin-Cummings, 1988.

[Qia94]   Jessica Qian, Satyamurthy Pullela, and Lawrence Pillage, "Modeling the 'effective capacitance' for the RC interconnect of CMOS gates," *IEEE Transactions on CAD/ICAS*, 13(12), December 1994, pp. 1526-1535.

[Rab75]   Lawrence R. Rabiner and Bernard Gold, *Theory and Application of Digital Signal Processing*, Prentice Hall, 1975.

[Rab96]   Jan M. Rabaey, *Digital Integrated Circuits: A Design Perspective*, Prentice Hall, 1996.

[Ram65]   Simon Ramo, John R. Whinnery, and Theodore van Duzer, *Fields and Waves in Communication Electronics*, New York: John Wiley and Sons, 1965.

[Raz98]   Behzad Razavi, *RF Microelectronics*, Upper Saddle River NJ: Prentice Hall PTR, 1998.

[Reg70]   William M. Regitz and Joel A. Karp, "Three-transistor-cell 1024-bit 500-ns MOS RAM," *IEEE Journal of Solid-State Circuits*, SC-5(5), October, 1970, pp. 181-186.

[Riv82]   Ronald L. Rivest and Charles M. Fiduccia, "A ``greedy" channel router," *Proceedings, 19th Design Automation Conference*, ACM/IEEE, 1982, pp. 418-424.

[Rot66]   J. P. Roth, "Diagnosis of automata failures: a calculus and a method," *IBM Journal of Research and Development*, 10(4), July, 1966, pp. 278-291.

[Roy93]   Kaushik Roy and Sharat C. Prasad, "Circuit activity based logic synthesis for low power reliable operations," *IEEE Transactions on VLSI Systems*,

1(4), December, 1993, pp. 503-513.

[Roy00]   Kaushik Roy and Sharat C. Prasad, *Low-Power CMOS VLSI Circuit Design*, New York:Wiley Interscience, 2000.

[Rub83]   J. Rubinstein, P. Penfield, Jr., and M. A. Horowitz, "Signal Delay in RC Tree Networks," *IEEE Transactions on CAD/ICAS*, CAD-2(3), July, 1983, pp. 202-211.

[Sai93]   A. Saini, "Design of the Intel Pentium™ Processor," in *Proceedings, ICCD '93*, IEEE Computer Society Press, 1993, pp. 248-252.

[Sak92]   K. A. Sakallah, T. N. Mudge, and O. A. Olukotun, "Analysis and design of latch-controlled synchronous digital circuits," *IEEE Transactions on CAD/ICAS*, 11(3), March, 1992, pp. 322-333.

[Sak93]   Takayasu Sakurai, "Closed-form expressions for interconnect delay, coupling, and crosstalk in VLSI's," *IEEE Transactions on Electron Devices*, 40(1), January 1993, pp. 118-124.

[Sal89]   Alexander Saldanha, Albert R. Wang, Robert K. Brayton, and Alberto L. Sangiovanni-Vincentelli, "Multi-level logic simplification using don't-cares and filters," *Proceedings, 26th Design Automation Conference*, IEEE Computer Society Press, Los Alamitos, CA, 1989, pp. 277-282.

[Sal89]   Arturo Salz and Mark Horowitz, "IRSIM: an incremental MOS switch-level simulator," in *Proceedings, 26th ACM/IEEE Design Automation Conference*, ACM Press, 1989, pp. 173-178.

[San84]   Alberto Sangiovanni-Vincentelli, Maruo Santomauro, and Jim Reed, "A new gridless channel router: Yet Another Channel Router the Second (YACR-II)," *Proceedings, ICCAD-84* , IEEE Computer Society Press, Los Alamitos, CA, 1984, pp. 72-75.

[Sap00]   Sachin S. Sapatnekar, "A timing model incorporating the effects of crosstalk on delay and its application to optimal channel routing," *IEEE Transactions on CAD/ICAS*, 19(3), May 2000, pp. 550-559.

[Sax83]   Tim Saxe, "CLL—A Chip Layout Language (version 4) ," Stanford University Technical Report, September, 1983.

[Sec85]   Carl Sechen and Alberto Sangiovanni-Vincentelli, "The TimberWolf placement and routing package," *IEEE Journal of Solid State Circuits*, SC-20(2), April, 1985, pp. 510-522.

[Ser89]   Donald P. Seraphim, Ronald Lasky, and Che-Yu Li, *Principles of Electronic Packaging*, McGraw-Hill, 1989.

[Sha38]   Claude E. Shannon, "A symbolic analysis of relay and switching circuits," *Transactions AIEE*, 57, 1938, pp. 713-723.

[Shi93]   Katsuhiro Shimonhigashi and Koichi Seki, "Low-voltage ULSI design," *IEEE Journal of Solid-State Circuits*, 28(4), April, 1993, pp. 408-413.

[Sho81] M. Shoji, "Electrical design of the BELLMAC-32A" microprocessor," in *Proceedings, IEEE International Conference on Circuits and Computers*, IEEE Computer Society Press, 1982, pp. 1132-115.

[Sho88] Masakazu Shoji, *CMOS Digital Circuit Technology*, Prentice Hall, 1988.

[Shu91] C. B. Shung, R. Jain, K. Rimey, E. Wang, M. B. Srinistava, B. C. Richards, E. Lettang, S. K Azim, L. Thon, P. N. Hilfinger, J. M. Rabaey, and R. W. Brodersen, "An integrated CAD system for algorithm-specific IC design," *IEEE Transactions on CAD/ICAS*, 10(4), April, 1991, pp. 447-463.

[Sin88] Kanwar Jit Singh, Albert R. Wang, Robert K. Brayton, and Alberto Sangiovanni-Vincentelli, "Timing optimization of combinational logic," *Proceedings, ICCAD-88* , IEEE Computer Society Press, Los Alamitos, CA, 1988, pp. 282-285.

[Smi00] David R. Smith and Paul D. Franzon, *Verilog Styles for Synthesis of Digital Systems*, Upper Saddle River NJ: Prentice Hall, 2000.

[Sno78] Edward A. Snow, *Automation of Module Set Independent Register-Transfer Level Design*, Ph.D. Thesis, Carnegie-Mellon University, April, 1978.

[Sta98] J. Staunstrup and W. Wolf, eds., Hardware/Software Co-Design: Principles and Practice, Norwell MA: Kluwer Academic Publishers, 1998.

[Sun84] Hideo Sunami, Tokuo Kure, Norikazu Hashimoto, Kiyoo Itoh, Toru Toyabe, and Shojiro Asai, "A corrugated capacitor cell (CCC)," IEEE Transactions on Electron Devices, ED-31(6), June, 1984, pp. 746-753.

[Sut99] Ivan Sutherland, Bob Sproull, and David Harris, *Logical Effort: Designing Fast CMOS Circuits*, San Francisco: Morgan Kaufman, 1999.

[Sye82] Zahir A. Syed and Abbas El Gamal, "Single layer routing of power and ground networks in integrated circuits," *Journal of Digital Systems*, 6(1), 1982, pp. 53-63.

[Sze85] S. M. Sze, *Semiconductor Devices: Physics and Technology*, New York: John Wiley and Sons, 1985.

[Szy85] T. G. Szymanski, "Dogleg channel routing is NP-complete," *IEEE Transactions on CAD/ICAS*, CAD-4(1), January, 1985, pp. 31-41.

[Szy88] Thomas G. Szymanski and Christopher J. Van Wyck, "Layout Analysis and Verification," Chapter 8, *Physical Design Automation of VLSI Systems*, Bryan T. Preas and Michael J. Lorenzetti, eds., Benjamin-Cummings, 1988.

[Tak85] Yoshihiro Takemae, Taiji Ema, Masao Nakano, Fumio Baba, Takashi Yabu, Kiyoshi Miyakasa, and Kazunari Shirai, "A 1 Mb DRAM with 3-dimensional stacked capacitor cells," in *Digest of Technical Papers, 1985*

IEEE International Solid-State Circuits Conference, IEEE, 1985, pp. 250-251.

[Tho91]   Donald E. Thomas and Philip Moorby, *The Verilog Hardware Description Language*, Kluwer Academic Publishers, Norwell, MA, 1991.

[Tho98]   Donald E. Thomas and Philip R. Moorby, The Verilog Hardware Description Language, Fourth Edition, Boston: Kluwer, 1998.

[Tri94]   Stephen M. Trimberger, ed., *Field-Programmable Gate Array Technology*, Kluwer Academic Publishers, Norwell, MA, 1994.

[Tse86]   C. J. Tseng and D. P. Siewiorek, "Automated synthesis of data paths in digital systems," *IEEE Transactions on CAD/ICAS*, CAD-5(3), July, 1986, pp. 379-395.

[Ung69]   Stephen H. Unger, *Asynchronous Sequential Switching Circuits*, Wiley-Interscience, 1969.

[Uye92]   John P. Uyemura, *Circuit Design for CMOS VLSI*, Kluwer Academic Publishers, 1992.

[VHD88]   VHDL Standards Committee, *IEEE Standard VHDL Language Reference Manual*, IEEE Std 1076-1977, 1988.

[van90]   Lukas P. P. P. van Ginneken, "Buffer placement in distributed RC-tree networks for minimal Elmore delay," in *Proceedings, ISCAS*, IEEE, 1990, pp. 865-868.

[Vin98]   James E. Vinson and Juin J. Liou, "Electrostatic discharge in semiconductor devices: an overview," *Proceedings of the IEEE*, 86(2), February 1998, pp. 399-418.

[Wak89A] Kazutoshi Wakabayashi and Takeshi Yoshimura, "A resource sharing and control synthesis method for conditional branches," *Proceedings, ICCAD-89* , IEEE Computer Society Press, Los Alamitos, CA, 198, pp. 62-65.

[Wal64]   C. S. Wallace, "A suggestion for a fast multiplier," *IEEE Transactions on Electronic Computers*, EC-13(1), February, 1964, pp. 14-17.

[Was78]   R. L. Wasdak, "Fault modeling and logic simulation of CMOS and MOS integrated circuits," *Bell System Technical Journal*, 57, May-June, 1978, pp. 1474-1499.

[Wil81]   T. W. Williams and N. C. Brown, "Defect level as a function of fault coverage," *IEEE Transactions on Computers*, C-30(12), December, 1981, pp. 987-988.

[Wil83]   Thomas W. Williams and Kenneth P. Parker, "Design for testability—a survey," *Proceedings of the IEEE*, 71(1), January, 1983, pp. 98-112. Reprinted in Vishwani D. Agrawal and Sharad C. Seth, *Test Generation for VLSI Chips*, IEEE Computer Society Press, 1988.

[Wol89]   Wayne Wolf, "How to build a hardware description and measurement system on an object-oriented programming language," *IEEE Transactions on CAD/ICAS*, 8(3), March, 1989, pp. 288-301.

[Wol97]   Wayne Wolf, "An architectural co-synthesis algorithm for distributed, embedded computing systems," *IEEE Transactions on VLSI Systems*, 5(2), June, 1997, pp. 218-229.

[Wol00]   Wayne Wolf, *Computers as Components: Principles of Embedded Computing System Design*, San Francisco: Morgan Kaufman, 2000.

[Xue96]   T. Xue, E. S. Kuh, and D. Wang, "Post global routing crosstalk risk estimation and reduction," in *ICCAD-96 Digest of Technical Papers*, IEEE Computer Society Press, 1996, pp. 302-309.

[Yan78]   Edward S. Yang, *Fundamentals of Semiconductor Devices*, McGraw-Hill, 1978.

[You92]   Ian A. Young, Jeffrey K. Greason, and Keng L. Wong, "A PLL clock generator with 5 to 110 MHz of lock range for microprocessors," *IEEE Journal of Solid-State Circuits*, 27(11), November, 1992, pp. 1599-1607.

Index

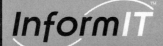
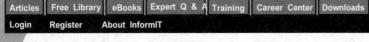